TANK WASTE RETRIEVAL, PROCESSING, AND ON-SITE DISPOSAL AT THREE DEPARTMENT OF ENERGY SITES

FINAL REPORT

Committee on the Management of Certain Radioactive Waste Streams Stored in Tanks at Three Department of Energy Sites

Nuclear and Radiation Studies Board

Division on Earth and Life Studies

NATIONAL RESEARCH COUNCIL
OF THE NATIONAL ACADEMIES

THE NATIONAL ACADEMIES PRESS
Washington, D.C.
www.nap.edu

THE NATIONAL ACADEMIES PRESS 500 FIFTH STREET, N.W. Washington, DC 20001

NOTICE: The project that is the subject of this report was approved by the Governing Board of the National Research Council, whose members are drawn from the councils of the National Academy of Sciences, the National Academy of Engineering, and the Institute of Medicine. The members of the committee responsible for the report were chosen for their special competences and with regard for appropriate balance.

Support for this study was provided by the U.S. Department of Energy under cooperative agreement number DE-FC01-04-EW07022. Any opinions, findings, conclusions, or recommendations expressed in this publication are those of the author(s) and do not necessarily reflect the views of the organizations or agencies that provided support for the project.

International Standard Book Number 0-309-10170-0

Additional copies of this report are available from the National Academies Press, 500 Fifth Street, N.W., Lockbox 285, Washington, DC 20055; (800) 624-6242 or (202) 334-3313 (in the Washington metropolitan area); Internet, http://www.nap.edu

Cover: Images courtesy of the U.S. Department of Energy.

Printed in the United States of America.

THE NATIONAL ACADEMIES

Advisers to the Nation on Science, Engineering, and Medicine

The **National Academy of Sciences** is a private, nonprofit, self-perpetuating society of distinguished scholars engaged in scientific and engineering research, dedicated to the furtherance of science and technology and to their use for the general welfare. Upon the authority of the charter granted to it by the Congress in 1863, the Academy has a mandate that requires it to advise the federal government on scientific and technical matters. Dr. Ralph J. Cicerone is president of the National Academy of Sciences.

The **National Academy of Engineering** was established in 1964, under the charter of the National Academy of Sciences, as a parallel organization of outstanding engineers. It is autonomous in its administration and in the selection of its members, sharing with the National Academy of Sciences the responsibility for advising the federal government. The National Academy of Engineering also sponsors engineering programs aimed at meeting national needs, encourages education and research, and recognizes the superior achievements of engineers. Dr. Wm. A. Wulf is president of the National Academy of Engineering.

The **Institute of Medicine** was established in 1970 by the National Academy of Sciences to secure the services of eminent members of appropriate professions in the examination of policy matters pertaining to the health of the public. The Institute acts under the responsibility given to the National Academy of Sciences by its congressional charter to be an adviser to the federal government and, upon its own initiative, to identify issues of medical care, research, and education. Dr. Harvey V. Fineberg is president of the Institute of Medicine.

The **National Research Council** was organized by the National Academy of Sciences in 1916 to associate the broad community of science and technology with the Academy's purposes of furthering knowledge and advising the federal government. Functioning in accordance with general policies determined by the Academy, the Council has become the principal operating agency of both the National Academy of Sciences and the National Academy of Engineering in providing services to the government, the public, and the scientific and engineering communities. The Council is administered jointly by both Academies and the Institute of Medicine. Dr. Ralph J. Cicerone and Dr. Wm. A. Wulf are chair and vice chair, respectively, of the National Research Council.

www.national-academies.org

COMMITTEE ON THE MANAGEMENT OF CERTAIN RADIOACTIVE WASTE STREAMS STORED IN TANKS AT THREE DEPARTMENT OF ENERGY SITES

FRANK L. PARKER, *Chair*, Vanderbilt University, Nashville, Tennessee
HADI ABU-AKEEL, AMTENG Corp., Vienna, Virginia
JOHN S. APPLEGATE, Indiana University School of Law, Bloomington
HOWIE CHOSET, Carnegie Mellon University, Pittsburgh, Pennsylvania
PAUL C. CRAIG,* University of California, Davis
ALLEN G. CROFF, Oak Ridge National Laboratory (retired), Tennessee
PATRICIA J. CULLIGAN, Columbia University, New York City, New York
KEN CZERWINSKI, University of Nevada, Las Vegas
RACHEL DETWILER, Braun Intertec Corp, Minneapolis, Minnesota
EDWIN E. HERRICKS, University of Illinois at Urbana-Champaign
TISSA ILLANGASEKARE, Colorado School of Mines, Golden
MILTON LEVENSON, Bechtel International (retired), Menlo Park, California
PAUL A. LOCKE, Johns Hopkins University, Baltimore, Maryland
MICHAEL H. MOBLEY, Mobley Radiation Consulting, Clarksville, Tennessee
DIANNE R. NIELSON, Utah Department of Environmental Quality, Salt Lake City
KEN E. PHILIPOSE, AECL/Chalk River Laboratories, Ontario, Canada
ALFRED P. SATTELBERGER, Argonne National Laboratory, Illinois
ANNE E. SMITH, CRA International, Washington, D.C.
LESLIE SMITH, University of British Columbia, Canada
DONALD W. STEEPLES, University of Kansas, Lawrence

Consultant

RODNEY C. EWING, University of Michigan, Ann Arbor

Staff

MICAH D. LOWENTHAL, Study Director
BARBARA PASTINA, Study Director
JOHN R. WILEY, Senior Program Officer
DARLA J. THOMPSON, Research Associate
LAURA D. LLANOS, Senior Program Assistant
MARILI ULLOA, Senior Program Assistant

---

* Dr. Paul Craig resigned from the committee on August 22, 2005.

List of Report Reviewers

This report has been reviewed in draft form by individuals chosen for their diverse perspectives and technical expertise, in accordance with procedures approved by the National Research Council's Report Review Committee. The purpose of this independent review is to provide candid and critical comments that will assist the institution in making its published report as sound as possible and to ensure that the report meets institutional standards for objectivity, evidence, and responsiveness to the study charge. The content of the review comments and draft manuscript remain confidential to protect the integrity of the deliberative process. We wish to thank the following individuals for their review of this report:

David Adelman, University of Arizona
Robert Bernero, U.S. Nuclear Regulatory Commission (retired)
Barry Burks, TPG Applied Technology
Sue Clark, Washington State University
Bill Echo, Idaho National Engineering and Environmental Laboratory (retired)
Roy E. Gephart, Pacific Northwest National Laboratory
Linn Hobbs, Massachusetts Institute of Technology
Terence C. Holland, Consultant
David C. Kocher, SENES Oak Ridge, Inc.
Leonard Konikow, U.S. Geological Survey
Edward Lahoda, Westinghouse Science and Technology Center
Jane C. S. Long, Lawrence Livermore National Laboratory
Mal McKibben, Citizens for Nuclear Technology Awareness
Richard A. Meserve, Carnegie Institution

Although the reviewers listed above have provided many constructive comments and suggestions, they were not asked to endorse the report's conclusions or recommendations, nor did they see the final draft of the report before its release. The review of this report was overseen by Hyla Napadensky, Napadensky Energetics Inc. (retired), and John Ahearne, Sigma Xi. Appointed by the National Research Council, they were responsible for making certain that an independent examination of this report was carried out in accordance with institutional procedures and that all review comments were carefully considered. Responsibility for the final content of this report rests entirely with the authoring committee and the institution.

Preface

During World War II and the Cold War that followed, the United States built large production capabilities for and stockpiles of nuclear weapons. This involved the chemical processing (reprocessing) of the spent nuclear fuel from plutonium production reactors. The highly radioactive wastes from reprocessing were put in underground tanks as neutralized liquor at the Hanford and Savannah River Sites in Washington State and South Carolina, respectively. A separate, but similar, program at the Idaho National Laboratory reprocessed spent nuclear fuel from naval nuclear reactors and experimental and test reactors; most of the waste from these reprocessing activities was converted to a granular form and stored in above-ground bins (tanks inside silos), although some highly acidic waste was left in the facility's underground stainless steel storage tanks.

At the Hanford and Savannah River Sites, the chemical processes used for extracting key radionuclides changed over time and were applied to different nuclear fuels and targets. As a result, the wastes produced by reprocessing were somewhat different at different times. The liquid wastes were subjected to chemical treatments to inhibit corrosion of the carbon steel tanks and to precipitate key radionuclides. Wastes from different chemical processes or from processing different materials were mixed together; some wastes were reprocessed multiple times for additional separations; and materials such as zeolites and diatomaceous earth were discarded in some tanks. The tanks themselves vary in design, even at one site and within one tank farm. Some waste was intentionally discharged into the ground at Hanford, and there were leaks in the tank farms at each site. Most of the tank wastes now are highly heterogeneous and the conditions in which they are stored vary.

Studies of means to solidify the highly radioactive tank wastes began roughly 50 years ago. Progress has been slow. This is an indication of the complexity and difficulty of the problem. Now the United States is entering a new phase, under very different social, political, and legal conditions than were in place at the time the wastes were created. The Department of Energy (DOE), which is responsible for managing the wastes and their consequent risks, has been operating one facility for immobilizing high-level radioactive waste through vitrification at the Savannah River Site and has finished operation of another at the West Valley Demonstration Project. DOE is building a processing and immobilization facility at Hanford and has recently selected a technology for immobilizing the liquid waste in tanks in Idaho. DOE has cleaned out about 12 of the 246 tanks at the sites, and 2 of the cleaned tanks have been closed by filling them with grout.

Ideally, all wastes would be removed from the tanks. However, it is widely recognized that it is prohibitive in terms of worker risk and economic cost to exhume the tanks or remove all of the wastes from all of the tanks. The debate now is over how much removal of the wastes from the tanks is enough, how much removal of radionuclides from the retrieved waste is enough, and whether grout is an adequate form of residual waste immobilization onsite. In 2004, Congress passed legislation (the Ronald Reagan National Defense Authorization Act for Fiscal Year 2005) to facilitate progress toward tank waste disposal in South Carolina and Idaho, providing a framework for determining what wastes may remain on-site. In the same act, Congress asked the National Academies to evaluate DOE's plans for some of these wastes at the Savannah River Site, the Hanford Site, and the Idaho National Laboratory.

A committee of 21 volunteers with diverse expertise was charged to fulfill the congressional request, issuing an interim report in 6 months and the final report in one year (see Appendix A for biographical sketches of the committee members). This proved to be a difficult challenge. Due to the complexity of DOE's tasks, there are many important details in several hundred substantial documents, and some details were not constant throughout the course of the study. Decisions and reviews are continuing even as this report is being published. The committee carried out its task as DOE was working with South Carolina, Idaho, and the U.S.

Nuclear Regulatory Commission on the first waste determinations to be made under the new legislation. DOE and others worked hard to get the committee the information it needed. However, an unfortunate feature of the timing is that some key documents were issued late in the course of the study (after the beginning of September 2005 and through the committee's review process).

Because of the long time frames in which some of the waste remains hazardous, the committee recognized that the future is uncertain and circumstances, capabilities, and knowledge may continue to change as dramatically in the future as they have in the past. We also recognize that given such limited experience with some of the chemical processes to treat wastes, some being used in these applications for the first time, there is a high degree of technical uncertainty added to the considerable uncertainties about costs and what the society values. We have tried to avoid scientific and engineering hubris in viewing our capabilities to predict what will occur in the future and how man-made objects will perform over these long time frames.

We have refrained from looking holistically at the problem of environmental releases of radioactive and hazardous chemical materials from the sites over time because it was not in our charter. However, we would be remiss if we did not call attention to the other radioactive and hazardous chemicals at the sites that also can pose risks to human health and the environment. As noted in the report, the trade-off between the cost and risk of retrieving and processing tank and bin wastes must take into consideration risks from other waste and contamination already committed to the site.

Time constraints, new developments, and the absence of some information even in the vast quantity of documentation provided necessarily limit the extent to which the committee could answer Congress's questions. Some related issues, such as seismic concerns, could not be considered. Nevertheless, the report contains important messages that we hope Congress, DOE, and others will find informative and helpful. Some of these messages are reiterated and elaborated from the committee's interim report. The final report does not reproduce the committee's finding and recommendation concerning "Class C" concentration limits. The committee stands by them, as indicated in Appendix E, but has nothing further to add on that subject.

The committee thanks the people who provided input to this study, including the many presenters listed in the Appendix D, as well as members of the public who spoke at those meetings. We specifically acknowledge several individuals who helped to coordinate meetings and respond to special requests by the committee: Mark Gilbertson, Randy Kaltreider, and Ken Picha at DOE headquarters; Tom Caldwell, Bill Clark, Ginger Dickert, Peter Hill, Christine Langton, Bill Pearson, Sherri Ross, Terry Spears, and Steve Thomas at the Savannah River Site; Shelly Sherritt and David Wilson at the South Carolina Department of Health and Environmental Control; Anna Bradford, David Esh, and Scott Flanders at the U.S. Nuclear Regulatory Commission; Lorie Cahn, Keith Lockie, and Keith Quigley at the Idaho National Lab; Kathleen Trever at the Idaho Department of Environmental Quality; Ryan Dodd, Bill Hewitt, Fred Mann, Roger Quintero, and Roy Schepens at the Hanford Site; Suzanne Dahl, Jane Hedges, and Michael Wilson at the Washington Department of Ecology; Russell Jim of the Yakama Indian Nation; and Ken Niles at State of Oregon Department of Energy.

Finally, the committee thanks the members of its staff—Laura Llanos, Micah Lowenthal, Barbara Pastina, Darla Thompson, Marili Ulloa, and John Wiley—all of whom worked long hours to support the committee throughout the study. In particular, we thank Micah and Barbara for their work in directing the study and Laura for ensuring that meetings went smoothly. We also want to acknowledge Kevin Crowley, director of the Nuclear and Radiation Studies Board, for his assistance at several points. This study was a large and challenging task that the committee could not have completed without their support.

Frank L. Parker, *Chair*
Committee on the Management of
Certain Radioactive Waste Streams
Stored in Tanks at Three Department of
Energy Sites

Contents

Executive Summary[1]

The U.S. Congress asked the National Academies to evaluate the Department of Energy's (DOE's) plans for retrieval and on-site disposal of certain radioactive wastes[2] stored in underground tanks at three DOE sites[3] and to make recommendations to improve those plans. The major results of this evaluation are summarized below. Readers are strongly encouraged to read the full report and particularly the findings and recommendations for further details.

- DOE's overall approach for management and disposal of tank wastes is workable, but important technical and programmatic challenges remain. In particular, the essential question, How clean is clean enough? applies to all cleanup activities and does not have a unique, numerical solution. The amount of waste to be retrieved from these tanks and how much of that should be disposed on-site is a decision in which DOE must consider a range of technical and nontechnical factors, including technical capabilities for waste retrieval and radionuclide separation from the removed wastes; cost, both in terms of dollars spent and worker doses incurred per increment of risk reduction achieved; and the potential risks from other wastes to be left on-site. **DOE should pursue a more risk informed, consistent, participatory, and transparent process for making decisions about how much waste to retrieve from each of its tanks or group of tanks, and how much of that waste to dispose at each of the three sites.**
- Only 2 of the 246 tanks at the three sites have been cleaned out and backfilled with grout, and none has had a permanent cover installed. At this early stage in the process, there is still time to develop tools and processes to address problems described here and in the full report. **DOE should initiate a targeted, aggressive, collaborative research program to develop and deploy needed innovative technologies for tank waste retrieval, treatment, closure, and disposal.**
- **DOE's current knowledge of tank waste characteristics is adequate for retrieving waste from tanks at all three sites.** DOE needs to know the waste composition in greater detail for processing purposes and to confirm compliance with performance objectives, but this must be done after waste retrieval when mixing makes representative sampling of the retrieved waste possible and when samples of the tank heels can be taken.
- **DOE should decouple its schedule for tank waste retrieval from its schedule for tank closure for those tanks that still contain significant amounts of radioactive material after initial waste retrieval is completed.** More broadly, because decisions about planned disposal activities require multiple inputs, DOE should not make decisions based solely on schedule conformance. Decoupling will enhance future opportunities to remove additional radioactive material from these tanks as retrieval technologies are improved. If implemented properly, decoupling for individual tanks need not delay the final closure of the tank farms. There is little technical advantage in the accelerated closure of the tanks.
- DOE plans to make waste determinations for individual tanks and small groups of tanks. Documents demonstrating compliance with performance objectives will be generated for each draft waste determination. The ongoing review of draft determinations by the U.S. Nuclear Regulatory Commission and host

[1]A detailed summary of the committee's report is presented in the next chapter.

[2]These wastes are the result of reprocessing spent fuel and targets from defense reactors and contain radionuclide concentrations above Class C quantities as defined in Title 10, Part 61 of the Code of Federal Regulations (10 CFR 61).

[3]The Savannah River Site, South Carolina; Hanford Site, Washington; and Idaho National Laboratory.

states, as required by the Ronald Reagan National Defense Authorization Act of 2005 (NDAA) and the state-approved closure plans it demands, is improving the technical quality and public transparency of DOE's planning efforts. **DOE should continue to seek transparent, independent peer review of critical data and analyses used to support decisions about tank waste retrieval, processing, and disposal even if review is not required under the NDAA.**

- DOE is just beginning to develop plans for post-closure monitoring of closed tank farms and associated disposal sites. The main objective of this monitoring is to verify compliance with the performance objectives in 10 CFR 61. However, some of the assumptions made in DOE's waste determinations need to be confirmed. Therefore, **DOE should develop plans now for a post-closure monitoring program and begin to build provision for monitoring into its tank closures and disposal facilities.**

The report provides several site-specific findings and recommendations; these are summarized below:

- *Savannah River Site*: The committee has serious reservations about aspects of DOE's plans for tank closure, including the point of compliance and assumptions about exposure scenarios and waste inventories remaining after tank cleanup. The committee is also concerned about DOE's plans to dispose of large inventories of radionuclides in the Saltstone Vaults on-site, and that the tank space crisis may lead DOE to dispose of additional radioactive material on-site. **To reduce the quantities of radionuclides to be disposed of on-site, DOE should develop alternates or enhancements to the deliquification, dissolution, and adjustment treatment process to solve its tank space problems.**
- *Hanford Site*: The committee also has reservations about DOE's plans to use bulk vitrification as a secondary process for treating low-activity waste for on-site disposal. **DOE should arrange for a transparent, independent, technical review of the bulk vitrification process to assess its performance and safety.**
- *Idaho National Laboratory*: DOE is making good progress in tank cleanup and closure.

A number of other significant issues will have to be resolved by DOE. The committee did not examine these issues in depth because DOE has not developed detailed plans for them, as yet, but **DOE should review and resolve these issues with deliberate speed.**

- These include remediation of plugged and leaking underground pipes and interwall spaces in double-walled tanks; the disposition of calcine bin waste at the Idaho site; regulatory approvals for the off-site disposal of some Hanford tank waste and Idaho sodium-bearing tank waste; the philosophy and methodology for post-closure monitoring; and plans for carrying out long-term stewardship, including how the federal government will maintain control "in perpetuity" at sites unsuitable for unrestricted release.

Summary

Waste from reprocessing of defense spent nuclear fuel is currently stored in large underground tanks at the Hanford Site in Washington State; the Savannah River Site in South Carolina; and the Idaho National Laboratory in Idaho. Overall, there are 246 waste tanks relevant to this study: 51 tanks containing 426 million curies (MCi; 15.8×10^{18} exabecquerels, (Bq) in 138,000 cubic meters (m^3) of waste at the Savannah River Site; 177 tanks containing 193 MCi (7.14×10^{18} Bq) in 204,000 m^3 of waste at the Hanford Site;[1] and 11 tanks and 7 bin sets currently containing about 41 MCi (1.52×10^{18} Bq) in 5,000 m^3 of waste at the Idaho National Laboratory.

In the Ronald Reagan National Defense Authorization Act of 2005 (NDAA, Section 3146 of Public Law 108-375), Congress asked the National Academies[2] to evaluate the Department of Energy's (DOE's) plans to manage radioactive waste streams from reprocessed spent fuel that (1) exceed the concentration limits for Class C low-level waste; (2) are stored in tanks at the sites mentioned above; and (3) DOE plans to dispose on-site rather than in a repository for spent nuclear fuel and high-level waste. The full statement of task can be found in Appendix B. At the request of Congress, the committee issued an interim report about the Savannah River Site (NRC, 2005a) in the summer of 2005, and this final report addresses the statement of task for all three sites. This summary describes the study and the committee's findings and recommendations, and provides abbreviated answers to Congress's questions.

The committee acquired a large number of documents and held five public meetings to obtain information from experts, affected parties, and interested members of the public. DOE and the other participants were responsive to the committee's requests; however, some data and analyses were not available to the committee (not yet collected, not yet performed, or not yet made public), and some plans had not yet been formulated or finalized by the time this report entered the National Research Council report review process in January 2006.[3] DOE issued most of the documents supporting its proposed tank waste disposition decisions in September and October 2005. Although DOE furnished the committee with hundreds of documents, some containing thousands of pages, and even though these recent documents provided quantitative examinations of many questions relevant to the committee's charge, there are no clear, definitive answers to some of the questions that Congress posed to the committee. As a result, the committee was unable to evaluate fully DOE's plans to manage its tank waste. The committee also had to operate within congressionally mandated schedule constraints, which limited the extent to which individual documents could be evaluated, particularly where DOE provided limited evaluation of cost, worker safety, and long-term human and environmental health consequences of technology alternatives for tank waste management.

Section 3146 of the NDAA, which contains the request for this study, is related to Section 3116 of the same act. Section 3116 explicitly enables DOE to determine that some tank waste from reprocessing of spent fuel is not "high-level waste" and can be disposed of on-site at the Savannah River

[1]Capsules of radioactive cesium and strontium and the so-called German logs contain another 136 MCi (5×10^{18} Bq) of radioactivity. While DOE considers the cesium and strontium capsules to be nuclear materials rather than tank wastes, the committee included them in the study because they are highly radioactive materials extracted from the tank wastes and DOE says it plans either to dispose of them on-site or to combine them with the high-activity waste stream to be vitrified and sent to geologic disposal. In essence, DOE faces the same decision about the capsules that it faces when deciding what to do with radioactive material separated in the Waste Treatment Plant. The main difference is when the separations were carried out.

[2]The operating arm of the National Academies, the National Research Council, appointed a committee to undertake this study under the auspices of the Nuclear and Radiation Studies Board. Biographical sketches of committee members can be found in Appendix A.

[3]With few exceptions, the committee's report is based on information received before January 1, 2006.

Site (SRS) and the Idaho National Laboratory (INL), provided that requirements in that section are met.[4] Section 3116 provides waste determination criteria that are applicable to South Carolina and Idaho but do not apply in Washington. The Hanford Site is governed by DOE's orders, including Order 435.1, and the federal and state laws in effect before the passage of the NDAA.[5]

In this final report the committee recommends actions that it believes could

- Reduce the quantity of radioactive material left on-site;
- Increase DOE's understanding of other factors that affect dose and risk—namely, the long-term performance of waste forms and other barriers to the release of radionuclides to the environment;
- Improve DOE's analyses of the choices it must make as it moves to retrieve reprocessing wastes from the tanks at the sites and dispose of some of them under the provisions of Section 3116 of the NDAA and DOE's waste management Order 435.1;
- Improve the likelihood that these wastes will be processed and immobilized in an efficient manner; and
- Improve DOE's decision making through a more risk-informed process.

The committee judges that these actions will increase DOE's ability to comply with the performance objectives in 10 CFR 61[6] and other applicable regulations and will help DOE fulfill its requirement to take actions to ensure that releases of radioactivity to the environment are as low as reasonably achievable (ALARA), with economic and social considerations taken into account.

BACKGROUND

If waste retrieval and processing facilities worked perfectly and at an acceptable cost, the objectives of DOE's tank cleanup program would be to remove all of the waste from the tanks; separate all of the retrieved radioactive constituents from the bulk of the waste; immobilize the radioactive waste for off-site disposal in a way that minimizes residual hazards; and minimize overall operational hazards from residual waste remaining on-site. No real waste retrieval system, however, will retrieve all of the waste; neither will a real separation process completely separate radioactive constituents from nonradioactive components. In the real world, some waste will be left in the tanks, some tank waste that leaked or was intentionally released to the soil will be left on-site, and some radioactive constituents will remain after treatment in the waste that is disposed of on-site.

FINDINGS AND RECOMMENDATIONS

Summarized below are the committee's general and site-specific findings and recommendations (see Chapters III through IX for details).

1. DOE's overall approach is workable but there are technical and programmatic challenges in reaching the goals of the tank remediation program.

DOE's overall approach for managing its tank wastes and the framework in which this must be done is workable: to the maximum extent practical, retrieve the waste from the tanks, separate the recovered waste into high- and low-activity fractions, and dispose of both waste remaining in tanks and recovered low-activity waste on-site in a manner that protects human health and the environment. Nonetheless, DOE faces technical and programmatic challenges in implementing this approach. Examples of technical challenges include retrieving waste from tanks with significant obstructions at the Savannah River Site and from tanks with leaks at the Hanford Site; and assessing the uncertainties in the performance of planned waste processing approaches, such as the deliquification, dissolution, and adjustment (DDA) process at the Savannah River Site and the bulk vitrification process at Hanford.[7] Programmatic challenges are those affecting the success of the tank cleanup program, such as budgetary challenges and regulatory challenges (see Chapter VIII).

2. Decisions about planned disposal activities require multiple inputs and should not be dictated solely by schedule conformance.

Basing tank management and waste disposition decisions only on performance assessments to demonstrate compliance with performance objectives is inadequate because such assessments do not take into account all of the various factors

[4]The legal definition of high-level radioactive waste HLW, as set out in the Nuclear Waste Policy Act (42 U.S.C. Section 10101), is waste that is "(A) the highly radioactive waste material resulting from the reprocessing of spent nuclear fuel, including liquid waste produced directly in reprocessing and any solid material derived from such liquid waste that contains fission products in sufficient concentrations; and (B) other highly radioactive material that the Commission, consistent with existing law, determines by rule to require permanent isolation." There is no particular concentration of radioactive material or dose limit associated with this definition.

[5]DOE Order 435.1 governs the management of radioactive waste at DOE sites. It includes waste criteria for determining that certain wastes are not high-level waste.

[6]The performance objectives of land disposal facilities for radioactive waste are defined in 10 CFR 61 Subpart C: (1) protect the general public from environmental releases and make releases as low as reasonably achievable, (2) protect inadvertent intruders, (3) protect individuals during operations, and (4) provide stability of the site after closure (see Appendix C).

[7]Other technical challenges concerning waste retrieval, processing, immobilization, and monitoring are described in Chapters III, IV, V, and VII.

that could be important to decisions such as the evolution of the full risk profile (risks across the site under different exposure scenarios) over time; compliance with federal facilities agreements; changes in costs and changes in how people value health and the environment; progress to build confidence in the program; and other site risks, among others. All of these factors are increasingly uncertain the further into the future that people attempt to anticipate conditions.

DOE uses conformance to a schedule (i.e., meeting milestones) as one of three criteria in determining what constitutes removal of waste to the maximum extent practical in the Section 3116 draft waste determination for Tanks 19 and 18 at the Savannah River Site. Although meeting agreed-upon schedules and milestones is important and in many cases legally enforceable, a schedule-driven approach could lead to retrieval and closure actions that may later be judged insufficient.

In addition to DOE, several other parties play an important role in the ultimate success of DOE's program: Congress, the Environmental Protection Agency, the Nuclear Regulatory Commission (USNRC), host states, American Indian nations, local governments, and other stakeholders. A risk-informed, transparent, participatory, and consistent decision-making process facilitates the involvement of these parties and enhances the effectiveness of the process.

3. Tank cleanup is a multidecade project allowing opportunities to improve the efforts to retrieve waste from the tanks

The milestones to close all tanks are at least a decade away (ranging from 2016 to 2032).[8] Of all 246 high-level waste tanks, only 2 have been closed (i.e., grouted) so far, and about 14 are being prepared for closure.[9] However, the cleanup of each tank will provide DOE with experiences that can be used to improve cleanup of the others. Following closure, modification of stabilized tank waste that is left on-site will be difficult, meaning that the form, concentration, and mass of tank waste left on-site will basically be fixed. Retrieval of waste from each tank will be somewhat different because each tank's specific combination of waste type, tank design and construction, and operation history is unique. Complete closure of all tanks will involve many one-of-a-kind and first-of-a-kind endeavors that would be carried out more effectively by building on experience with tank waste remediation. DOE is learning from its experience to date, but there still are substantial opportunities to continue to improve its program with respect to waste retrieval, processing, immobilization, and disposal, monitoring compliance and performance assessment, decision making, and research and development. Each of these opportunities is discussed in greater detail below.

Waste Retrieval (see Chapter III)

Depending on the particular requirements of individual tanks, DOE should use the most effective sequence and combination of waste retrieval tools to ensure that waste is removed to the maximum extent practical. When the limit of a given technology is reached, DOE should utilize, when necessary, other waste retrieval tools (already available or to be developed) so that the maximum extent practical is not contingent solely on what technology has already been deployed or on the proposed tank cleanup schedule. Reaching the limit of a given technology does not in itself demonstrate that all practical efforts for retrieval have been made.

The committee continues to believe that DOE should decouple the schedule for tank waste removal from the schedule for tank closure on a case-by-case basis, particularly in the case of tanks with significant heels (radioactive material remaining after planned retrieval operations are complete), as is likely in tanks with obstructions and/or with recalcitrant waste. In these tanks, more time may be needed to implement additional waste retrieval methods. Decoupling will enhance future opportunities to remove additional radioactive material from these tanks as retrieval technologies are improved. If implemented properly, decoupling for individual tanks need not delay the final closure of the tank farms. There is little technical advantage in the accelerated closure of the tanks.

Waste Processing (see Chapter IV)

When selecting a waste processing technology, DOE should take into account the impacts, flexibility, and robustness of processing facilities and waste forms. The Savannah River Site plans to use the DDA process to free up tank space, but the committee has concerns about the amount of radioactive material that the DDA process would allow to be disposed as low-activity waste on-site. The committee also has concerns about the bulk vitrification option for Hanford's supplemental low-activity waste treatment (see site-specific concerns, below). The cost and risks to workers, members of the public, and the environment if the processes should fail to perform acceptably, along with schedule uncertainties, need to be taken into account in making decisions among alternatives.

[8]Currently, the closure schedule for Hanford is 2024 for single-shell tanks and 2032 for double-shell tanks; at the Savannah River Site, the closure milestones are 2022 for Type I, II, and IV tanks; and 2028 for Type III tanks; at Idaho, the six-phase tank closure process began in 2005 and will reach completion in 2016. No milestone has been selected for closing the Idaho bins.

[9]DOE has officially submitted a waste determination to close the following tanks: Tanks 18 and 19 at the Savannah River Site, Tanks WM-180 through 186, and tanks WM-103 through WM-106 at the Idaho site (see Table F-1 in Appendix F). A separate state and USNRC review required under the Hanford Federal Facility Agreement is underway for Hanford's Tank C-106.

Waste Immobilization *(see Chapters V, VI, and IX)*

The committee agrees with DOE's selection of cementitious material (grout) as the most appropriate material for tank closure and does not foresee the development of better alternatives. However, the committee has concerns about DOE's understanding of the very-long-term performance of the grout used to inhibit water flow and immobilize waste in closed tanks. As a result, the committee recommends further short- and long-term research and development on the performance of cementitious materials. These efforts should be tailored to the formulations of grout planned for use in tank closures and waste immobilization and to the demands DOE places on their long-term performance in its performance assessments.

Performance Assessments and Monitoring Compliance *(see Chapters VI and VII)*

The committee views monitoring programs and performance assessments as iterative, interrelated, evolutionary activities that require updating as new information becomes available and as changes occur at the sites. The sites have not yet, for the most part, developed plans for post-closure monitoring so the committee is not able to comment on them. DOE Order 435.1 requires that plans for closure of high-level waste facilities include a monitoring plan and further requires that iterations of performance assessments for low-level waste disposal facilities continue through facility closure and beyond, as needed (DOE, 2001a). It is understandable that post-closure monitoring is not DOE's highest priority right now, given that closure of the tank farms is still decades away. Plans are needed, however, before closure because some of the components of monitoring systems should be built into the closure system. DOE should begin to build provision for monitoring into its tank closures and disposal facilities and develop plans for a post-closure monitoring program, ensuring that post-closure monitoring and the updating of performance assessments are given appropriate attention as the site progresses toward closure and beyond.

External and independent peer review of DOE's draft waste determinations and performance assessments introduced by Section 3116 of the NDAA has led to demonstrable improvement in DOE's analyses (such as incorporating sensitivity studies) and the technical documents that are being prepared. This in turn has sharpened the understanding of DOE's rationale, assumptions, analyses, and conclusions. DOE should continue to seek transparent, independent peer review of critical data and analyses used to support decisions about tank waste retrieval, processing, and disposal even if review is not required under the NDAA.

Decision Making *(see Chapter VIII)*

Determining how clean is clean enough for tank waste retrieval, separation, and disposal is a decision in which DOE must consider a range of technical and nontechnical factors. The question does not have a unique, numerical solution. In such decisions, DOE should take into account, in addition to the performance assessments results for specific locations at specific times, how the risks from the materials left on-site vary over space and time; technical capabilities for waste retrieval and radionuclide separation from the removed wastes; cost, both in terms of dollars spent and worker doses incurred per increment of risk reduction achieved; and the potential risks from other wastes to be left on-site. Given the technical and programmatic challenges in DOE's waste management environment, one way to improve decision making is to adopt a more risk-informed, participatory, transparent, and consistent decision-making process. Such a process, as recommended in a previous National Research Council report (NRC, 2005b), would give regulators, Congress, the public, and especially DOE a clearer idea of the challenges and choices that DOE faces. It also will make DOE's planning more robust, in the sense that it is more likely to succeed in its mission. DOE has taken steps to improve its transparency in its most recent draft waste determinations and performance objectives demonstration documents, which describe how DOE reached its decisions and provide supporting data and analyses for understanding the rationale for its decisions (Sams, 2004; Buice et al., 2005; DOE-ID, 2005a; DOE-SRS, 2005a; Rosenberger et al., 2005). The committee commends this improvement and encourages DOE to continue to increase transparency, accessibility, participation, and peer review in all aspects of its tank waste management program.

Research and Development *(see Chapter IX)*

As DOE is in the initial stages of retrieval and closure, and as the committee continues to see delays in key pieces of the tank program (e.g., Salt Waste Processing Facility at the Savannah River Site and Waste Treatment Plant at Hanford; see below), it is increasingly clear that there is more time for implementing a research and development program that could improve waste retrieval, tank stabilization, and low-activity waste immobilization. DOE should initiate a targeted, aggressive, collaborative research and development program focused on (1) options for chemical cleaning of tanks; (2) emerging technologies to assist in tank waste removal, including robotic enhancements to current waste retrieval technologies; and (3) near- and long-term performance and monitoring of tank fill materials as they interact with the environment. Based on experience with the Environmental Management Science Program,[10] Tanks Focus

[10]DOE's Environmental Management Science Program (EMSP) was created by the 104th Congress to stimulate basic research and technology development for environmental cleanup of the nation's nuclear weapons complex (NRC, 1997a).

Area,[11] and similar programs, a 10-year program on the order of $50 million per year would seem appropriate to generate the technological know-how needed for continuous improvement of tank waste management.

Site-Specific Findings and Recommendations

Savannah River Site

Compliant Tank Volume for Processing Needs

Tank wastes at the Savannah River Site are found in three different physical forms: a salt solution, a water-soluble saltcake, and an insoluble sludge. All phases contain radioactive materials. For the last 10 years, the site's Defense Waste Processing Facility has immobilized sludge in glass and poured the glass into steel canisters, which are stored pending shipment off-site for disposal in a high-level radioactive waste repository. DOE has stated that it needs open volume in compliant waste tanks[12] for secondary wastes from sludge treatment to ensure that sludge removal from noncompliant tanks continues apace and the Defense Waste Processing Facility continues to operate at full capacity. DOE ultimately plans to use the Salt Waste Processing Facility to remove radionuclides from most of the salt solution and saltcake phases of the tank waste, which occupy most of the volume in compliant waste tanks. To obtain open tank volume for waste inputs before the Salt Waste Processing Facility is operational and for efficient operation of that facility, DOE plans to use two interim processes: DDA and a separate, low-throughput chemical processing unit. Because of the recently announced 26-month delay in start-up of the Salt Waste Processing Facility, DOE has been forced to reexamine its alternatives in obtaining that open tank volume.

The committee reemphasizes its concern[13] that too much waste will be processed through the DDA if it is a standalone process. There are two principal reasons for this concern. First, as described in DOE's plans as of 2005 (when the committee's interim report was prepared), DDA would send large amounts of radioactive material to the Saltstone Vaults, orders of magnitude more than was originally envisioned. Second, because of the 26-month delay in operation of the Salt Waste Processing Facility, it is possible that additional radioactive material could be disposed in saltstone if DDA has to operate longer than previously expected. In its salt waste determination (DOE, 2006), DOE said it intends to put no more radioactive material in the saltstone than described in the draft salt waste determination (DOE-SRS, 2005b); but DOE has not yet proposed any solutions to the tank space problem arising from the 26-month delay. Thus, the committee could not evaluate the problem further. The committee reiterates, however, that DOE should seek alternatives to the DDA process, either by slowing waste inputs (slowing operations or gaining efficiencies) or by finding storage alternatives for the least hazardous of the tank wastes to free up storage space.

Point of Compliance

DOE's point of compliance (the location where compliance with performance objectives is determined) for its F Tank Farm is 1.6 kilometers (1 mile) away from the facility boundary, rather than the standard 100 meters (109 yards) away. When DOE uses a nonstandard point of compliance, it should state clearly the potential exposures closer to the disposal facility in case assumptions about human behavior and institutions do not turn out to be true. The selection of the point of compliance has both policy and technical dimensions. The committee believes that those technical dimensions should be stated clearly and prominently so that the policy decision is well informed.

Estimated Doses from the Predicted Waste Residuals in the F Tank Farm

In estimating the residual tank inventories for its performance assessment calculations for the F tank farm, DOE assumes that future efforts to clean out tanks will be much more effective than they were for most of the tanks that have already been cleaned out. The committee views this assumption as both optimistic and unsupported. Without a technical basis for the inventory estimates, the committee does not have confidence in the results of the performance assessment for the F Tank Farm.[14]

Hanford

The challenges DOE faces at Hanford are significant and varied, but given that the revised performance assessment for the single-shell tank farms at Hanford had not been issued by the end of 2005, the committee was unable to evaluate DOE's plans with respect to several elements of the charge from Congress. However, the committee has concerns about the technical performance and safety features of the bulk vitrification option for supplemental low-activity waste treat-

[11] The Department of Energy's Tanks Focus Area (TFA) was funded by DOE's Office of Science and Technology to provide technical assistance with issues related to tank wastes at the Savannah River Site, Hanford Site, Idaho National Laboratory, Fernald Site, and West Valley Demonstration Project.

[12] Compliant waste tanks meet the Resource Conservation and Recovery Act requirements for storage of hazardous waste. Noncompliant tanks do not meet those requirements (e.g., full secondary containment).

[13] See the committee's interim report (NRC, 2005a). The summary of that report can be found in Appendix E.

[14] No tank closures have been proposed yet for the H Tank Farm.

ment. Each site is pursuing different technologies for immobilizing its non-high-level tank waste—grout at the Savannah River Site, steam reforming at the Idaho site, and vitrification in the Waste Treatment Plant and bulk vitrification[15] at Hanford. The Hanford bulk vitrification process is currently less well developed technically than either the Savannah River Site saltstone or the Idaho National Laboratory steam reforming. Before selecting an immobilization technology at Hanford, DOE should sponsor a detailed, transparent, independent, technical review of bulk vitrification versus other options, focusing on process risks and uncertainties.

Idaho National Laboratory

Idaho faces smaller, simpler challenges than either the Savannah River or Hanford Sites in cleaning out the tanks. Less spent fuel was reprocessed at the site—all of it with the same process; most of the tank waste was calcined and resides as granular solids in bins; and what liquid waste was stored in the tanks did not separate substantially into sludge and salt because the waste was left in its acidic state. The Idaho National Laboratory is making good progress in dealing with its liquid wastes. However, it remains to be seen whether the solidified waste (the calcine) stored in bins will be as easy to remove as projected by site personnel.

CONGRESSIONAL CHARGE TO THE COMMITTEE

The committee's charge from Congress contains six specific topics. Each topic is presented below and is followed by the committee's response.

Topic 1: The "Department's understanding of the physical, chemical, and radiological characteristics of the waste referred to above, including an assessment of data uncertainties."

The committee believes that DOE has reached a point in its analysis of the physical, chemical, and radiological characteristics of the waste in the tanks where further understanding would not change its overall approach substantially. DOE's knowledge of the waste in the tanks is sufficient for waste retrieval. DOE needs to know the waste composition in greater detail for processing purposes and to confirm compliance with performance objectives, but this must be done after waste retrieval when mixing makes representative sampling of the retrieved waste possible and when samples of the tank heels can be taken. Even then, the waste composition need only be known sufficiently for reliable and efficient processing to take place. Some processing methodologies, such as steam reforming and grouting, do not require detailed feed characterization, whereas others, such as vitrification, may require greater knowledge and control of the waste characteristics. When these requirements are very stringent, it may be necessary to look for processing and immobilization technologies that accommodate a wider range of feed characteristics. The costs and the risks to workers, members of the public, and the environment if the processes should fail to perform acceptably have to be taken into account for each processing option. For different processes and different locations, the knowledge required may be different.

Topic 2: "Any actions additional to those contained in current plans that [DOE] should consider to ensure that the plans will comply with the performance objectives of Part 61 of Title 10, Code of Federal Regulations".

10 CFR 61.41 states: "Reasonable effort should be made to maintain releases of radioactivity in effluents to the general environment as low as is reasonably achievable." After DOE shows that its plans meet the dose limits, DOE should further demonstrate how its plans for waste retrieval and immobilization meet ALARA requirements to protect workers, the public, and the environment now and in the future. In Section 3116(a)(2) of the NDAA, a criterion for on-site disposal is that the waste "has had highly radioactive radionuclides removed to the maximum extent practical." The risks posed depend on the assumed location and time at which the performance criteria must be met. DOE has issued only a few documents detailing how it determined what amounts of material left in a tank would be acceptable. There is not sufficient information available to evaluate whether all of the components of importance to such decisions have been taken into account. However, it would be advantageous to have a common process that illustrates how risks under each option are likely to change over space and time, which would be useful in determining what wastes can be disposed on-site. An illustrative and extremely simplified example of such a process is given in Chapter X.

Topic 3: "The adequacy of the Department's plans for monitoring disposal sites and the surrounding environment to verify compliance with those performance objectives."

Monitoring is important in performance assessment model validation exercises and for early detection of failures. It also allows remedial actions to be taken at the earliest possible time, thereby minimizing human and environmental impact and cost. The committee judges that monitoring within the disposal facility is the most desirable location for the early detection of problems, followed by detection in the vadose zone, and finally by detection in the nearest aquifer. The committee's overall impressions are that the sites' monitoring programs are satisfactory at fulfilling their current goals, which in most cases are site characterization and operations monitoring to assess regulatory compliance. However, the committee believes that the DOE plans for

[15]Bulk vitrification is the lead candidate for the Hanford supplemental low activity waste process; however, the preferred technology has not yet been officially selected.

monitoring should go beyond the requirement of verifying compliance with performance objectives. DOE should start planning its post-closure monitoring programs so that provision for monitoring can be built into closure plans and designs.

Topics 4 and 5: "Existing technology alternatives to the current management plan for the waste streams mentioned above and, for each such alternative, an assessment of the cost, consequences for worker safety, and long-term consequences for environmental and human health"; and any "technology gaps that exist to effect improved efficiency in removal and treatment of waste from the tanks at the Hanford, Savannah River, and Idaho sites."

The committee had to operate within schedule constraints, which limited the extent to which it could evaluate technology alternatives. Such evaluation was particularly difficult because DOE was able to provide only limited information on cost, worker safety, and long-term human and environmental health consequences of technology alternatives for tank waste management. However, it is apparent that DOE should continue to adapt existing and develop new technologies for effective waste retrieval, with emphasis on tanks with obstructions and recalcitrant waste. Additionally, the committee recommends a targeted aggressive, collaborative research and development program on chemical cleaning of the tanks, mechanical waste retrieval, and tank filling materials for tank stabilization. The committee recommends support at approximately $50 million per year to focus on technologies that could become available in the near-term (within 10 years) in time to be implemented during the tank cleanup program.

Topic 6: "Any other matters that the committee considers appropriate and directly related to the subject matter of the study."

Following are issues that the committee believes are important, but either DOE's plans are not detailed enough at this time to make specific recommendations or the issues are independent of tank management plans:

- Remediation of pipelines, leaking underground pipes and interwall spaces in double-walled tanks, and other auxiliary equipment in the tank farms could be challenging, particularly at Hanford where there are about 100 plugged pipelines (see Chapter III).
- Although the Idaho National Laboratory should focus on tank wastes first, some consideration should be given to the calcine waste and bins and their disposition (see Chapter III).
- DOE needs regulatory approvals for the off-site disposal of some Hanford tank waste and Idaho sodium-bearing tank waste.
- The philosophy and methodology for post-closure monitoring needs to be developed and articulated.
- Attention should be paid to long-term stewardship, post-closure monitoring, and the meaning of institutional control "in perpetuity," and rigorous (within the limits of long-term prediction) planning for these activities should commence. A focus on these issues is important and would incorporate, but extend beyond, DOE's tank waste management plans (see Chapters VII and VIII).

1

Introduction

In the Ronald Reagan National Defense Authorization Act (NDAA) of 2005 (Section 3146 of Public Law 108-375), Congress directed the Department of Energy (DOE) to request a study from the National Academies[1] evaluating DOE's plans to manage radioactive waste streams from reprocessed spent fuel that

> exceed the concentration limits for Class C low-level waste as set out in Section 61.55 of Title 10, Code of Federal Regulations;[2,3] DOE plans to dispose of on the sites specified below rather than in a repository for spent nuclear fuel and high-level waste; and are stored in tanks at the Savannah River Site, South Carolina; Idaho National Engineering and Environmental Laboratory, Idaho; and the Hanford Reservation, Washington.

Congress asked the National Academies to assess the following:

1. DOE's knowledge of the physical, chemical, and radiological characteristics of the waste in the tanks;
2. Actions that DOE should consider to ensure that management plans comply with the performance objectives for land disposal facilities; [4]
3. DOE's monitoring plans to verify compliance with the aforementioned performance objectives;
4. Existing technology alternatives for waste management;
5. Technology gaps for waste retrieval and management; and
6. Any other matters that the committee considers appropriate and directly related to the subject matter of the study.

Task element (6) was reinforced by Representative John Spratt (5th District of South Carolina) and House Armed Services Committee staff, who presented the charge to the committee at its first meeting in March 2005 (see Appendix D). The congressman and staffers asked the committee to interpret the charge broadly to include any relevant matters of importance, with emphasis on any portion of the tank waste that would be disposed at the sites.[5] Congress asked for an interim report on the Savannah River Site within six months and a final report on all of the sites within twelve months from the beginning of the study (January 2005). The interim report was issued in August 2005 (NRC, 2005a). The summary[6] of that report is reproduced in Appendix E along with a section based on developments since the interim report and

[1]The operating arm of the National Academies, the National Research Council, appointed a committee to undertake this study under the auspices of the Nuclear and Radiation Studies Board. Biographical sketches of the committee members can be found in Appendix A.

[2]Through Part 61 of Title 10, Code of Federal Regulations (10 CFR 61) titled "Licensing Requirements for Land Disposal of Radioactive Waste," the U.S. Nuclear Regulatory Commission regulates near-surface disposal of commercial low-level waste. Subpart 10 CFR 61.55 classifies low-level radioactive waste as Class A, B, or C, according to the concentrations of key radionuclides in the waste. Class C waste must meet more rigorous waste form requirements to ensure stability and requires additional measures at the disposal facility to protect against inadvertent intrusion. The regulation states that low-level waste that exceeds Class C limits is not generally suitable for near-surface disposal.

[3]For the purpose of this study, the committee interprets the concentration criterion to apply to the waste streams stored in tanks prior to waste processing or immobilization.

[4]The performance objectives of land disposal facilities for radioactive waste are defined in 10 CFR 61 Subpart C: (1) protect the general public from environmental releases and make releases as low as reasonably achievable, (2) protect individuals from inadvertent intrusion, (3) protect individuals during operations, and (4) provide stability of the site after closure (see Appendix C).

[5]Specifically, Representative Spratt said "I thought it imperative that the scientific experts we were calling upon not be narrowly scoped by Congress, but rather have the authority and the latitude to look into matters unforeseen or unknown by Congress that may have a bearing on the subject."

[6]The committee's interim report is available online at *http://www.nap.edu/catalog/11415.html*.

feedback that DOE provided to the committee. This report is the final report addressing the statement of task in full.

The committee worked with DOE, the South Carolina Department of Health and Environmental Control (DHEC), the Idaho Department of Environmental Quality, the Washington State Department of Ecology, the Oregon Department of Energy, the U.S. Environmental Protection Agency (EPA) Region 10, the U.S. Nuclear Regulatory Commission (USNRC), Native American tribal nations, DOE's contractors, and others to obtain the information needed for the study. To this end, the committee obtained a large number of documents and held five public meetings to obtain information from experts and interested members of the public. DOE staff and contractors were responsive to the committee's requests; however, some data and analyses were not available to the committee (not yet collected, not yet calculated, or not yet made public[7]), and some plans had not yet been formulated or finalized by the time this report entered the National Research Council's report review process in late January 2006. Therefore, because the reports were not available for most of the tanks, the committee could not evaluate DOE's intentions for complying with the performance objectives of 10 CFR 61, as required in Section 3116 of the 2005 NDAA.

Section 3116, which is closely related to Section 3146 requesting this study, explicitly enables DOE to determine that some tank waste is from reprocessing of spent fuel is not high-level waste and can be disposed of on-site at the Savannah River Site and the Idaho National Laboratory, provided the requirements in that section are met.[8] Section 3116 provides new waste determination criteria at these two sites but not at the Hanford Site, where DOE Order 435.1 is still in effect (albeit under the threat of litigation, see Chapter VIII).[9]

[7]Under the Federal Advisory Committee Act Amendments of 1997 (Public Law 105-153), any document provided to the committee from outside the National Academies must be made available to the public, unless the document is exempt from disclosure under the Freedom of Information Act (Public Law 89-554) and its amendments. As a result, DOE and others could not provide any document that was undergoing security review, internal scientific review, or legal and policy review and was therefore not ready for public release. The information-gathering phase for the interim report lasted from March through October 2005. Some of the information that the committee had requested was not available because it did not exist, was still being calculated, or was under review.

[8]Before this law was signed, the full legal definition of high-level radioactive waste was waste that is "(A) the highly radioactive waste material resulting from the reprocessing of spent nuclear fuel, including liquid waste produced directly in reprocessing and any solid material derived from such liquid waste that contains fission products in sufficient concentrations; and (B) other highly radioactive material that the [Nuclear Regulatory] Commission, consistent with existing law, determines by rule to require permanent isolation" (U.S. Code, Title 42, Chapter 108, Nuclear Waste Policy, Section 10101). There is no particular concentration of radioactivity or dose limit associated with this definition.

[9]Order 435.1 is an internal DOE order governing the management of radioactive waste at DOE sites. Order 435.1 includes waste determination criteria for different types of waste, including high-level waste.

In this final report the committee evaluates and recommends actions that it believes could

1. Reduce the quantity of radioactive material left on-site;
2. Increase DOE's understanding of other factors that could reduce dose and risk—namely, the long-term performance of waste forms and other barriers to the release of radionuclides to the environment;
3. Improve DOE's analyses of the choices it must make as it moves to retrieve reprocessing wastes from the tanks at the Savannah River Site, Hanford, and Idaho National Laboratory and dispose of them under the provisions of Section 3116 and Order 435.1;
4. Improve the likelihood of processing and immobilizing these wastes in an efficient manner; and
5. Improve DOE's decision making through a risk-informed process.

The committee judges that these actions will increase DOE's ability to comply with the performance objectives in 10 CFR 61 and other applicable regulations, and will help DOE fulfill its requirement to take actions to make releases of radioactivity to the environment as low as reasonably achievable (ALARA), with economic and social considerations taken into account.

The material in this report builds on the interim report; the committee reviewed new material from the Savannah River Site and held meetings at the Idaho National Laboratory and the Hanford Site in Washington, which broadened the committee's perspective.

OBJECTIVES FOR THE TANK WASTES

The committee believes that as a starting point for analyzing options, developing plans, and making waste management decisions, it is useful to examine the life cycle of the wastes, identifying both their current conditions and the desired objectives, or "end states."[10] The choices and decisions that represent different paths from the current condition to the desired end states can then be delineated. In this section, the committee examines what the ideal objectives would be for the tank wastes and what real-world obstacles make those objectives difficult to achieve.

Ideal Objectives for the Tank Wastes

In an ideal world—that is, if waste retrieval and processing facilities worked perfectly and at an acceptable cost—the objectives would be the following:

[10]A previous National Research Council report addressed end states for DOE's Environmental Management sites with an application to Hanford tank waste (NRC, 1999a).

1. *Remove tank waste.* The first priority for most, if not all, of the interested parties is to retrieve the waste from the tanks,[11] particularly tanks that lack full secondary containment. If all of the waste could be retrieved, tank closure would be a minor concern because there would be no residual radiological hazards. No real waste retrieval system, however, will retrieve all of the waste. It is also impractical to exhume the tanks and dispose of them off-site.[12]
2. *Separate the radioactive constituents from the salt solutions and bulk chemicals from the sludge.* Separations are carried out primarily to reduce the volume of high-activity waste that must be immobilized, stored, and ultimately shipped to a deep geologic repository for disposal. The radionuclides constitute only a small portion of the waste volume. However, in any real separation system, complete separation of the radioactive components is not possible, which results in a low-activity waste that contains a small fraction of the radionuclides but most of the chemicals in the original tank waste. The low-activity waste would be disposed in a manner appropriate for the lower hazard it poses.
3. *Immobilize radioactive waste for disposal.* The more hazardous radioactive wastes must be immobilized in a manner suitable for acceptance in a deep geologic repository (for high-level waste or transuranic waste) and the less hazardous bulk wastes must be immobilized in a form suitable for land disposal in a manner that prevents unacceptable accidental direct exposures and inhibits leaching of contaminants. The reality is that there is no perfect immobilization matrix for waste; some leaching with time is inevitable.
4. *Minimize operational hazards.* The ideal objective is to eliminate operational hazards. However, some hazards always remain due to the impracticality of completely eliminating unexpected conditions, accidents, and human error. Operational hazards apply mainly to workers. Worker safety is the top priority in the near term before radionuclides and hazardous chemicals move into the accessible environment. There are also operational hazards to the public, for example, during waste transportation from the sites to a repository. Regulations require DOE to keep exposures to the public and workers as low as reasonably achievable, with economic and social considerations taken into account.
5. *Minimize residual hazards to reduce long-term maintenance.* The ideal objective, again, is to eliminate the hazard of waste staying on-site, whether immobilized waste residues in the tanks (or even the tank structure, piping, and other internals) or immobilized low-activity waste. In this case, the wastes would be left in a condition that would not require institutional controls or long-term monitoring and maintenance to prevent unacceptable exposure of workers, the public, and the environment. Because it is impossible to clean the tanks perfectly, completely separate the highly hazardous radionuclides, and totally immobilize the waste, there will always be some residual hazard left on-site and therefore long-term maintenance is likely to be needed.

Real-World Challenges: Difficulties in Achieving Ideal Objectives

While DOE and others may strive toward these ideal outcomes, reality makes it quite difficult to achieve them, as discussed above. Examination of the disposition options, including technological capabilities, relative risks, costs, and other trade-offs, is necessary to select an appropriate plan of action. Chapter II provides background information that illustrates the challenges and difficulties at each site. Key aspects of the difficulties are examined in later chapters, and recommended strategies for addressing those challenges are offered.

[11]Throughout this report, "tank waste removal" refers to removal of waste from the tank and secondary containment system, if applicable.

[12]Exhuming and disposing these extremely large and contaminated waste tanks off-site would pose a tenfold increase in risks to workers, according to DOE estimates. DOE's former assistant secretary for environmental management, Jesse Roberson, estimated in 2004 that "it would cost as much as $50 billion more over the life cycle of the department's cleanup program and extend that life cycle by decades to have to process all of our tank waste as high-level waste for disposal in a geologic repository, including exhuming the tanks themselves, cutting them up and packaging them for disposal" (Roberson, 2004).

II

Background and Overview of Current Situation

In this chapter the committee provides factual background information in support of its analyses, findings, and recommendations presented in this and subsequent chapters.

OVERALL APPROACH

The Department of Energy's (DOE's) overall approach for managing its tank wastes is the following: To the maximum extent practical, retrieve the waste from the tanks (and bins in Idaho, see below); separate (process) the recovered waste into high- and low-activity fractions; and dispose of both remaining tank heels and recovered low-activity waste on-site in a manner that protects human health and the environment. Figure II-1 is a simplified illustration of such an approach. The details of this approach are discussed in the following chapters: waste retrieval in Chapter III, waste processing plans in Chapter IV, and tank closure plans in Chapter V.

THE THREE SITES

This section provides background information on the Savannah River Site, the Hanford Site, and the Idaho National Laboratory and their tank waste.

Savannah River Site

The Savannah River Site has 51 underground tanks[1] that are used for storing 138,000 m^3 (36.4 million gallons)[2] of hazardous and radioactive waste.

The Savannah River Site started generating tank waste in 1954 when a large chemical processing facility, called the F Canyon, was brought into service to separate uranium and plutonium from irradiated targets and spent nuclear fuel from on-site reactors to support the U.S. nuclear weapons program. A second chemical processing facility, the H Canyon, was brought on line in 1955.

Each canyon facility piped highly radioactive liquid waste from the chemical processing operations to a set of tanks located in its area: the F Area Tank Farm has 22 tanks and the H Area Tank Farm has 29 tanks.[3] The tanks range in size from about 2,850 to 4,900 m^3 (750,000 to 1.3 million gallons). They are vertical cylinders, approximately 23 to 26 m (75 to 85 feet) in inner diameter and 7.5 to 11 m (24.5 to 35 feet) in height from the inner tank floor to the ceiling. They are buried at a shallow depth (1 to 3 m below the land surface), mostly above the water table, although four tanks in the H Area Tank Farm are nearly submerged in the saturated zone (that is, the water table reaches nearly to the top of the tanks).

Access to the interior of the tanks is gained through portals, called risers, which rise from the top of the tank through the ground cover to the land surface. The number of risers in each tank ranges from 7 to 40, and the diameters of most of the apertures range from 58 to 107 cm (23 to 42 inches) depending on tank type. Some risers are larger: The center riser of a Type IV tank is approximately 2.7 m (9 feet) (Fogle, 2002). Annulus ports on Type III tanks are as small as 20 cm (8 inches, see below).

Most of the tanks have a carbon steel inner wall and a cylindrical outer vault wall constructed of concrete, with a space between them called the "annulus."[4] Tanks that have

[1]Two tanks (Tanks 17 and 20) were filled with grout and closed in 1997, and three tanks (16, 18, and 19) were cleaned and taken out of service, so there are currently 46 tanks in service. DOE plans to close Tanks 18 and 19 next.

[2]This report presents quantities in SI units and the equivalent value in English units in parentheses, e.g., 11 liters (2.9 gallons). The only exception is radioactivity, which is reported in curies first with the quantity in becquerels after, because becquerels are very rarely used in discussion of tank wastes.

[3]A map of the Savannah River Site and the General Separations Area where the tanks are located can be found in Appendix J.

[4]The gap between the primary and secondary containment is about 60 cm (2 feet).

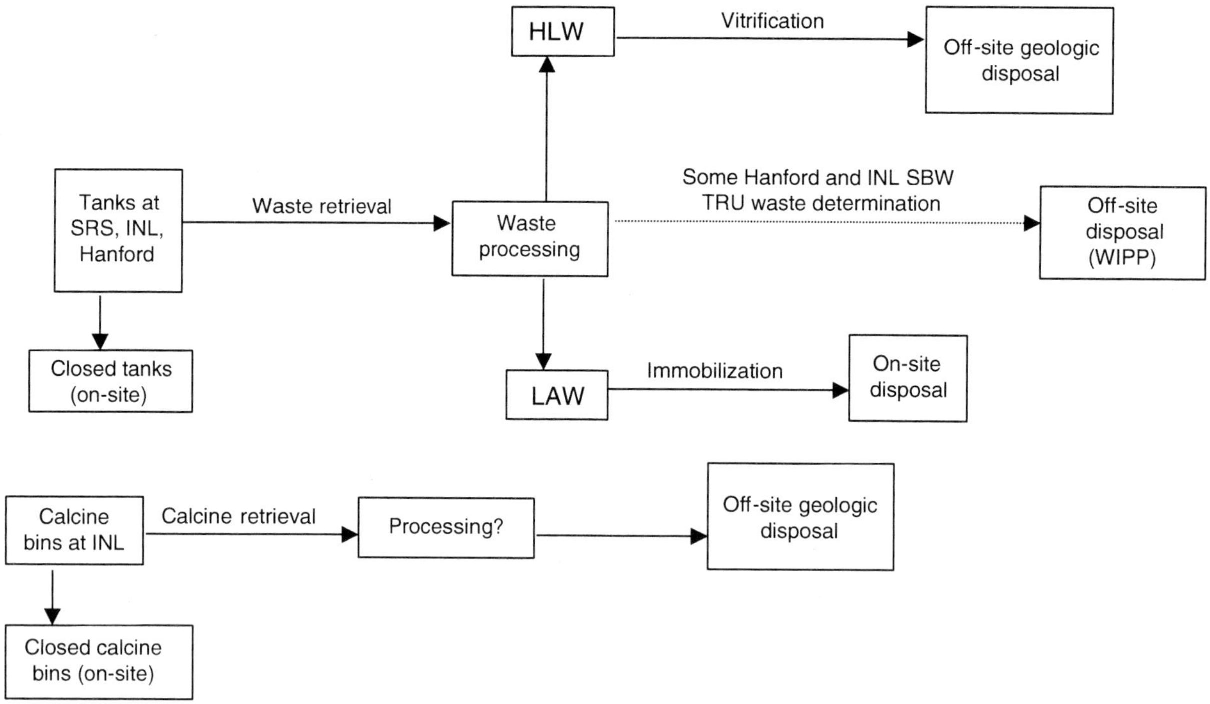

FIGURE II-1 Simplified flowsheet of DOE's waste management plan for its tank and bin wastes at the Hanford, Idaho, and Savannah River sites. The waste processing box corresponds to the Waste Treatment and Immobilization Plant and supplemental treatment plant for Hanford; the steam reforming plant for the Idaho sodium-bearing waste; and sludge washing, and the Salt Waste Processing Facility at the Savannah River Site. It is not clear right now whether the Idaho calcine will be accepted in a deep geologic disposal as is or if it requires further processing. Some Hanford tank waste and the steam-reformed waste from the sodium-bearing waste at Idaho will undergo a waste determination to declare it defense transuranic (TRU) waste and ship it to the Waste Isolation Pilot Plant (WIPP). NOTE: INL = Idaho National Laboratory; HLW = high-level waste; LAW = low-activity waste; SBW = sodium-bearing waste; WTP = Hanford Waste Treatment Plant

a metal liner on the outer wall are said to have a "secondary containment" (i.e., a tank inside a tank). If the outer liner rises only partway up the outer wall, it provides only partial secondary containment. Eight of the tanks have no annulus or secondary containment (Type IV tanks), 16 have partial secondary containment (Types I and II tanks), and 27 have full secondary containment (Types III and IIIA tanks).

Figure II-2 illustrates the four general tank types. Only the Type III and IIIA tanks with full secondary containment are considered "compliant tanks" under the site's Federal Facility Compliance Agreement and Consent Order (Federal Facility Agreement), the agreement regulating waste under the Resource Conservation and Recovery Act (RCRA) requirements for wastes stored in tanks (40 CFR 264.193 (b)). The "noncompliant" tanks are generally past their 30-year design life, and many (13, at last report; DOE-SRS, 2005c) have a history of cracks or leakage (either from the tank into the annular secondary region or from the surrounding media into the tank or annulus),[5] although only one tank is believed to have leaked a small quantity of waste to the environment. Waste levels in the tanks have been lowered below the location of known leaks, and at present DOE believes that there are no active leaks except in Tank 5 from which waste is currently being retrieved.

All but the Type IV tanks contain dense networks of vertical and horizontal "cooling coils," pipes that circulate cooling water. The cooling water removes heat produced from radioactive decay in the waste.

Savannah River Site Tanks Inventories

Based on information provided by site personnel, the global inventory of chemicals and radionuclides in the tank farms is reasonably well understood. The information is based on analytical data, reactor fuel burnup and discharge records, reprocessing plant processes, flowsheets, and records of chemical purchases and operations.

Waste from the canyons contains acids and other chemicals used in the separation processes, chemicals added to neutralize and alkalinize the waste, and radionuclides (fission products, such as cesium-137, and actinides, such as neptunium-237) not separated during recovery of plutonium

[5]Leaks are detected by visual inspection or by conductivity probes in the annulus.

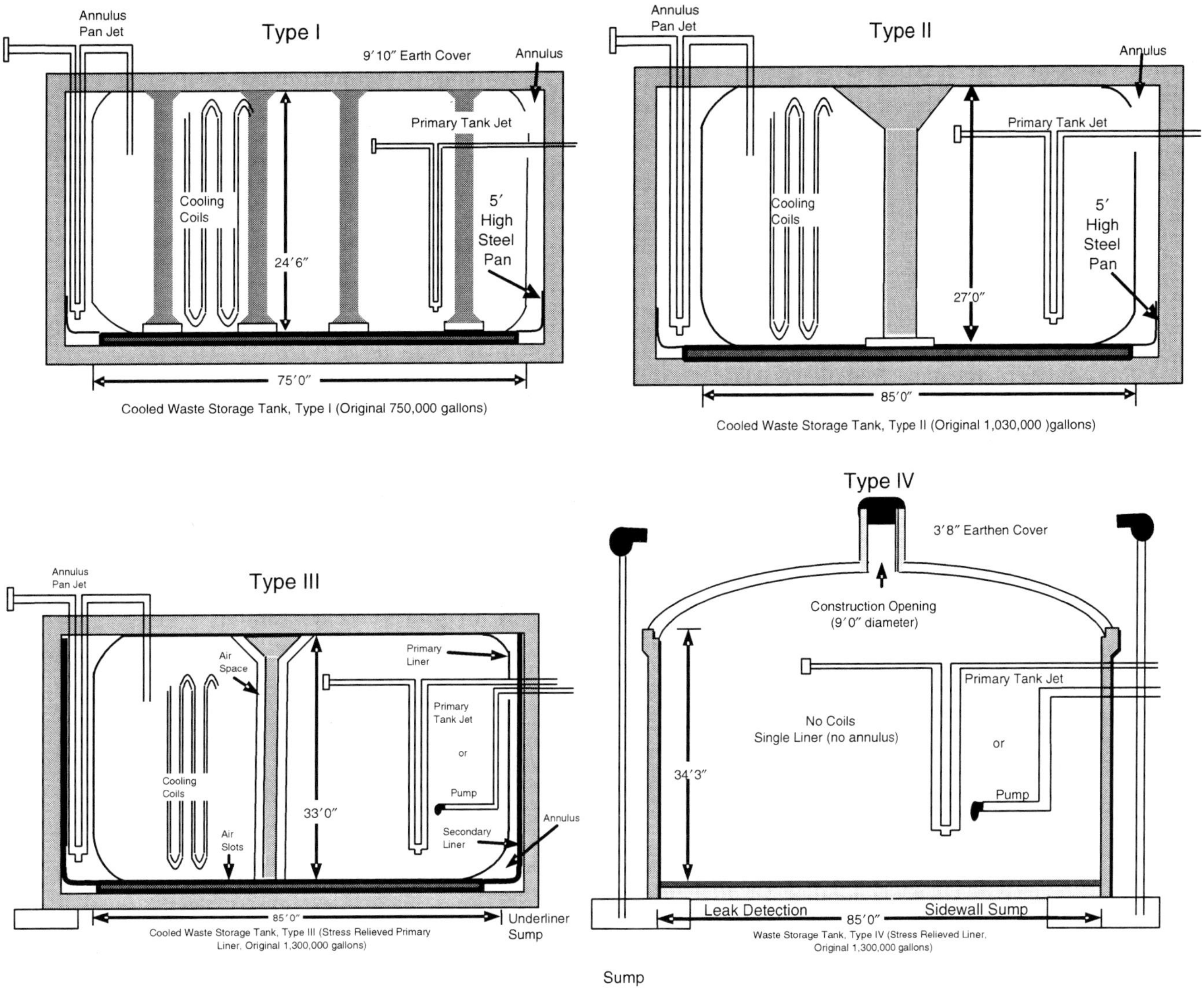

FIGURE II-2 Diagrams of tank types at the Savannah River Site (not drawn to scale). There are twelve Type I tanks, four Type II tanks, twenty-seven Type III (and IIIA) tanks, and eight Type IV tanks in the tank farms. Risers are not depicted in most of these diagrams but are present on each tank. SOURCE: Adapted from Mahoney, 2005.

and uranium. To prevent corrosion of the carbon steel tanks, sodium hydroxide was added to neutralize the acid and make the waste alkaline before it was pumped to the tanks. This caused metals and most radionuclides to precipitate as an insoluble sludge,[6] which settled to the bottom of the tanks (see Figure II-3). The liquid remainder, or unconcentrated supernate, contains soluble salts and is referred to as a salt solution. If concentrated by evaporation, much of the salts initially in solution will crystallize to form a solid saltcake. Thus, the wastes in the tanks exist mainly in three physical forms: sludge, supernatant liquid ("supernate"), and saltcake. Together, the supernate and saltcake are referred to as salt waste.

To conserve tank space, most of the salt solutions have been processed through an evaporator (a heated tank that evaporates water from waste) to produce saltcake, leaving relatively small volumes of concentrated supernate solution. The total estimated radioactivity in each physical form is shown in Figure II-4 and Table II-1, which also lists the radioactivities of other wastes on the site. Further details are provided below.

The supernate contains more than 90 percent of the inventory of soluble radioactive species, mainly cesium-137.

[6]The terms "insoluble" and "soluble" are used here to describe chemical species that exist preferentially in the solid phase or the liquid phase, respectively, in the larger medium (the waste in a tank). No species will exist exclusively in one phase. In the cases discussed here, however, all but a very small fraction of the chemical mentioned exists in the preferred phase in that medium. Also, the sludge entrains some soluble radioactive species.

FIGURE II-3 Photograph of a Savannah River Site tank sludge sample. SOURCE: Caldwell, 2005a.

The saltcake is a solid material composed of more than 99 percent salts, such as sodium nitrate, that contains concentrations of soluble and insoluble radioactive constituents lower by approximately a factor of 10 to 20 than what is in the sludge. The waste in the tanks (see Figure II-4) contains approximately 426 million curies (MCi; 1.58×10^{19} becquerels) of radioactivity; approximately half of the radioactivity is in the sludge and half in the salt waste. Most of the volume is in the salt waste, approximately 128,000 m^3 (33.8 million gallons), whereas the sludge represents approximately 9,800 m^3 (2.6 million gallons).

More than 95 percent of the radioactivity in the salt waste comes from cesium-137 (and its short-lived decay product, barium-137^{m}) and strontium-90 (and its short-lived decay product, yttrium-90). Both the cesium and the strontium isotopes have half-lives of approximately 30 years. The cesium poses a particular hazard for people working near the waste because it emits penetrating radiation (gamma rays).

Other radioactive constituents in the waste are of concern for other reasons: DOE has concluded that carbon-14, selenium-79, technetium-99, iodine-129, tin-126, and neptunium-237 dominate the long-term risk to the public from disposed waste because of their long half-lives and their mobility in the environment (Cook, 2005). The actinide isotopes, including isotopes of plutonium and americium, decay into a series of other radioactive substances (together referred to as a decay chain) and also constitute long-term hazards, particularly for inadvertent intruders.

Although the global tank farm inventory is reasonably understood, individual tank inventories have greater uncertainties (Reboul and Hill, 2005). Sampling individual tanks is difficult because the waste is highly radioactive, and furthermore it is heterogeneous and segregated into different forms and compounds in different portions of the tanks.

DOE reported to the committee (Hill, 2006) on three sampling studies:[7] a 2002 statistical comparison of slurried sludge samples from eight individual tanks versus characterization predictions for those tanks, a review of seven supernate samples, and a review of six short (3 feet or 1 meter) salt core samples. These reviews found the following:

- For significant radionuclides,[8] the predicted inventory was on average a factor of 1.6 greater than the measured inventory and 95 percent of the predicted inventories were within a factor of 2.5 less than predicted and a factor of 8 more than predicted. All predictions were between a factor of 10 of the measured value.
- For all elements considered to be significant (i.e., at least 1 weight percent of the total dried solids), the predicted concentration was within a factor of 1.12 less than the measured concentration and 95 percent of the predicted inventories were between a factor of 4 less than predicted and a factor of 2.5 more than predicted. All of the differences were within a factor of 10.

DOE believes that uncertainties in the saltcake radionuclide constituents (i.e., all nuclides that are characterized for saltcake predictions) are within a factor of 2. DOE describes the error in predictions of minor chemical constituent concentrations (those for sodium phosphate, sodium chloride, sodium fluoride, sulphate, etc.) in the saltcake as plus or minus 50 percent. Uncertainties in supernate radio-

[7]These studies were not yet public when the committee completed its report, so the reviews were only described to the committee.

[8]At the time of the sampling study DOE described, significant radionuclides were those that contributed to inhalation dose potential.

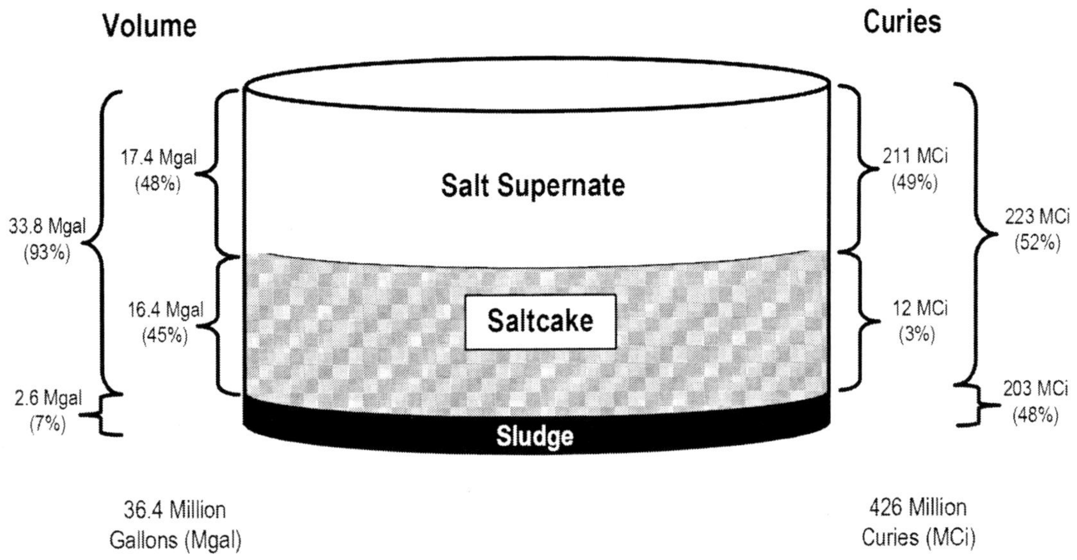

FIGURE II-4 Aggregated volume and radioactivity distributions among the tank waste phases in all tanks at the Savannah River Site as of December 2004. SOURCE: DOE-SRS, 2005b.

TABLE II-1 Inventory of Radioactive Waste by Type at the Savannah River Site.

Type of Waste	Volume (m^3)	Radioactivity (Ci)
Total waste in the tanks comprising[a]	138,000	426 million
Sludge`	9,800	203 million
Saltcake	62,000	12 million
Supernate	66,000	211 million
Vitrified high-level waste[b]	1,500	10 million
Stored transuranic waste[c,d]	11,000	490,000
Buried transuranic-contaminated waste and soil[c]	4,500	18,500
Low-level waste stored[c]	15,276	1.3 million[d]
Low-level radioactive waste in disposal cells[c]	698,000	11 million[e]
Saltstone as of 2005	25,000[f]	225[f]
Saltstone (DOE projected)	410,000[a]	3-5 million[a]
E Area vaults	117,000[g]	10 million[g]
Old Burial Ground	Unknown[h]	4.5 million[h]
Tanks at closure (DOE projected)	140[i]	0.72 million[i]
TOTAL	> 867,000	446.6 million

NOTE: Shaded area lists wastes that are expected to remain on-site. The data are from different sources, are measured or estimated at different times, and did not indicate quantified uncertainties. This table does not include spent fuel from research reactors. 1 Ci = 3.7×10^{10} Bq.

[a] DOE-SRS, 2005b.
[b] As of January 31, 2006, 2044 canisters containing approximately 0.74 m^3 per can.
[c] DOE, 2001b.
[d] Assumes same concentration as E Area Vaults (85 Ci/m^3).
[e] Not decay corrected, hence an overestimate.
[f] WSRC, 2004a.
[g] E Area Vault Waste Information Tracking System (WITS) provided by DOE (Clark, 2005a).
[h] WSRC, 1997.
[i] Based on assumed residuals in the tanks (DOE-SRS, 2002) and average concentrations (Buice et al., 2005). See Chapter VI of this report for discussion of this inventory.

nuclide predictions for cesium-137, technetium-99, and iodine-129 are within 10 percent, other fission products are within a factor of 2, and actinides are within a factor of 10. DOE believes that predictions of minor chemical constituent concentrations in the supernate are within 50 percent (plus or minus 50 percent).

The low-activity waste disposed as saltstone will be characterized through sample analysis, and therefore, the uncertainty in this waste stream inventory will be small. The impetuses for sampling waste that is retrieved from the tanks are the Salt Waste Processing Facility and the Defense Waste Processing Facility (DWPF) feed delivery specifications. The Savannah River Site also plans to sample waste tank heels to demonstrate how the tank heels meet the performance objectives in 10 CFR 61. Detailed discussion of tank heel sampling can be found in the performance objective demonstration document for Tanks 18 and 19 (Buice et al., 2005).

Savannah River Site Waste Processing

DOE's plan to manage the waste retrieved from the tanks is to separate the radioactive from the nonradioactive components, the latter of which make up nearly the entirety of the waste volume. This processing generates two waste streams: (1) a high-activity, low-volume waste stream, which will be immobilized and disposed off-site in a deep geologic repository, and (2) a low-activity waste stream, which is to be disposed in near-surface vaults on-site. Figure II-5 illustrates the waste flows that DOE has described for tank wastes at the Savannah River Site. The wastes planned for repository disposition are not the subject of this study because they are not planned for on-site disposal. They are included here because the management of tank wastes must be considered as a system of interconnected parts.

Sludge Processing

For nearly 10 years, DOE has been retrieving sludge from tanks at the Savannah River Site for immobilization in glass. After retrieval from the tank, the sludge is transferred to a dedicated waste tank where it is "washed" to remove soluble salt constituents that will interfere with the glass-forming process and to reduce the volume of material that is sent to the DWPF for vitrification into logs of waste glass. The logs are to be disposed off-site in a high-level waste repository. The wash water and a low-activity liquid waste stream from the DWPF are sent back to the tanks (see Figure II-5).

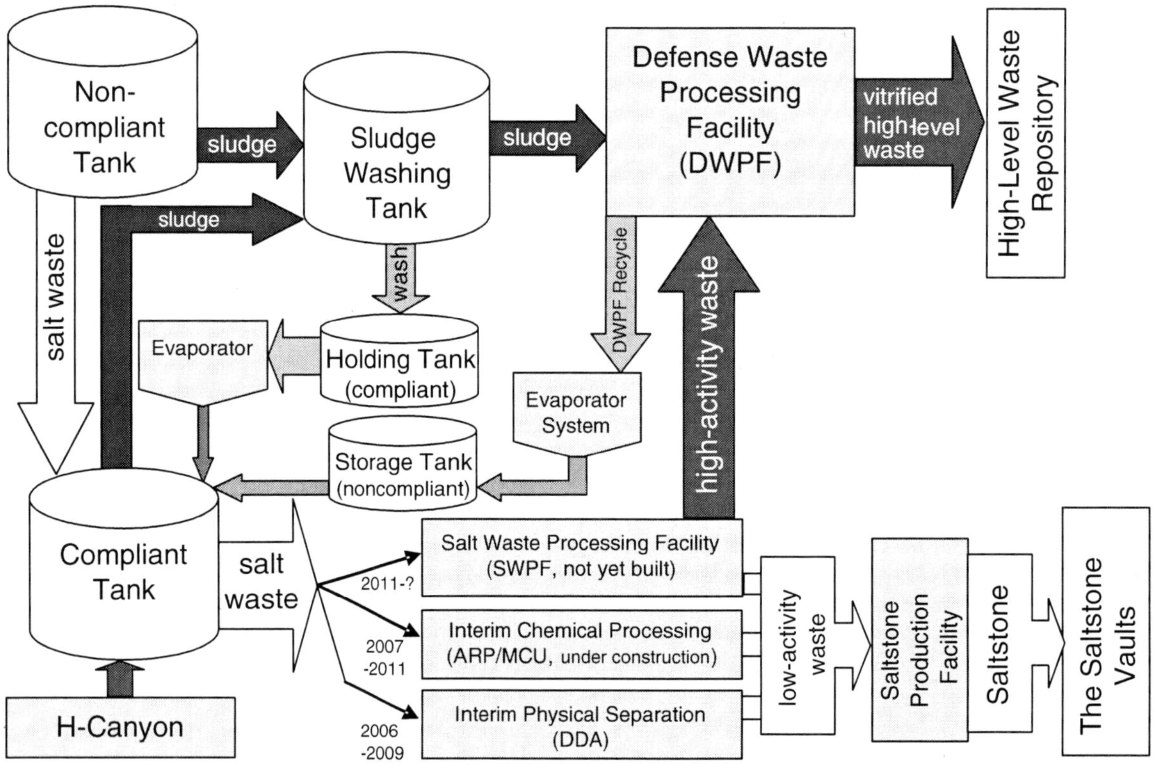

FIGURE II-5 Waste flows in the Savannah River Site waste management plans. Note that the sizes do not necessarily scale with the sizes of the waste flows. Noncompliant tanks are those that do not have full secondary containment (i.e., a tank inside of a tank). NOTE: ARP = Actinide Removal Process; DDA = deliquification, dissolution, and adjustment; MCU = Modular Caustic Side Solvent Extraction Unit.

Salt Waste Processing

DOE indicated to the committee that the Savannah River Site is facing a "tank space crisis" because of net waste inputs from current waste processing and waste removal operations. To ensure that sludge removal from noncompliant tanks continues apace and DWPF continues to operate at full capacity DOE is proposing to begin processing salt waste as soon as possible. DOE is still developing facilities to process the salt waste (supernate and saltcake) at the Savannah River Site. Three progressively more sophisticated and effective separation processes are to be brought into service for processing different batches of salt wastes: DOE proposes to use two "interim" processes (described in Chapter IV) for what it calls "low-activity salt," that is, salt waste that contains what DOE considers to be "low concentrations of radionuclides," relative to the average until the Salt Waste Processing Facility (SWPF) begins operations. The SWPF was scheduled to begin operation in 2009; however, DOE recently announced an estimated 26-month delay in startup operations because of seismic concerns in the building design (Terhune and Kasper, 2005).

When the SWPF comes online, it will process the majority of the salt waste, "augmented as necessary by ARP" (DOE, 2006). Tank wastes are to be processed to concentrate the radionuclides into a high-activity waste stream that will be vitrified at the DWPF. The other separated fraction, consisting mainly of the nonradioactive salts and other constituents with low concentrations of radionuclides that make up the less contaminated, low-activity waste stream, is to be immobilized in the Saltstone Production Facility—an operation that mixes liquid waste with grout[9] to create a waste form referred to as saltstone, which is disposed on-site as a monolith in concrete vaults. Until now, the Saltstone Production Facility has handled very low activity waste. The higher radioactivity anticipated in the liquid waste that DOE plans to send to the facility prior to SWPF startup has required DOE to reconfigure the equipment and facility as well as add shielding in certain areas.

Tank Closure

After waste is retrieved from a tank, DOE plans to operationally "close" the tank, i.e., fill the tank with tailored layers of grout, and sever and seal external penetrations. At some point in the future, groups of tanks will be formally closed and engineered barriers (e.g., a cap) will be emplaced. The plans for waste retrieval and tank closure are discussed in Chapters III and V, respectively.

[9]Except where otherwise indicated, the term "grout" is used here to mean a cementitious material used for waste immobilization or tank fill; see Chapter V.

The Hanford Site

The Hanford Site was the world's first plutonium production factory. It was designed, constructed, and operated by E.I. du Pont de Nemours and Company to provide the Manhattan Project with material for the cores of some of the first nuclear weapons.[10] The Hanford Site occupies 1,517 km^2 (585 square miles) of land on the Columbia River in south central Washington State. The plutonium production reactors were built in the "100 Area," widely spaced along the Columbia River so that the reactors could use river water as coolant. The chemical processing plants were built 10-20 km south of the river in the middle of the site, called the "200 Area," or the Central Plateau (see Figure II-6).

Nine plutonium production reactors were built and operated at Hanford, and all of them have been shut down. When the production facilities were operational, irradiated reactor fuel was transported by rail from the 100 Area reactors to chemical separation plants in the 200 Area. Five chemical reprocessing facilities (T-Plant, B-Plant, U-Plant, REDOX Plant, and PUREX Plant) and a plutonium finishing plant operated over the history of the site. The waste from these facilities was stored in large underground tanks, and during some of the early years of operation, lower-activity (compared to tank waste) liquid waste was discharged into the ground. The last reprocessing facility ceased operations in 1990.

The Hanford Site has 149 single-shell tanks and 28 double-shell tanks (Figure II-7). Tanks are grouped into 12 single-shell tank farms and 6 double-shell tank farms in the 200 East and 200 West Areas of the site. The tanks are interconnected by underground pipes and served the five chemical processing facilities mentioned above. The tank farms also have ancillary equipment used to divert and direct waste within each tank farm, such as valve boxes and pump pits, and between tank farms.

The 149 single-shell tanks were constructed between 1943 and 1964. These are vertical cylindrical structures that range in size from approximately 200 m^3 to nearly 3,800 m^3 (55,000 to 1 million gallons)—133 of the 149 are 2,000 m^3 (500,000 gallons) or larger—and were constructed of a concrete shell lined with a single layer of carbon steel. A typical tank is 23 m (75 feet) in diameter and 9 to 16 m tall (30 to 54 feet). The top dome is unlined concrete with 2 to 3 m (6 to 10 feet) of earthen cover (Elmore and Henderson, 2001a). Most of the tank bottoms are slightly concave (bowl shaped, with the low point in the center). Access to the tank interiors is achieved through risers that range from 10 cm to 1.1 m (4 to 42 inches) in diameter (Elmore and Henderson, 2001a). The number of risers in a single-shell tank ranges from 9 to more than 20.

[10]Du Pont turned operation of Hanford over to General Electric in 1946. Subsequently, du Pont became involved in the design and construction of the facilities at the Savannah River Site in 1950, and it continued to operate that site until 1989.

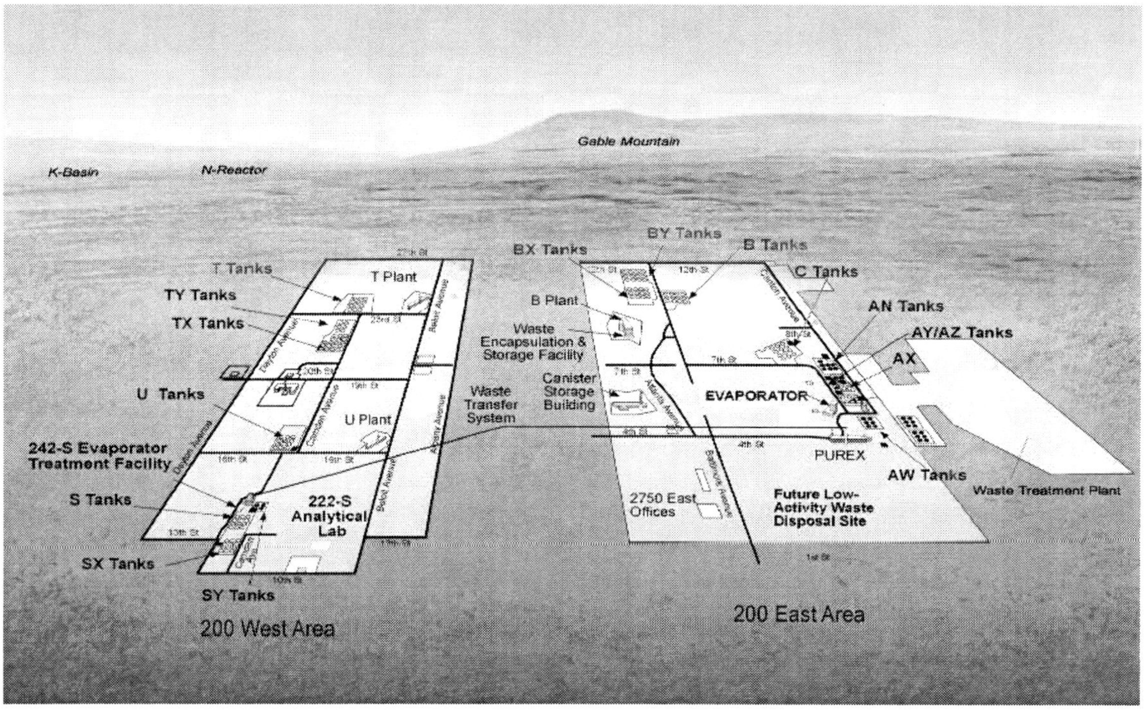

FIGURE II-6 Map of the Hanford Site indicating the areas related to plutonium production and storage of high-level radioactive waste. NOTE: Figure does not include every facility in the 200 Area. SOURCE: Mann, 2005.

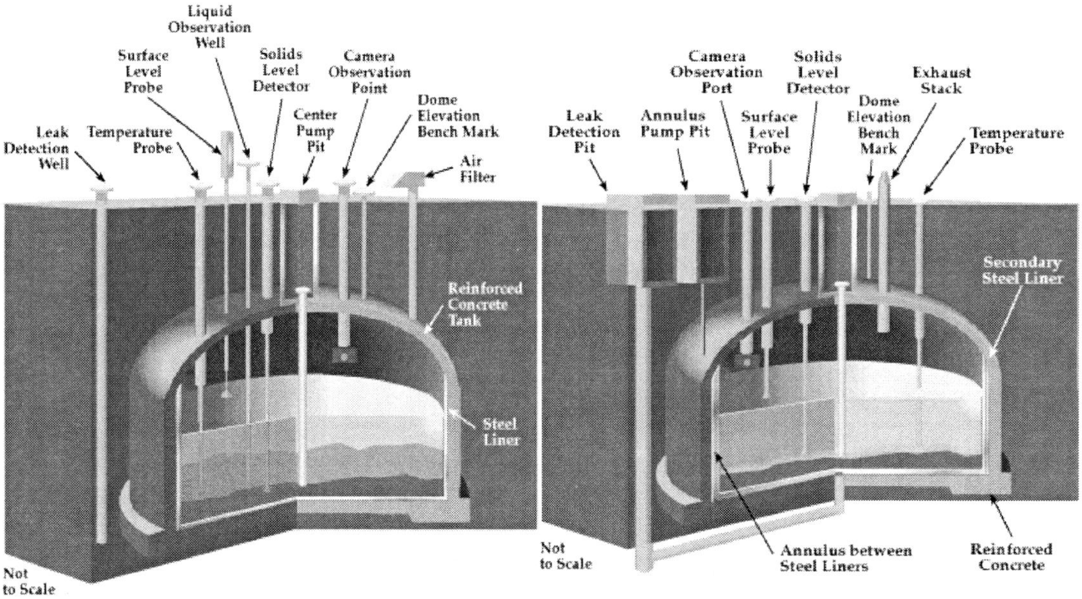

FIGURE II-7 Hanford single-shell tank (on the left) and double-shell tank diagram (on the right). SOURCE: *www.Hanford.org*.

Many of the tank farms do not have evaporators, and early volume reductions were achieved by allowing the waste to boil in the tanks (see Footnote 9). None of the tanks has cooling coils. None of the welds on the carbon steel liners of the single-shell tanks was stress relieved. The combination of thermal stresses and exposure to hot, aggressive solutions has resulted in stress-corrosion cracking at the welds of some tanks. All single-shell tanks are regarded as beyond their design lives and sixty-seven of the single-shell tanks have leaked or are assumed to have leaked about 3,800 m^3 (1 Mgal) of waste into the environment (see below).

Starting in 1968, the site built 28 double-shell tanks. All of the double-shell tanks are similar in size and design to the largest of the single-shell tanks, but they enclose the waste entirely in a double layer of carbon steel separated by an annular space to collect and monitor leaks. The last was built in 1986. All double-shell tanks were stress-relieved during construction to substantially reduce the likelihood of stress-corrosion cracking. None of the double-shell tanks has leaked.

Hanford Tank Inventories

The form of Hanford tank wastes is similar to that at the Savannah River Site: highly alkaline waste in supernatant, saltcake, and sludge phases. The sodium hydroxide and sodium nitrite used for corrosion control in the tanks formed sodium nitrate cakes and hydrated oxides of radionuclides and other chemicals in the waste, creating a sludge on the floor of the tanks. However, Hanford dealt with a greater variety of fuels and at different times used more chemical processes than did the Savannah River Site, which is reflected in the diversity of tank waste compositions at Hanford. Table II-2 lists the quantities of waste stored, disposed, or discharged to the environment at the Hanford Site.

Beginning in the late 1960s, cesium was separated from the supernatant liquid in waste tanks and strontium was removed from the sludge that was dissolved and processed. These separations, which removed approximately 90 percent of the strontium and cesium from the processed waste, produced intensely radioactive halide salts, now stored on-site in capsules (see Sidebar II-1).[11]

Hanford processed 97,000 metric tons of irradiated uranium. Between 1944 and 1980, approximately 700,000 m^3 (185 million gallons) of liquid radioactive waste was pumped into 149 single-shell tanks. Releases from the tanks and the piping system were first reported in 1956 and 67 tanks now are estimated to have released between 2,200 and 3,800 m^3 (580,000 and 1.0 million gallons) of tank waste into the ground (Honeyman, 2005). Prompted by leaks in some of the single-shell tanks, DOE pumped the free and drainable liquids from those tanks into double-shell tanks and implemented a remedial investigation program to determine the nature and extent of past leaks and various interim corrective measures to reduce groundwater impacts. Now, approximately 121,000 m^3 (about 32 million gallons) of saltcake, sludge, and interstitial liquid waste remains in the single-shell tanks (Honeyman, 2005). The site inventory in the 177 Hanford tanks today consists of approximately 204,000 m^3 (54 million gallons) of radioactive waste containing 193 MCi (7.14 EBq) of radioactivity. In 1989, the defense-related plutonium production mission at Hanford ended and all production reactors and processing plants were shut down. Most of the work at the site now supports the mission of managing the waste and environmental problems at the site.

The 204,000 m^3 (54 million gallons) of waste is what remains in the tanks from the roughly 2 million cubic meters (525 million gallons) of tank wastes generated between 1944 and 1988. The balance of the waste was evaporated (71.3 percent),[12] disposed to the ground after some radionuclide removal (28.5 percent), or leaked directly to the ground (about 0.25 percent). The saltcake and sludge in the single shell-tanks contain a little over 98 MCi (3.6×10^{18} Bq) of radioactivity. The double-shell tanks have about 95 MCi (3.5×10^{18} Bq) contained in waste consisting mostly (80 percent) of liquids, but also of sludges and salts (Wiegman, 2004; Honeyman, 2005). A visual inventory of the radioactivity in Hanford tank waste is shown in Figure II-8.

Aside from concerns about leaks, safety concerns related to the tanks arose as a result of hypothesized and observed chemical reactions, excessive heating, and the possibility of nuclear reactions (criticality) in some of the waste. Of greatest concern was the observed buildup of hydrogen gas generated by chemical and radiolytic reactions in one tank (later found in others). The potential for a flammable mixture to be ignited prompted DOE to install a mixing pump in Tank SY-101, which is located in the 200 West Area, to prevent local buildup of flammable concentrations and to institute a set of operational controls utilizing flammable-gas monitors. Heat generation due to radioactive decay and

[11]While DOE considers the cesium and strontium capsules to be nuclear materials rather than tank wastes, the committee has examined them because they are highly radioactive materials extracted from the tank wastes and DOE says it plans either to dispose of them on-site or to combine them with the high-activity waste stream to be vitrified and sent to geologic disposal. In essence, DOE faces the same decision about the capsules that it faces when deciding what to do with radioactive material separated in the Waste Treatment and Immobilization Plant. The main difference is when the separations were carried out.

[12]Different techniques have been used to reduce the volume of wastes through evaporation. Early techniques included allowing the wastes to self-boil because of the decay heat they generated. Temperatures in many of the tanks routinely were above 150°C (300°F), and one tank got above 310°C (590°F). Other early techniques included in-tank evaporation, either by inserting an electric heater into the waste or by circulating hot air into the tanks. Large-scale evaporation began in the 1970s by operating Evaporator-Crystallizers in the 200 West and 200 East Areas (Gephart, 2003).

TABLE II-2 Inventory of Radioactive Waste by Type at the Hanford Site.

Type of Waste	Volume (m^3)	Radioactivity (Ci)
Total tank wastes[a,b]	204,000	193 million
Single-shell tanks	121,000	98 million
Double-shell tanks	83,000	95 million
Radionuclides separated from tank wastes		
Cs and Sr capsules	~4	125 million
German logs	34 logs	11 million
Waste leaked into environment from tanks[c]	2,200-3,800	0.3 million
Early tank waste intentionally discharged to soil[c]	454,000-492,000	65,000-4.7 million
Evaporator condensate released to ground[c]	1 million	~3,000
Stored transuranic (TRU) waste[d,e]	48,000[d]	246,000[d]
	46,000[g]	300,000[g]
Buried TRU-contaminated waste and soil[d,e]	107,000	92,000
Low-level waste (including mixed) stored	187,000[g]	5.5 million[g]
	9,300[d]	Not available[d]
Low-level radioactive waste in disposal cells	1.2 million[d]	12 million[d]
	283,000[g]	11 million[g]
Low-level radioactive waste at U.S. Ecology commercial disposal site (not DOE)[h]	382,000	3.9 million
TOTAL	>3.58 million	>350 million

NOTE: Shaded entries are wastes that ultimately are expected to remain on-site. The data are from different sources, are measured or estimated at different times, and did not indicate quantified uncertainties. 1 Ci = 3.7×10^{10} Bq.

[a] Schepens, 2005. For more information on cesium and strontium capsules and the German logs, see Appendix K.
[b] This table does not include spent fuel and sludge from the K-Basins or spent nuclear fuel from the Fast Flux Test Facility (FFTF) and the Shippingport Pressurized Water Reactor.
[c] NRC, 2001a, decay corrected to mid-1990s. The lower-bound number for intentionally discharged radioactivity accounts only for cesium-137 and strontium-90.
[d] DOE, 2001b.
[e] As of 1996.
[f] Not decay corrected, hence an overestimate.
[g] DOE-RL, 2004a.
[h] Quantities as of January 1, 2000 (Washington State, 2000).

SIDEBAR II-1
Cesium and Strontium Capsules

Some of the tank waste at Hanford was processed to remove cesium and strontium. Cesium was removed by ion-exchange columns. Strontium was removed by a solvent extraction method. This method produced liquid and solid alkaline waste containing high concentrations of organic complexants that retain some radioactive elements in solution. These separations, which removed approximately 90 percent of the strontium and cesium from the waste processed, were performed in the B Plant and produced intensely radioactive halide salts (cesium chloride and strontium fluoride). These salts were encapsulated in 2,217 metal cylinders, some of which were used both on-site and off-site as radiation sources. The cylinders are approximately 7 cm (2.75 inches) in diameter and 50 cm (approximately 20 inches) long. The off-site applications never developed as expected and ceased entirely in 1988 after one capsule being used in the commercial sector was found to be leaking (USNRC, 1989). All capsules were returned to Hanford by 1996. As of 1997, nearly 300 capsules had been dismantled and their contents repackaged. Currently, 1,936 capsules[a] are stored at the Waste Encapsulation and Storage Facility at Hanford. Although 23 of these had to be overpacked (i.e., sealed in a larger stainless steel container) due to swelling, the capsules are generally considered to be in good condition (DOE-RL, 2002).

The total volume occupied by these capsules is about 4 m^3 (150 cubic feet), but they account for more than 40 percent of the tank waste radioactivity at Hanford. Off-site disposal at a geologic repository by 2020 is the reference disposition option for these materials. A previous National Research Council report describes the technical challenges that these cesium and strontium capsules present for continued storage and eventual disposal (NRC, 2003a).

[a] 1,335 containing about 83.5 MCi (3.1 EBq) of cesium and its barium decay product, and 601 containing about 36.5 MCi (1.4 EBq) of strontium and its decay product. These are decay corrected to July 2005 from DOE-RL (2003).

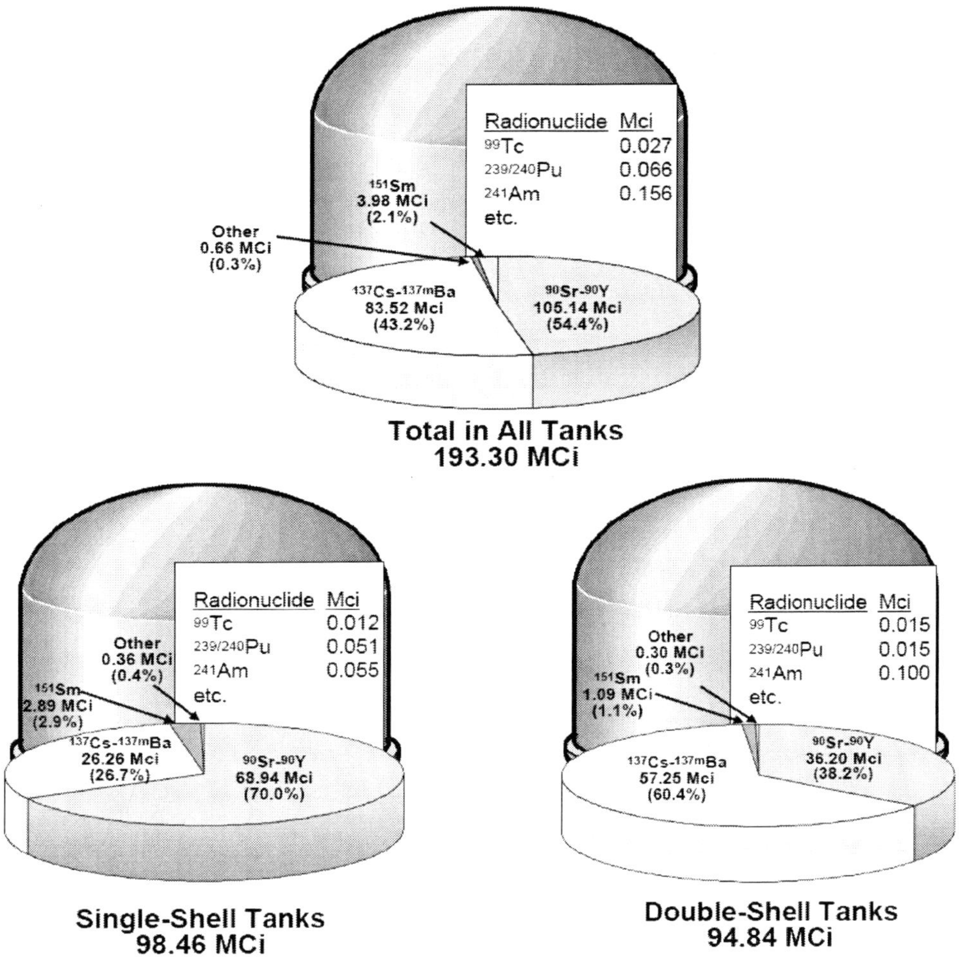

FIGURE II-8 Radioactivity in Hanford tank wastes, decay corrected to January 2004. SOURCE: Honeyman, 2005.

exothermic chemical reactions in the waste (ferrocyanide reacting with nitrate and nitrite mixtures; organic complexants and organic solvents) raised concerns that temperatures in some single-shell tanks could exceed the tanks' structural limits. One tank was termed a "high-heat tank" due to heat from radioactive decay. DOE determined that the waste should be transferred to a double-shell tank, which is better equipped to take the heat load. Investigation of the chemical component of this potential problem revealed that radiolysis had diminished the concentrations of the relevant species and that they were diluted enough by other components of the waste to prevent significant further reaction rates.

Based on information provided by site personnel, the committee judges that the global inventory of chemicals and radionuclides in the tank farms at Hanford is reasonably well understood and is derived in a manner similar to that employed at the Savannah River Site. Individual tank waste inventories have greater uncertainty (Honeyman, 2005). The composition of the waste in Hanford tanks is not fully known because of poor record keeping concerning waste inputs to particular tanks and transfers among the tanks, and the difficulty and high cost of sampling and assay of samples. As of the end of 2005, 86 single-shell tanks and 17 double-shell tanks had been core-sampled. Most of the remaining 74 tanks have been sampled via grab samples[13] or auger samples. There are 32 single-shell tanks that have not been sampled since 1986.

Like the tanks at the Savannah River Site, many Hanford tanks have a bottom layer of sludge containing strontium

[13]The preferred method to estimate inventory is sampling, supplemented by process knowledge. Core, liquid "grab" (using the "bottle-on-a-string" method), and vapor-phase sampling are the methods currently used to characterize waste in the tanks prior to waste retrieval. Auger samples are samples obtained with an "auger tip" (similar to a drill bit), a solid or tubular drill rod, and a "T" handle. The auger tip drills into the waste as the handle is rotated, and material retained on the auger tip is brought to the surface.

and transuranic elements along with hydroxides, oxides, and phosphates of the various nonradioactive metals present in the waste. Above the sludge layer sits one of two different salt media: in the single-shell tanks there are sodium nitrates, nitrites, phosphates, aluminates, carbonates, and sulfates in the form of a saltcake with interstitial liquids containing radioisotopes of cesium, technetium, and iodine. Double-shell tanks have a slurry of the aforementioned salts topped by supernatant liquid composed of the same materials, but bearing trace concentrations of strontium and transuranic radionuclides. Finally, a vapor resides above the liquid and solid contents of the tanks. The vapor is mostly air with small amounts of hydrogen, nitrous oxide, ammonia, trace organic chemicals, and water vapor. Some of the tanks contain other materials, such as debris, cement, diatomaceous earth, and broken or obsolete contaminated equipment (see Chapter III).

Because of the risks to workers and the high cost of data acquisition, the tanks are not sampled for general characterization purposes (i.e., just to know everything in a tank); however, they are sampled for specific data needs in accordance with the site's Data Quality Objectives or if requested by oversight groups. For example, the Defense Nuclear Facilities Safety Board (DNFSB, 1993) directed the Hanford Site to sample the tanks for flammable gas. Sampling at Hanford is currently driven by waste compatibility and chemistry control for corrosion mitigation, Waste Treatment and Immobilization Plant feed delivery needs, and single-shell tank retrieval actions to support tank closure.

Hanford Tank Waste Processing

As mentioned above, Hanford tank waste consists of highly alkaline sludge, saltcake, and supernate. Cesium and strontium isotopes and their decay products comprise most of the radioactivity in the waste. Current planning is to retrieve all waste in the single- and double-shell tanks, separate low-activity waste from the high-level waste, operate the Waste Treatment and Immobilization Plant and supplemental treatment systems to immobilize in glass most of the waste, and package remaining transuranic waste determined not to be from reprocessing of spent fuel for shipment to WIPP for disposal. Waste in the 28 double-shell tanks is to be retrieved and immobilized in glass in the Waste Processing and Immobilization Plant. Plans for the highly radioactive waste processed from the contents of the 149 single-shell tanks are still being developed, but it too is expected to be treated and immobilized in the proposed vitrification facility. Details and status of waste retrieval, processing, and tank closure plans are described in Chapters III, IV, and V, respectively.

The Idaho National Laboratory

The Idaho National Laboratory was established in 1949 as the National Reactor Testing Station on what had previously been a bombing and artillery range for the U.S. Navy. The site occupies 2,303 km^2 (890 square miles) of land in southeast Idaho, approximately 40 km (25 miles) west of Idaho Falls. Nine primary facility areas scattered mostly across the southern half of the site support missions related to naval nuclear propulsion and civilian and military nuclear applications. Some of these missions are ongoing, but others have ended. The spent nuclear fuel processing facilities are located at the Idaho Nuclear Technology and Engineering Center (INTEC), formerly the Idaho Chemical Processing Plant.

Between 1953 and 1992, the Idaho Chemical Processing Plant reprocessed 44 metric tons of heavy metal of U.S. government spent nuclear fuel primarily to recover highly enriched uranium (NRC, 1999b). The processing of spent fuel at the Idaho National Laboratory was similar to processing carried out at the Savannah River Site and the PUREX facility at Hanford. Spent nuclear fuel was dissolved in nitric acid and other strong mineral acids and then sent through further processing steps to recover uranium, neptunium, krypton, barium, and xenon. The highly radioactive waste from the first cycle of the solvent extraction system, containing most of the fission products, was piped to the underground tanks. The 11 stainless steel tanks, each with a typical capacity of 1,136 m^3 (300,000 gallons), are located within concrete vaults. There are annular spaces between the outside of the tanks and the vault walls. Some of the tanks have cooling coils along their bottoms and walls. Three of the tanks were designed for use with less radioactive wastes and, thus, did not receive first-cycle waste from the reprocessing facilities.

Unlike waste at Hanford and the Savannah River Site, sodium hydroxide was not added to the liquid tank waste, thus reducing the volume of storage space needed. Because the tank farm components were made of stainless steel that was compatible with the waste, there was no need to neutralize the waste streams. In fact, the waste streams sent to the tank farm were purposely kept acidic to minimize waste precipitation, to simplify later waste retrieval, transfers, and processing. Maintaining acidic waste streams also reduces the possibility of accidental nuclear criticality in the tank farm. Most of these wastes were removed from the tanks and sent to the calciner for processing.

Calcine Waste in Bins

The Waste Calcining Facility, which operated from 1963 until 1981, and the New Waste Calcining Facility, which operated between 1982 and 2000, were designed and constructed to calcine (i.e., rapidly evaporate and decompose anions such as nitrate and carbonate to yield a granular solid) aqueous wastes generated from spent fuel reprocessing. The calciners are fluidized-bed units: the liquid waste is sprayed into a vessel containing an air-fluidized bed of granular (200-500 μm diameter) particles and heated to 400-600°C. The

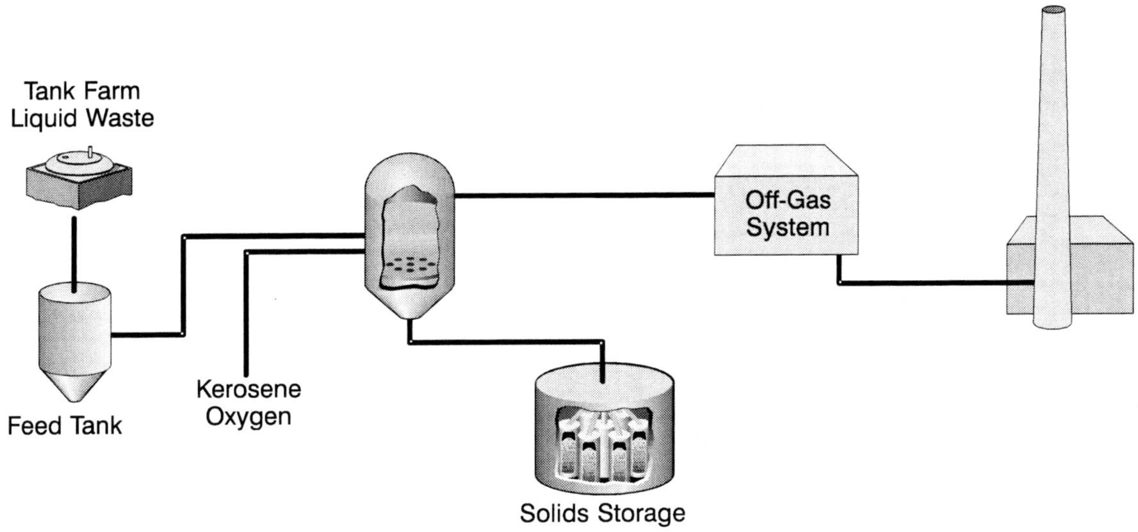

FIGURE II-9 Calcination process (top) and calcine simulant (bottom).

original design used an internal heat exchanger to heat the bed. Later, the system was modified, and kerosene and oxygen were injected into the vessel (see Figure II-9). Kerosene burns in the hot vessel, and its combustion provides the necessary process heat to vaporize the waste. The liquid and volatile portion of the anions evaporate, and most of the remaining constituents of the waste adhere to the granular bed particles. This process reduces the volume of the liquid waste by a factor of between 2 and 10.

The granular calcine waste was piped pneumatically into tall, stainless steel bins, (see Figure II-10). The bins are contained in concrete vaults called Calcined Solids Storage Facilities (also known as "bin sets"). The bin sets were built from the late 1950s to late 1980s and were designed to last 500 years. There are seven bin sets that contain from three to twelve bins each. The bin sets are located below (or partially below) the ground surface. The bins are different in size, sometimes even within the same bin set. Bin heights range from 6 m (20 feet) to 21 m (68 feet); their (outer) diameters range from 0.9 meters (3 feet) to 4.1 m (13.5 feet). In the case of annular bins, the space between the outer cylinder and the inner cylinder varies from 0.6 m (2 feet) to 1.9 m (6.25 feet). Storage volumes in each bin set range from approximately 226 m^3 (8,000 cubic feet) to approximately 1,506 m^3 (53,200 cubic feet).

All of the bins can be accessed from the top through installed risers, except for those in bin set I, which has no access risers. The number of risers varies from one to five per bin. In general, the annular bins have more access risers than do the cylindrical bins. Bin set I is expected to be the most challenging for waste retrieval because there is no installed retrieval access. The bins also contain numerous internal obstructions, such as internally mounted wall stiffeners and bottom braces, which could hinder waste retrieval operations (Steiger and Swenson, 2005). The largest bin set (bin set VII) is empty. Bin set I consists of four sets of three concentric units; bin sets II and III are composed of seven cylindrical units, while bin set IV is composed of three cylindrical units; bin sets V, VI, and VII are composed of seven annular units (see Figure II-10).

Calcine Bin Inventory

Approximately 41 MCi (1.52×10^{18} Bq) of waste,[14] nearly all of the liquid waste from reprocessing of spent fuel at the Idaho National Laboratory, had been calcined by May 2000, when the calciner was shut down to comply with a 1999 modification to a notice of noncompliance consent order with the State of Idaho. The composition of the calcine varies depending on the composition of the fuel and its cladding, as well as any chemicals added during reprocessing and calcination. Aluminum- and zirconium-clad fuels yield calcines containing alumina (Al_2O_3) and zirconia (ZrO_2) as major constituents. Hydrofluoric acid was used to dissolve zirconium-clad fuel. Aluminum nitrate was added to the liquid waste to complex the fluoride. Calcium nitrate was added to the waste at the calcining facility (not in the tank farm) to prevent fluoride volatility in the calcination process. Thus, calcium fluoride (CaF_2) and alumina are also major constituents of zirconia calcine. Boron, sodium, chromium, iron, lead, mercury, and other trace metals are present in the calcine waste as oxides; some sodium and potassium are present as nitrates; and some magnesium and calcium carbonate from the dolomite startup bed is also present.

[14]This value is decay corrected to 2006.

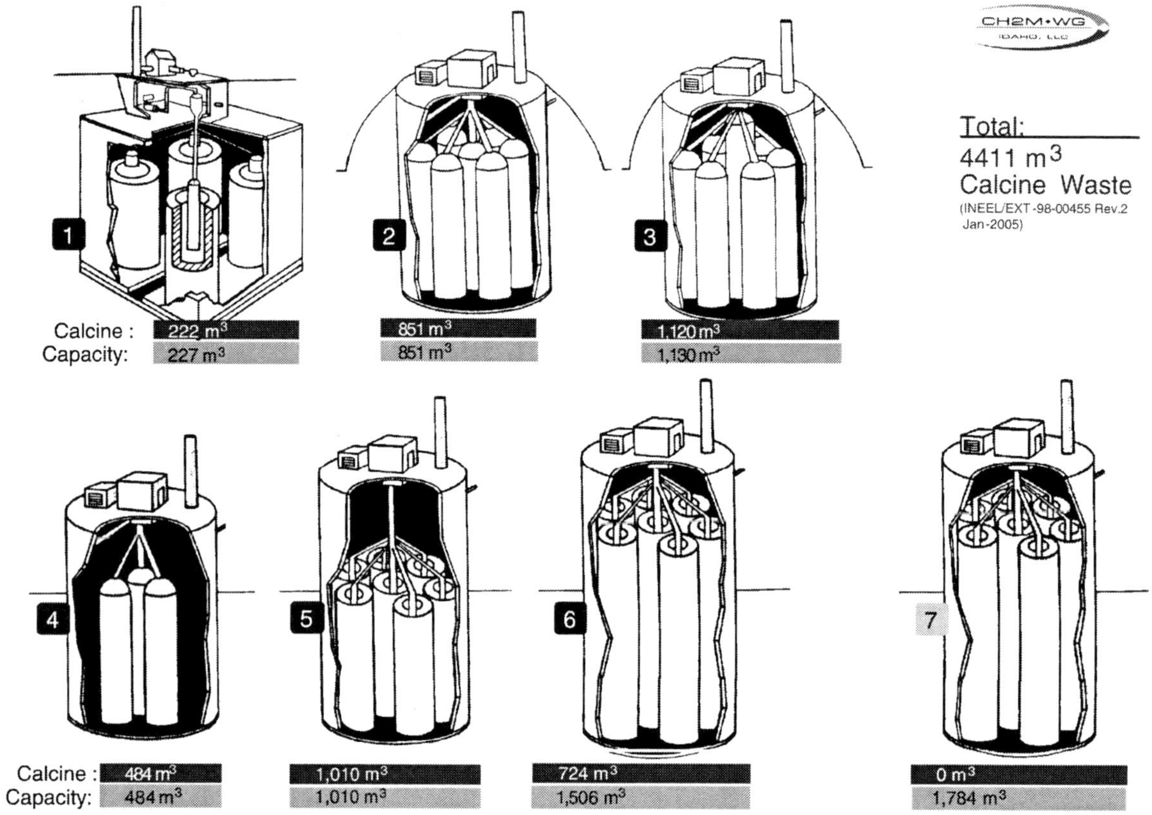

FIGURE II-10 Calcine solids storage facilities. The horizontal line outside the bin shows the location of the land surface. SOURCE: Patterson, 2005.

Very few characterization data are available on the calcine at the Idaho National Laboratory because of the high dose levels and the difficulty of reaching the material in the bins. Some of the information that is of current interest, particularly the concentration of long-lived radioactive nuclides and RCRA metals, was not routinely collected at the time of waste generation. Information gaps were filled using process knowledge. The relative error bound for calcine inventory is 14 percent at a 95 percent confidence level (Steiger and Swenson, 2005).

A previous National Research Council report (NRC, 1999b) discusses calcine characterization in Idaho bins and points out both the discrepancies among characterization information and the heterogeneity of calcine properties potentially existing in the bins. Moreover, what is known today about the calcine appears to be based on pilot tests with cold surrogates and not on sampling information. For example, the only samples of actual calcine that have been retrieved from bins consist of two core samples collected from the second bin set in 1979. Another calcine sample from the output of the calciner was collected in 1993. The calcination heat source was changed from indirect liquid-metal heating to in-bed combustion just around the time this waste was calcined. In-bed combustion generates both oxidizing and reducing chemical environments in different regions, which could affect calcine properties. To date, no samples of calcine produced by in-bed combustion have been retrieved from bins.

Sodium-Bearing Waste in Tanks

The roughly 500 kCi (1.85×10^{16} Bq) of radioactivity in the liquid radioactive waste remaining at the INTEC tank farm is called sodium-bearing waste. DOE describes sodium-bearing waste as "a liquid mixed radioactive waste produced from the second and third cycles of spent nuclear fuel reprocessing and waste calcination, liquid wastes from INTEC closure activities stored in the Tank Farm, solids in the bottom of the tanks, and trace contamination from first cycle reprocessing extraction waste" (DOE-ID, 2002).

Sodium bearing waste has high concentrations (more than 2 moles per liter) of sodium nitrate salts, resulting from the addition of sodium hydroxide to the washing solution to enhance its effectiveness in removing some residues. Some

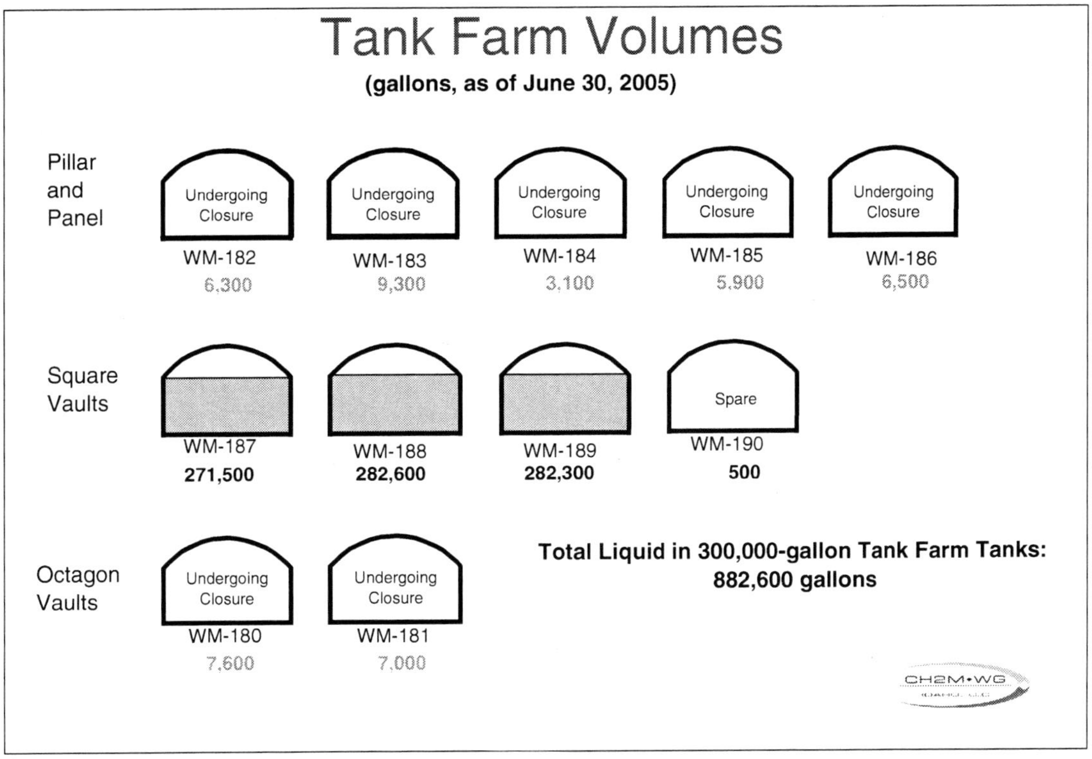

FIGURE II-11 Idaho National Laboratory tank farm volumes. SOURCE: Lockie et al., 2005.

of this liquid waste was sent through the calciner to produce calcine waste. Figure II-11 shows how much waste is stored in each of the roughly 1,100 m^3 (300,000 gallon) tanks. DOE is now cleaning seven tanks, and one tank is a clean spare.

Idaho Tank Inventories

The vast majority of Idaho tank waste is in a liquid form, but a small amount of insoluble solids can be found at the bottoms of the tanks. DOE has sampled waste from several tanks. In a typical tank, the cesium-137 and strontium-90 (and their short-lived decay products borium-137m and yttrium-90) account for most of the radioactivity in both the solids and the liquids. In tank WM-187, for example, strontium and cesium together contribute 97 percent of the total radioactivity of 0.22 Ci per liter (8.1 × 10^9 Bq per liter). Isotopes of plutonium (mostly plutonium-241 and plutonium-238) constitute 1.9 percent (Barnes et al., 2004). The inventory of radioactive waste in the INTEC Tank Farm is listed in Table II-3.

More than 187 samples have been retrieved from the tanks at the Idaho National Laboratory since 1987. Most of the samples are obtained via steam jet pump from the tanks. Some samples were obtained directly via a Light Duty Utility Arm (Olson, 2005). The uncertainty in quantities is generally within 10 percent for most chemical constituents; less than 20 percent for most radionuclides; and about 30 percent for solids; however, many organics were not able to be detected during sample analysis. The tanks are currently sampled for steam reforming[15] processing needs.

Idaho Tank Waste Processing and Tank Closure

Waste processing at the Idaho National Laboratory is different from that at the other two sites: There are no plans to perform chemical separations on the liquid waste or calcine to generate a high-activity fraction and a low-activity fraction. DOE has recently selected steam reforming as the technology to convert sodium-bearing waste into a solid form (DOE-ID, 2005b). The calcine waste will be put in a form suitable for disposal in a monitored geologic repository and will be ready for shipment out of Idaho by 2035. DOE plans to ship its solidified sodium-bearing waste to the WIPP

[15]In a typical steam reforming process, superheated steam, along with the material to be treated and co-reactants, is introduced into a fluidized bed reactor where water evaporates, organic materials are destroyed, and the waste constituents are converted to a granular, leach-resistant solid (NRC, 2005b; see also, DOE-EM, 2005 and DOE-ID, 2002).

TABLE II-3 Inventory of Radioactive Waste by Type at the Idaho National Laboratory

Type of Waste	Volume (m^3)	Radioactivity (Ci)
Total tank and bin waste[a]	~5,000	35-36 million
Comprising		
Treated Sodium-bearing waste in tanks	~ 500-800	~520,000
Calcine waste in bins	4,400	35 million
Waste leaked into environment from pipes and valves[b]	107	37,000
Service wastewater injected to aquifer[c]	45 million	22,000
Stored transuranic waste[d,e]	65,000	343,000
Buried transuranic-contaminated waste and soil[d,e]	37,000	297,000
Low-level waste (including mixed) stored	2,200	Not available
Low-level radioactive waste in disposal cells [d]	158,000	12 million
TOTAL	>45 million	>49 million

NOTE: Shaded entries are wastes that ultimately are expected to remain on-site. These data are from different sources, are measured or estimated at different times, and did not indicate quantified uncertainties. This table does not include spent nuclear fuel stored on-site (i.e., from naval and test reactors as well as from Fort St. Vrain and Three Mile Island) or contaminated soil at the evaporation ponds, which have been remediated. 1 Ci = 3.7×10^{10} Bq.

[a] Lockie, 2005a. Decay corrected to 2012.
[b] Cahn, 2005. Not decay corrected, therefore an overestimate.
[c] DOE-ID, 2003a. Radioactivity not decay corrected. 99.8% of the radioactivity is tritium.
[d] DOE, 2001b.
[e] As of 1996.

facility in New Mexico. DOE's plans for and status of waste retrieval, processing, and tank closure at Idaho are described in Chapters III, IV, and V respectively.

POLICY BACKGROUND

The 1954 Atomic Energy Act (AEA) gave the Atomic Energy Commission (the predecessor agency of both DOE and USNRC) the authority to manage nuclear waste generated from both defense and commercial nuclear fuel cycle activities. The 1982 Nuclear Waste Policy Act (NWPA) defined the term "high-level waste" (HLW) and officially adopted deep geologic disposal as the nation's long-term strategy for managing this waste.

The definition of HLW, as set out in the Nuclear Waste Policy Act (42 U.S.C. Section 10101), is:

> (A) the highly radioactive waste material resulting from the reprocessing of spent nuclear fuel, including liquid waste produced directly in reprocessing and any solid material derived from such liquid waste that contains fission products in sufficient concentrations; and (B) other highly radioactive material that the [Nuclear Regulatory] Commission, consistent with existing law, determines by rule to require permanent isolation.

It is apparent from this text that Congress defined HLW in the AEA and the NWPA in terms of its source. Section 3116 of the NDAA provides an exception to this definition at the sites in South Carolina and Idaho. DOE Order 435.1 still applies to waste determinations at Hanford and potentially to other wastes at the Savannah River and Idaho sites to which Sect. 3116 does not apply.

In 1993, the USNRC first set out criteria to determine which portions of certain Hanford nuclear fuel reprocessing waste are not HLW (the waste so determined is also called "waste incidental to reprocessing" in some documents).[16] DOE, which regulates itself on most matters related to radioactive waste, developed Order 435.1 which contains provisions for determining that some wastes are not HLW and, thus, can be managed as low-level waste or transuranic waste (DOE, 1999a; 1990b; 2001b). According to DOE Order 435.1, waste can be determined to be incidental to reprocessing by two methods, "citation" or "evaluation." The citation method simply lists certain wastes, such as resins and clothing, that DOE identifies as incidental to reprocessing. The evaluation method is based on three criteria provided to DOE by USNRC in 1993 in its denial of a petition for proposed rulemaking concerning the definition of HLW (Bernero, 1993).

> The Commission . . . has indicated . . . it would regard the residual fraction as "incidental" waste, based on the

[16] The first official document referring to "waste incidental to reprocessing" is the provisions of DOE Manual 435.1 concerning determining whether DOE tank waste is not HLW. "Incidental" waste is mentioned in a March 4, 1993 Federal Register Notice in which the USNRC set forth criteria for determining that waste from Hanford double-shell tanks disposed of in a grout facility would not be HLW USNRC found that the principles for waste classification are well established, endorsing the criteria DOE later used in Order 435.1 (NRC, 2005b; see also Bernero, 1993).

> Commission's understanding that DOE will assure that the waste: (1) has been processed (or will be further processed) to remove key radionuclides to the maximum extent that is technically and economically practical; (2) will be incorporated in a solid physical form at a concentration that does not exceed the applicable concentration limits for Class C low-level waste as set out in 10 CFR Part 61; and (3) will be managed, pursuant to the Atomic Energy Act, so that safety requirements comparable to the performance objectives set out in 10 CFR Part 61 are satisfied. (Bernero, 1993)

Thus, the tank waste that USNRC reviewed and that was destined for disposal on-site would not be considered HLW if it met the criteria. On this basis, DOE Manual 435.1 created in effect three implicit subcategories of waste: (1) high-level waste, (2) non-high-level waste that is managed as low-level waste, and (3) non-high-level waste that is managed as transuranic waste (see Appendix C, Table C-1).

Using the provisions of DOE Manual 435.1, DOE proposed to determine that certain wastes at the three DOE sites that are the subject of this report are not HLW, a step needed for DOE to carry out its separation strategy (high-activity and low-activity) for the tank wastes. This process came to an abrupt halt in 2003 when DOE was sued in Idaho by the Natural Resources Defense Council, Snake River Alliance, Confederated Tribes and Bands of the Yakama Nation, and the Shoshone Bannock Tribes. The plaintiffs argued that Order 435.1 exceeded DOE's authority under the AEA and the NWPA. In 2004, the court found that the standards DOE established by rule were too discretionary and offered no effective limitation on the agency's ability to determine which waste could be managed as low-level waste and disposed on-site. The federal district court in Idaho ruled in favor of the plaintiffs, finding that DOE could not continue with its management activities in reliance on Order 435.1.[17]

DOE appealed the district court's decision. The U.S. Court of Appeals for the Ninth Circuit did not rule on the legal merits of the district court's ruling. It reversed the district court on the procedural ground that the case was not yet "ripe" for judicial determination.[18] In other words, the Ninth Circuit expressed no opinion on the legality of Order 435.1, but put off the question for a later time, when DOE actually takes action under the authority of Order 435.1. Although the decision that struck down Order 435.1 was vacated, the Order could be contested at its first use. This leaves Order 435.1 in some degree of legal limbo in Idaho, where the only existing opinion (albeit vacated) is negative and in Washington state, which is also in the Ninth Circuit.

DOE saw the rulings as a major impediment to its pursuit of a separation strategy at the Hanford and Savannah River Sites and to tank closure at all three sites. So, even before the Ninth Circuit rendered its decision on the appeal, DOE sought a statutory remedy from Congress. In Section 3116 of the Ronald Reagan National Defense Authorization Act of 2005, Congress established criteria for determining that some waste from spent fuel reprocessing is not high-level waste and may be disposed of on-site at the Savannah River Site and the Idaho National Laboratory. The Hanford Site, however, was not included in the provisions of Section 3116 because the state of Washington explicitly is not covered or bound by the section. In its criteria, Congress implicitly divided the non-high-level waste from spent fuel reprocessing destined for on-site disposal into two subclasses, depending on the concentrations of radionuclides in the waste in relation to Class C concentration limits in 10 CFR 61.55 (see Appendix C) although the differences are only procedural (NRC, 2005a). Therefore, under Section 3116, at the Savannah River Site and the Idaho National Laboratory (but not Hanford), there are essentially three subclasses or categories of tank waste from reprocessing: HLW, non-HLW Class C or less, and non-HLW greater than Class C.

Section 3116 is similar to Order 435.1 in many ways, most importantly in the standard for removal of radionuclides to the maximum extent practical and in the use of the performance objectives in 10 CFR 61 as benchmark criteria for on-site disposal. However, there are some critical differences. First, Section 3116 addresses only wastes that are to be disposed of on-site and which are subject to a state compliance agreement whereas the provisions of DOE Manual 435.1 could encompass any waste and its planned destination. Section 3116 does not say that waste disposed on-site is low-level waste, although it is implied that such wastes will be managed by near-surface disposal like other low-level waste disposed on-site. Section 3116 was intended to resolve the legality of the overall separation strategy at the Savannah River Site and of tank closures at the Savannah River Site and Idaho (but not, of course, at Hanford). Unlike Order 435.1, however, Section 3116 does not provide authority or guidance on tank waste determinations for retrieved non-high-level waste to be managed as transuranic waste, probably because defense transuranic waste is slated for geologic disposal at the Waste Isolation Pilot Plant in New Mexico and Section 3116 only applies to waste that stays on-site.[19]

Second, Section 3116 sets out roles for the host states and USNRC, which are absent from Order 435.1. DOE requested informal USNRC input on waste determinations performed before 2004 under Order 435.1 (Camper, 2005; Flanders, 2005). However, the USNRC did not have any official regulatory role in that capacity and provided general and nonbinding comments on DOE's waste determinations. The

[17]*NRDC, Inc. et al. v. Abraham*, 271 F. Supp. 2d 1260 (D. Idaho 2003).

[18]That is, DOE had not yet actually applied Order 435.1 in the Idaho case. *NRDC, Inc. et al. v. Abraham*, 388 F.3d 701 (9th Cir. 2004).

[19]See definition of transuranic waste in Appendix K.

situation at the Hanford Site is somewhat different because the Federal Facility Compliance Agreement and Consent Order for this site formally requires USNRC input on the effectiveness of DOE tank waste retrieval. Third, Section 3116 and Order 435.1 differ in their description of the degree of removal of the highly radioactive fraction:

- Section 3116(a)(2): "has had highly radioactive radionuclides removed to the maximum extent practical"
- Manual 435.1-1 (p. II-1): "has been processed, or will be processed, to remove key radionuclides to the maximum extent that is technically and economically practical"

The meaning of the unmodified term "practical" in Section 3116 requires some interpretation, i.e., whether it is the same as the "technically and economically practical" found in Manual 435.1-1 or something different (including more, fewer, or other considerations). Fourth, a time of compliance is not mentioned in 10 CFR Part 61. A time of compliance of 10,000 years has been recommended by Nuclear Regulatory Commission staff in its guidance on performance assessments, but this recommendation has not been approved by the commissioners, so it does not constitute official agency policy. However, DOE Order 435.1 specifies a time of compliance of 1,000 years for low-level waste disposal facilities, which complicates the issue. This difference may need to be resolved, however, DOE has made it a practice to carry performance-assessment calculations out to the peak dose within 10,000 years, possibly making the difference irrelevant. The time of compliance is discussed further in Chapters VI and VIII.

Multiple Legal Drivers and Decision-Making Authorities

The cornerstones of DOE's authority to manage radioactive waste are the AEA and the NWPA. However, the AEA and the NWPA are not the only applicable federal statutes. Federal legislation such as the Comprehensive Environmental Response, Compensation, and Liability Act (CERCLA), Resource Conservation and Recovery Act (RCRA), Clean Air Act, Clean Water Act, Safe Drinking Water Act, National Environmental Policy Act (NEPA), and correlative state laws all have a part to play. The relevant considerations under these statutes go well beyond, and often adopt different approaches than, the AEA, NWPA, and Section 3116 of the NDAA. Moreover, these other laws are not administered by DOE, but by the Environmental Protection Agency (EPA) and, through delegated authority, the states.

Through Order 435.1, DOE regulates storage and treatment as it relates to radiological components of waste at the three sites. Order 435.1 could be applied to waste determinations at Hanford, subject to resumption of the legal challenges brought previously in Idaho which were suspended on the basis of ripeness. Disposal actions for wastes that DOE determines under Section 3116 not to be high-level waste are to be monitored by the USNRC in coordination with the host state.

In addition, all sites have entered into Federal Facility Agreements and Consent Orders on behalf of DOE with each host state and EPA (and the U.S. Navy at Idaho) (Idaho FFACO, 1991; Hanford FFACO, 2003; SRS FFA, 1993; Idaho SACO, 1995). These agreements establish the operational goals and milestones for DOE's site cleanup operations. They provide authoritative interpretation of DOE's statutory and regulatory obligations, and they add requirements, such as milestones and technical performance specifications; however, Federal Facility Agreements cannot establish requirements that are contrary to existing federal laws.

DOE's plans for Savannah River Site tank waste disposition are subject to the approval of the South Carolina Department of Health and Environmental Control under the Savannah River Site Federal Facility Agreement (SRS FFA, 1993). The state regulates the hazardous component of the waste through the South Carolina Hazardous Waste Management Act while the tanks are closed under wastewater treatment and hazardous waste regulations (SCDHEC, 2004a, 2004b). The closure milestones in the Federal Facility Agreement for this site are 2022 for Type I, II, and IV and 2028 for Type III tanks.

Plans for the Idaho National Laboratory tank waste disposition are subject to the approval of the Idaho Department of Environmental Quality under the Idaho Federal Facility Agreement and Consent Order and the Settlement Agreement and Consent Order (Idaho FFACO, 1991; Idaho SACO, 1995). The 1995 court settlement (called the Settlement Agreement) among DOE, the U.S. Navy, and the State of Idaho requires that sodium-bearing waste be solidified and made ready for disposal outside Idaho by 2009. The 1995 court settlement also requires all high-level waste to be prepared for removal from Idaho by 2035. Through the Idaho Hazardous Waste Management Act (HWMA, modeled on RCRA), the state regulates the treatment and storage of the hazardous components of the waste (sodium-bearing waste, tanks, and calcine). The waste remaining in the tanks following closure may be subject to continued RCRA-HWMA regulation. A 1991 Notice of Noncompliance consent order signed by both the EPA and Idaho established a schedule for DOE to cease use of the tank farm and perform closure activities. The tanks should be closed in six phases from 2005 to 2016. DOE Order 435.1 regulates storage and treatment of radiological components of the waste (sodium-bearing waste, tanks, and calcine). Section 3116 does not apply to waste determination at Idaho if the waste is not slated for on-site disposal. According to a state regulator, potential disposal at the Waste Isolation Pilot Plant as transuranic waste would be regulated by EPA and the State of New Mexico; otherwise, the NWPA applies (Trever, 2005).

DOE's plans for Hanford Site tank waste disposition are subject to approval of the State of Washington, Department of Ecology under the Hanford Federal Facility Agreement and Consent Order, also known as the Tri-Party Agreement (Hanford FFACO, 2003). The RCRA closure plan (under the Washington Hazardous Waste Management Act) applies to tank farms, while the AEA and NWPA apply to tank residuals and waste left in pipes. Tank farm closure is performed under CERCLA. The Hanford closure schedule milestones in the Tri-Party Agreement are the year 2024 for single-shell tanks and 2032 for double-shell tanks.

Consideration of various other legal drivers for the tank wastes (such as the Clean Air Act and the Safe Drinking Water Act) illustrates another dimension of complexity. The numerous waste types related to the tanks are governed by many regulations, and in some cases the regulatory framework differs for the same type of waste stream depending on the site. Waste types in the tanks or related to tank waste are the following:

- Bulk tank waste;
- Tank residuals (including heel);
- Tank itself and interior equipment (e.g., cooling coils, nonretrievable cleaning equipment);
- Cesium and strontium capsule contents and containers (at Hanford only);
- Piping and valve boxes connecting tanks and between tank farms;
- Contaminated soil below tanks and pipes and in tank farm areas; and
- Used waste management equipment (e.g., HLW melters).

Different combinations of legal and regulatory standards apply to each of these. Chapter VIII contains findings and recommendations relevant to such a complex legal and regulatory framework.

DIFFERENCES AMONG SITES

Tables II-4 and II-5 show that there are major differences among the three sites in terms of the geology, hydrology, climate, physical and chemical composition of the waste, and tank-system designs. These differences are described in this section.

Differences in Natural and Man-Made Conditions at Each Site

The natural features at the sites differ, from elevation to rainfall, although the two western sites have more natural conditions in common with each other than with the Savannah River Site. It can also be seen that in the man-made features, particularly the fuel reprocessing methods

TABLE II-4 Differences in the Natural Features from Site to Site

	Savannah River Site	Hanford Site	Idaho National Laboratory
Average seasonal low/high temperature C (°F)	2/33 (36/92)	0/24 (32/76)	–7/18 (19/65)
Extremes low/high temperature C (°F)	–19/42 (–3/107)	–32/33 (–25/92[a])	–45/39 (–49/103)
Distance from tank farm to nearest surface water			
by land;	~0.9 km	15 km (downgradient)	61 m (ephemeral stream)
by groundwater	1.85 km	30 km	200 km
Natural flow of nearest surface water average, maximum (cubic meters per second)	0.5, 23.5 (rates measured in Four Mile Branch)	3,360, 19,500 (rates measured in the Columbia River)	Intermittent, 82.4 (estimated 100-year flood of Big Lost River)
Subsurface medium	Loamy sand, sandy clay, clay, silty clay	Unconsolidated glaciofluvial sands and gravels, fluvial-lacustrine sediments, basalt	Basalt overlain by alluvial deposits of gravel-sand-silt with silt and clay interbeds
Average depth from ground surface to water table	9.75 m	90-100 m	143 m
Average annual precipitation	124.4 cm	16 cm	22.1 cm
Average annual soil infiltration	40 cm	0.4-1.0 cm	0.36-1.1 cm
Complications for monitoring and modeling contaminant transport	Clay lenses; multiple aquifers with different flow rates and directions; large infiltration rate	Deep vadose zone, highly variable conductivity and sorption values, varied stratigraphy	Deep vadose zone, highly variable conductivity and sorption values (ranging several orders of magnitude); perched aquifers

[a] Hoitink et al., 2005.

TABLE II-5 Differences in the Man-Made Features from Site to Site

			Idaho National Laboratory	
Tank Specification	Hanford Site	Savannah River Site	SBW Tanks	Calcine
Number of tanks/areas to close	177/18 tank farms	51/2 tank farms	11[a]/1 tank farm	7 calcine bin sets
Tank types	2 (149 SST and 28 DST)	4 (Type I, II, III, IV)	1	2 (annular and cylindrical bins)
Tank sizes, 10^3 gal	55-1,160	750-1,300	30-318	60-471
Construction periods (or years when in service)	1943-1964 SSTs 1968-1986 DSTs	Type I: 1954-1965 Type II: 1956-1960 Type III: 1971-1992 Type IV: 1959-1965	1953-1966	1960s
Construction material	Carbon steel	Carbon steel	Stainless steel	Stainless steel
Tank maximum ages in years at closure	More than 75	More than 75	More than 60	More than 40
Tank conditions	67 confirmed and assumed leakers, estimated 1 million gallons to soil	11 leakers, 1 to soil	No leakers	No leakers
Tank depth relative to water table	Well above water table	Some tank bottoms in water table	Well above water table	Above surface
Extent of obstruction in tanks	Abandoned equipment, debris	Severe obstructions due to vertical cooling coils in most tanks	Little or no obstructions (cooling coils on the bottom and walls)	
Waste types	Viscous, alkaline liquid, sludge, saltcake, diatomaceous earth[a]	Viscous, alkaline liquid, sludge, saltcake, zeolite[b]	Acidic, liquid sodium waste, and small amount of sludge	Calcined powder
Waste volume, 10^6 gallons	54	33	1.4	1
Waste radioactivity, 10^6 Ci (3.7×10^{16} Bq)	193 in tanks 136 in capsules and German logs[c]	426	0.52	24
Retrieval schedule	SSTs complete by 2018[d] and DSTs by 2028[a]	2019 for Type I, II, and IV; 2024 for Type III	HLW retrieval complete by 1998; remaining liquid waste by 2012	Road-ready by 2035
Closure schedule	SSTs by 2024[d] and DSTs by 2032[a]	2022 for Type I, II, & IV; Type III by 2028;	In six phases from 2005 to 2016	Not yet determined

NOTE: DST = double-shell tank; HLW = high-level waste; SST = single-shell tank.

[a] Diatomaceous earth was used as waste sorbent material to immobilize residual supernatant liquid in tanks where liquid removal by pumping was not feasible (see Appendix K).

[b] Zeolites were used to remove cesium from the condensed steam recovered from an evaporator. Zeolite particles contain "trapped" cesium ions (along with other ions) and are difficult to retrieve by pumping because of their high settling rate (see Appendix K).

[c] Cs and Sr capsules = cesium and strontium capsules (see Appendix K).

[d] Currently reevaluating retrieval and closure schedules.

SOURCE: Adapted and elaborated from CH2M Hill Hanford Group, Inc., 2003.

and storage, the Savannah River and Hanford Sites have much more in common with each other than they do with Idaho.

The types of waste at the Idaho site are really distinct from those at Hanford and the Savannah River Site because only a single reprocessing technology was used and the wastes were kept in acidic condition and then converted into solids. This has made retrieval of the sodium-bearing wastes from their tanks much easier. DOE anticipates that the same will be true for the calcined solids.

Differences in Readiness at Each Site

Table F-1 in Appendix F shows that the Savannah River Site, the Hanford Site, and the Idaho National Laboratory are at different points in their removal and stabilization of tank wastes. As a result of these differences, the amount of detail in the committee's discussions of the processes and recommended changes at these sites varies greatly. The experience and progress toward final tank farm closures are different at each site. The Idaho National Laboratory, for example, has retrieved the waste from the majority of its tanks, 7 of the 11 1136 m^3 (300,000 gallon) tanks and all 4 of its 114 m^3 (30,000 gallon) tanks. Hanford has completed waste retrieval from only 4 of its 177 tanks, while the Savannah River Site has retrieved the waste from 4 of its 51 tanks—in each case a small percentage of the number of tanks to be emptied. Calcine retrieval has not yet been tested with the radioactive solids in the bins at the Idaho site.

In some cases, the tanks from which the wastes have been retrieved were chosen because it was thought that they would be the easiest to clean. DOE has put very few tanks (and their wastes) through the other major steps, namely separation, treatment, and disposal of the retrieved wastes and closure of the tanks. Therefore, there is little operational experience so far on tank remediation and closure.

To fill this lack of operational experience, move the program forward, and try to encompass all of the likely variants, some experimental work at the bench scale, less experimental work at a pilot scale, and computer programs used to model facility performance have been utilized at the sites. However, the sites acknowledge that there is no assurance that the tank remediation process will move forward seamlessly in the future with more than 240 tanks and bins to close.[20]

This section shows that there are varied geologic, topographic, and climatological differences among and within sites and that the design of the tanks and their waste contents vary widely, so unique solutions may be required for each tank or group of tanks.

However, given the uncertainties and the challenges ahead, the committee judges that it would be desirable to have general guidelines on technologies for tank waste remediation that apply to all tanks. There is no dispute that all snowflakes or fingerprints are unique, but that does not mean that there are no similarities or common features among them. The same is true of the solutions for tank waste retrieval. Tank waste retrieval is not like an assembly line in which everything is the same except the color. The distinctive design of the tank, the mixtures of the wastes in the tank, and the local geology, hydrogeology, and precipitation can influence the best methods for the removal of wastes to the maximum extent practical and how much needs to be removed.

The tanks also have many similarities: They are all steel, they are all located below ground, and most are above the water table. They have similar limitations in the size of the tools that can be inserted into the tank, and the techniques for removing the most difficult wastes from the tanks are similar. Therefore, many things can be learned from each waste removal operation, but the details of what removal operations to use in a particular tank cannot necessarily be determined in advance because conditions in the tank may be very different from earlier experience. However, it may be possible on the basis of earlier experience to identify some techniques that are more or less likely to work in that particular tank's environment.

FINDINGS AND RECOMMENDATIONS

Despite this being the background chapter, the first findings and recommendations are given here because they establish the framework for the report. They also explain why this report discusses processes (e.g., waste retrieval, waste processing, tank closure, monitoring, performance assessment, decision making) in a "global" way, while each site is treated individually at a more detailed level.

Finding II-1: There is great diversity in the natural and man-made conditions within and among sites.

Recommendation II-1: Each tank or group of tanks with similar problems should be addressed using an approach specifically tailored to best address the particular situation by taking into account site- and tank-specific conditions, previous experience, and advancements in technology.

Finding II-2: There has been limited operational experience acquired in tank remediation and closure so far at the Hanford Site, Idaho National Laboratory, and Savannah River Site.

Recommendation II-2: Given the early stage of the tank waste remediation program and the challenges ahead, the committee judges that it would be desirable to have general guidelines on applicable technologies for tank waste remediation that apply to all sites.

[20]As mentioned earlier, the Hanford Site has 177 tanks; the Idaho National Laboratory has 11 tanks and 7 calcine vaults; and the Savannah River Site has 51 tanks, two of which are already closed.

III

Tank Waste Retrieval

As shown in Chapter II (Figure II-1), the first major step in the cleanup of the Department of Energy's (DOE's) tank wastes is to retrieve wastes from the tank. The effectiveness of waste retrieval technology is central to the committee's task because it drives the cost, worker risk, and secondary waste production associated with the removal of waste from the tanks. These are key factors in deciding how much waste should be left behind in tanks for on-site disposal. The status, effectiveness, and challenges of waste retrieval technologies are the focus of this chapter.

DOE has developed and deployed a number of technologies for retrieving tank wastes. Because the three sites use similar technologies, this section begins with a generic description of the technologies and approaches used in various stages of the waste retrieval process. Most of this chapter addresses retrieval from the large underground storage tanks at the three DOE sites.

However, the Idaho National Laboratory also has six sets of large stainless steel bins containing reprocessing waste that was converted to a fine granular powder by a process called calcination. The calcined material is very different from the wastes stored in the tanks, and factors affecting its retrieval are discussed in the section concerning waste characterization. The chapter ends with the committee's findings and recommendations related to tank waste retrieval.

WASTE RETRIEVAL TECHNOLOGIES AND APPROACHES

As described in detail in Chapter II, the waste that is to be retrieved from DOE's large underground storage tanks can contain three distinct phases: (1) supernatant liquid, (2) precipitated saltcake containing mostly nonradioactive sodium compounds and soluble radionuclides such as cesium-137, and (3) viscous sludge.[1] The phases usually occur in this order from top to bottom although the sludge and saltcake may be layered to some extent and both contain interstitial liquids having a composition typical of the supernatant liquid. Other materials may be present in the tank, such as debris and abandoned equipment.

The objective of DOE's tank cleanup program is to retrieve waste to the maximum practical extent for subsequent processing. However, waste retrieval technologies suitable for removing large amounts of bulk waste are generally not suitable for removing small amounts of residual waste in a tank. Thus, tank waste retrieval technology is discussed in two parts, which address: (1) bulk waste retrieval (including supernatant liquid, saltcake, and sludge retrieval) and (2) retrieval of the residual remaining after bulk waste retrieval.

Bulk Waste Retrieval

Retrieval of the bulk of the tank waste necessarily begins with the supernatant liquid at the top and works down through the saltcake and sludge. The following sections describe the mobilization, collection, and removal of the bulk of these wastes.

Supernatant Liquid

Retrieving the bulk of the supernatant liquid is straightforward because liquid waste is inherently mobile and can be removed readily by a "transfer" pump, much like a sump pump in a basement (see Figure III-1). The transfer pump is lowered into the tank to remove the liquid through a pipe or hose into a doubly contained underground pipe. The pumps used for this purpose need relatively low power compared to the mixing pumps used to mobilize waste forms such as saltcake and sludge, as discussed below. The technology for retrieving supernatant liquid is efficient and well established, although minor improvements continue to be made as a result of field experience.

[1]The saltcake and the sludge are not present in the Idaho National Laboratory's tank waste because it is acidic.

FIGURE III-1 The SEEPEX transfer pump before installation. SOURCE: Reynolds, 2004.

Saltcake

Saltcake is composed mostly of readily soluble salts that precipitated because the saturation limit in the supernatant liquid was exceeded due to evaporation. The saltcake is dissolved to prepare it for removal by adding "fresh" water[2] to the tank. Although it would be preferable to avoid adding fresh water to the tanks by using already contaminated recycled fluids, these fluids, such as supernatant liquid, are less efficient in dissolving the saltcake because they already contain high concentrations of salts or solids. Both the Savannah River and the Hanford Sites use "inhibited water" or "raw water" to dissolve saltcake. At Hanford Tank S-112, approximately 95 percent of the tank waste inventory was retrieved using saltcake dissolution; additional retrieval using high-pressure (>5000 pounds per square inch [psi]) jets of water is being performed to break up and mobilize the remaining hard heel. The resulting solution is circulated through the tank until the desired specific gravity (usually 1.3 to 1.4) is achieved.[3] Tank S-102 is also currently undergoing saltcake dissolution.

Once the saltcake is dissolved, the resulting solution is then retrieved in the same manner as the supernatant liquid. As with supernatant liquid retrieval, this technology is efficient and well established with only minor improvements continuing to be made.

[2]Fresh water is usually filtered river or well water containing added chemicals (e.g., sodium nitrite) to inhibit corrosion of the carbon steel tanks. The solution is sometimes referred to as "inhibited water" or "raw water."

[3]In the present context, the term specific gravity refers to the ratio of the density of a liquid to the density of water at 4°C.

Sludge

Bulk retrieval of the sludge and other solid materials is considerably more challenging than supernatant liquid and saltcake retrieval because the sludge cannot readily be dissolved in water and pumped from the tank; it first has to be mobilized, suspended in the liquid, and collected near a transfer pump. At the Savannah River Site, the sludge is mobilized using large mixer pumps (also called "sluicing pumps" or "slurry pumps"; see Figure III-2) that mix it into a slurry by directing a jet of water into the waste layer.

At the Hanford Site, mixer pumps are planned for retrieval of waste from the newer double-shell tanks. However, the sludge waste in Hanford single-shell tanks is mobilized using different equipment. For structurally sound single-shell tanks, high-pressure jets of water and mixer pumps and transfer pumps, similar to the equipment used for saltcake retrieval, will be used to retrieve sludge. For single-shell tanks with questionable integrity, a vacuum retrieval system in combination with a mobile retrieval system (except for smaller-diameter 200-series single-shell tanks) will be used to retrieve sludge.

The Savannah River and Hanford Sites recognize that using recycled fluids, such as supernatant liquid from the same or surrounding tanks, to mobilize the sludge is better than using so-called fresh water because there is no net increase of waste volume and no subsequent need for evaporation. Moreover, the supernatant liquid has a viscosity 10 to 20 times higher than that of raw water because of its high salt content. The supernatant liquid has self-sealing properties (i.e., it will plug small leaks) and keeps the slurry better suspended. However, if only a small addition of fluid is

FIGURE III-2 Mixer pump used in Hanford Tank AZ-101.
SOURCE: Gasper, 2003.

needed to suspend solids, raw water is used to minimize equipment contamination, worker exposure, and risk of leaks.

The jets from the mixer pumps can be directed to impact various parts of the sludge layer and steer it toward the transfer pump intake. As with the supernatant liquid and dissolved saltcake, a transfer pump is then used to pump the slurry to a double-shell tank or compliant ("new-type") tank. Eventually, the sludge from different tanks will be sent to processing facilities at the Hanford Site to separate the high-activity fraction from the low-activity fraction of the waste before vitrification, or washed then vitrified at the Savannah River Site.

The primary challenge in sludge retrieval is creating a slurrying jet that has sufficient energy to mobilize and mix the sludge across distances on the order of half a tank diameter (10.7 to 12.2 m or 35 to 40 feet) amidst the internal structures and debris. The Savannah River Site approach has been to build a steel superstructure atop each tank that is sufficient to mount four large (200 to 300 horsepower [hp]) electric motors. Each motor rotates a 12-meter- (40-foot-) long drive shaft having a directional pump attached at the bottom to mobilize sludge into a slurry that can be removed from the tank with a transfer pump. At Hanford, single-shell tank retrieval from tanks with questionable integrity is performed using either installed sluicing jets, the Hydrolaser, the vacuum retrieval system, or a crawler system described later.[4]

The approach to bulk sludge retrieval described above suffers from two significant shortcomings, the first of which is cost. The typical cost to retrieve the bulk of the waste from a single large tank is about \$16 million to \$24 million depending on whether there is a need to build a massive superstructure to support the mixer pumps (DOE-SRS, 2005d). The second shortcoming is the occupational dose to workers during mixer pump removal and insertion and during maintenance to repair problems caused by the bearings along the pump shaft.

Calcine

DOE has not yet begun retrieval of the 4,411 m^3 of calcine from the bins at Idaho National Laboratory. Technologies being investigated for this purpose are discussed in Appendix G and issues concerning future retrieval are discussed in the section of this chapter dedicated to the adequacy of in-tank waste characterization for waste retrieval.

Technology Advances in Bulk Waste Retrieval

Advancements in technology for retrieval of bulk wastes are directed primarily at making engineering improvements in the large mixer pumps used to mobilize sludge to reduce cost and worker dose through incremental improvements in pump reliability. A notable recent technology at the Savannah River Site that may represent a step improvement in bulk waste retrieval is the "waste on wheels" (WOW) concept for bulk sludge removal. The WOW concept is centered on a specially designed submersible mixer (slurry) pump. This pump is similar to a conventional mixer pump except that a 305 hp drive motor is coupled to the pump and submerged, instead of being attached with a long shaft (see Figure III-3). The submersible mixer pumps are designed to be "self-supporting," with some tanks requiring "limited" structural steel supports, while others require a more robust structural steel supporting system similar to standard slurry pumps. The amount of structural steel required will depend on a technical evaluation of each tank.

Submersible mixer pumps are designed to be retractable, portable, and reusable. The infrastructure (power, instrumentation) that supports each of these is also designed to be movable; hence, the name WOW. These pumps address the shortcomings of traditional mixer pumps: They eliminate the troublesome 12-meter shaft, have sufficient power and directional control so that the number of pumps used in each tank can be reduced from four to two, and allow bulk waste retrieval in some tanks without having to build a costly steel support structure over the tank. These pumps have been successfully demonstrated in the cold test facility at the Savannah River Site and installed to retrieve waste in Savannah River Site Tank 5 as this report was being written. As a consequence, information on the effectiveness and/or shortcomings of the submersible mixer pump is only just

[4]Mixer pumps have been installed in Hanford double-shell tanks only to suspend, mix, and homogenize feed solutions and slurries for transfer to waste staging and ultimate transfer to the Waste Treatment Plant for treatment.

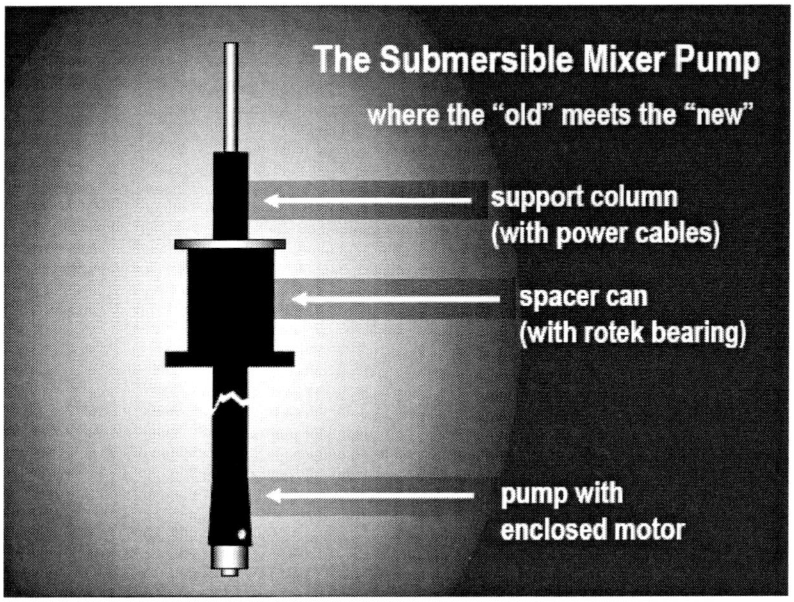

FIGURE III-3 Submersible mixer pump used at the Savannah River Site Tank 5. SOURCE: Clark, 2005b.

beginning to become available. Additional details are provided in a later section titled Retrieval Experience.

Residual Waste Retrieval

Residual waste is what remains after the application of bulk retrieval technologies described above. Residual waste can be composed of any or all of the following:

- Liquid and sludge in the bottom of the tank that could not efficiently be recovered using bulk retrieval technology;
- Radioactive material on the internal surfaces (e.g., walls, cooling coils) of the tank above the level of waste in the bottom of the tank; and
- Wastes composed of agglomerated materials that resist physical removal techniques to varying degrees. One example of such a material is the zeolite (see Appendix K) in some of the tanks at the Savannah River Site.

Waste remaining in the tank after the residual waste has been removed to the maximum extent practical is referred to as the "heel." The heel is the fraction of the waste that cannot be further mobilized and removed by practical means and, therefore, will be grouted inside the tank (see Chapter V). The heel is composed mostly of insoluble metal hydroxides, oxides, and (at Hanford) phosphates containing strontium-90 and transuranic isotopes that are often viscous or consolidated into solids that are difficult to remove or located in inaccessible locastions. To limit the amount of waste left in the tanks, the mobilization and collection steps of residual waste (after bulk waste retrieval) are critical to the success of the tank cleanup process.

Physical Technologies for Residual Waste Retrieval

The primary objective of residual waste retrieval technology is to mobilize and collect the waste at a point where it can be removed from the tank while minimizing secondary waste production. A variety of physical mobilization and collection technologies have been developed, and these have been implemented using a variety of deployment technologies. (Table F.2 in Appendix F summarizes the available technologies for residual waste retrieval.)

Physical technologies for mobilizing residual waste and collecting it at a point in the tank that is accessible by a transfer pump include using a combination of the following: hydraulic techniques involving the use of pressurized water; a vacuum; mechanical techniques to dislodge the waste for collection and mobilization. All physical techniques for retrieving residual wastes in tanks require some type of deployment technology. To date the dominant deployment technology for residual waste retrieval has been mechanical arms. The most popular type of arm used at DOE sites is a relatively simple mast inserted through a riser from which an arm containing one or two joints projects and some type of retrieval device extends (see Appendix G).

Chemical Technologies for Residual Waste Retrieval

When physical technologies for retrieving residual waste have not been deemed sufficient, DOE has employed chemical technologies to remove remaining tank wastes. Tests performed in the late 1970s and early 1980s (Bradley and Hill, 1977; Hill, 1978; West, 1980) proved oxalic acid solutions to be the most effective reagent for dissolving sludge while not corroding the carbon steel tank (see Appendix G).

One reason DOE has cited for not using oxalic acid is a concern that it may compromise nuclear criticality safety.[5] Since its interim report, the committee has received additional information from the Savannah River Site staff on criticality concerns. According to the Savannah River Site staff, sludge heels after bulk waste removal may contain enough fissile mass,[6] on average, to cause a criticality event (DOE-SRS, 2005e). As DOE recognizes, the available data are inconclusive with regard to the potential for criticality during oxalic acid dissolution because the bulk of the material, including the fissile material, may be removed before the ratio of fissile to poison falls below safe storage limits. Savannah River Site staff told the committee that each tank heel needs to be evaluated before attempting dissolution with oxalic acid to determine whether any potential for criticality exists and if so how to mitigate that risk. The committee has not seen any additional calculations regarding the criticality issues and cannot determine whether the concern is well founded. In Chapter IX, the committee recommends that DOE investigate the matter further.

According to DOE, there are additional downstream complications when using oxalic acid due to the large amounts of acid needed to dissolve the residual sludge. If oxalic acid is added to the sludge before washing, it goes to the evaporator and may precipitate in the feed tank. If oxalic acid is added to the salt tank, it may precipitate and form a hard salt layer that would require additional water to remove.

DOE has indicated that it is planning to use oxalic acid as a final cleaning step at the Savannah River Site on a tank-by-tank basis, taking into account the factors listed above (DOE-SRS, 2005d). Hanford is not planning at this time to use oxalic acid to clean its tanks.

Technology Advances for Mechanical Residual Waste Retrieval

DOE is considering the use of the Power Fluidic Technology™ developed by AEA Technology and a similar Russian Pulsating Mixer Pump technique for residual waste retrieval at the Hanford and the Savannah River Sites (Murray, 2005). This technology involves using two or more nozzles in a tank to establish a "back-and-forth" motion of the sludge and a vacuum induced by fluid flow through a "jet pump system," as shown in Figure G-6 in Appendix G.

[5]A criticality event is when a self-sustaining nuclear chain reaction occurs. Such an event can be a safety concern.

[6]Fissile mass is made of isotopes having a high probability of undergoing nuclear fission when struck by neutrons and, in the right quantities and configurations, can sustain a nuclear chain reaction.

This technology does not have any mechanical devices inside the tank and is supposed to use water more efficiently than conventional mixer pumps. Fluidic technology is claimed to be useful for bulk waste as well as residual waste retrieval and for mixing the grout with waste residuals that cannot be removed further. However, like other technologies, its effectiveness on waste in tanks with vertical cooling coils is uncertain.

Hanford tested a prototype using fluidic technology in a full scale cold test facility, and determined that the complex air and fluid controls and air handling systems would require significant maintenance. Concerns were also raised about radiation exposures to workers and meeting ALARA (as low as reasonably achievable) requirements for operations in the single-shell tank farm system. To the committee's knowledge, no further deployment of fluidic technology for waste retrieval from single-shell tanks is planned. Use of fluid jets for mixing grout with waste residuals is still under consideration.

GENERAL WASTE RETRIEVAL ISSUES

The discussion above has alluded to future difficulties that can be expected when retrieving waste from DOE's "complicated" tanks. Some of the important complications are discussed below. In addition to general issues, a number of site-specific issues may limit the extent to which wastes can be retrieved, as discussed in Appendix G.

- **Recalcitrant Waste Deposits:** Waste may be encrusted on internal tank surfaces or structures in a semidry form, or it can agglomerate in physical and chemical forms that resist physical removal technologies. These recalcitrant waste deposits must be mobilized before they can be collected and removed from the tanks.
- **Waste Accessibility:** To remediate a tank, the waste has to be accessible to waste removal streams or tools. One of the main challenges in waste removal is the number and type of physical obstacles in the tanks and tank design features that complicate waste retrieval operations. The most difficult issue to overcome is likely to be the vertical cooling coils in Savannah River Site tanks, which severely impede the ability to maneuver water jets, mechanical arms, and in-tank vehicles. The Hanford tanks, although they do not have cooling coils, have other internal obstructions such as the air-lift circulators that were used to stir and suspend the sludge in high-heat tanks.
- **In-tank Debris**: Most underground storage tanks, especially at Hanford and Savannah River Site, contain objects labeled as debris. These can include failed pumps, instrument trees, sluicers, hoses, and miscellaneous items. Some items have been left in place (e.g., suspended from the top of the tank), while others (such as tapes for measuring waste depth) have been dropped into the tanks. If heel retrieval systems that use vehicles or mobile in-tank systems

are to be deployed, debris may become a problem even if these systems are capable of removing or moving this debris around the tanks.

• **Residual Waste in Pipelines and Ancillary Equipment:** The large underground storage tanks that are the primary focus of retrieval are interconnected by myriad pipelines and ancillary equipment, such as small underground storage tanks, pumps, valve boxes used to transfer tank wastes among the large tanks, and other hardware used in waste retrieval operations. As a part of routine practice, the majority of pipelines and ancillary equipment has been or will be flushed with nonradioactive water. Consequently, the remaining waste is expected to contain amounts of radionuclides that are small relative to what is being left in the tanks. Foremost among the challenges associated with the pipelines, is the fact that some of them became plugged during use so they could not be flushed to remove most of the residual radionuclides. DOE is inclined to propose that most pipelines and ancillary equipment be flushed and grouted in situ, while it considers exhumation of plugged pipelines where necessary (Harbour et al., 2004; Schaus, 2005).

• **Leaks to the Environment:** To varying degrees, all three DOE sites have inadvertently released waste contained in tanks into the environment. Such releases raise two issues: (1) the degree of retrieval from the tanks that should be required, given the amount of radionuclides in the immediately surrounding environment, and (2) the application of removal to the "maximum extent practical" to leaked radionuclides.

EVALUATION OF RETRIEVAL TECHNOLOGY STATUS

The purpose of this section is to evaluate the status of retrieval technologies by drawing on the discussion in the foregoing sections.

Technology Availability

Table F-2 in Appendix F summarizes the main waste retrieval tools in DOE's "toolbox." The waste retrieval tools are grouped in broad classes (see column 1, Appendix F) because each site has developed or is using modified versions of the techniques. The table also indicates if and where a device was deployed or tested.

Retrieval Experience

All three DOE sites have operational experience retrieving liquid wastes from many tanks as a result of years of moving waste among the tanks as a part of day-to-day operations in the tank farms. There is no estimate of the amount of liquid waste that has been retrieved but such operations are considered routine. Table F-1 in Appendix F shows the status of tank waste retrieval operations at the three sites.

Both Hanford and Savannah River Site staff have some experience retrieving saltcake wastes (Idaho tanks have no saltcake) although not nearly as much as with liquids. In some cases, higher-pressure liquid jets and higher dissolution temperatures have been needed to overcome endothermic reactions and some complex waste chemistry solid-liquid equilibrium issues. However, DOE has substantial operating experience in dissolving saltcake and retrieving the resulting liquids.

The Savannah River Site has some operational experience with sludge removal. Sludge removal operations began in 1969 at the Savannah River Site for tank space management, to test waste retrieval technologies, or to retrieve sludge for processing and immobilization in the Defense Waste Processing Facility (Saldivar, 2002; DOE-SRS, 2005a). However, past sludge removal operations were not carried out with the purpose of removing waste from a tank "to the maximum extent practical." That is, there is much more experience with bulk retrieval of sludge than with residual waste retrieval. Savannah River Site tank retrieval experience is summarized as follows:

• Tank 16 was taken out of service because of leaks and became the first case in which DOE attempted to completely remove the bulk of the waste from a tank at the Savannah River Site. In 1979 254 m^3 (67,000 gallons) were removed with hydraulic techniques, and 5.3 m^3 (1,400 gallons) were removed in 1980 with chemical cleaning. There was minimal saltcake in this tank. The slurry pumps were extremely effective in removing the sludge. Oxalic acid was used to remove the residual waste. This tank has cooling coils.

• Waste has been retrieved from Tanks 17 and 20, and these tanks have been filled with grout. Tank 20 contained very little sludge, while Tank 17 contained a minimal amount of sludge and no zeolites. Neither of these tanks has cooling coils.

• Waste has been retrieved from Tanks 18 and 19. DOE has submitted a draft waste determination necessary to allow these tanks to be closed (DOE-SRS, 2005a). Tank 19 contained very little sludge and a substantial amount (49.2 m^3 [13,000 gallons]) of zeolites, while Tank 18 contained substantial amounts of sludge and minimal amounts (7.6 m^3 [2,000 gallons]) of zeolites. Neither of these tanks has cooling coils.

• Waste retrieval from Tank 5 began in October 2005 and is ongoing. Bulk waste retrieval ended mid-December 2005. The submersible mixer pump used on this tank is described above (Figure III-3). The volume of wet solids remaining after the first phase of bulk waste retrieval in Tank 5 is between approximately 68 m^3 (18,000 gallons) as calculated and 93 m^3 (24,500 gallons) as estimated from mapping the sludge mound at the bottom of the tank (Daily, 2005a). Most of the remaining solids were located in the northwest quadrant of the tank where an array of cooling coils partially obstructed the mixer discharge jets. This is the first tank with cooling coils that has undergone bulk retrieval operations since Tank 16. A second mixing campaign was initiated in late November 2005 using a new mixer orientation to focus on the remaining solids (Purohit, 2005). An additional 7.6 m^3 (2,000 gallons) of sludge were removed in the second sludge retrieval campaign (Daily, 2005b). The submersible mixer pumps have been taken out of Tank 5 and installed in Tank 6 for bulk waste removal. Waste removal activities in Tank 5 will resume following completion of waste removal from Tank 6. Heel removal and cleaning of Tank 5 will follow completion of the bulk waste removal phases.

• Tank 6 is also undergoing waste retrieval (February 2006). As noted above, the submersible mixer pumps have been removed from Tank 5 and installed in this tank. Waste removal activities in Tank 6 are anticipated to occur through April 2006. This tank also has cooling coils.

In summary, after Tank 16, Savannah River Site staff chose to begin waste retrieval from relatively uncomplicated tanks to obtain experience and is now proceeding to work on the more complicated Tank 5. Removal operations in Tank 5 and Tank 6 are DOE's first attempt to retrieve substantial quantities of sludge from tanks containing vertical cooling coils. Tank 5 is a leaking tank, with 13 known leak sites, although there is no significant waste accumulation in the tank annulus. A new leak appeared within two weeks of initiating sludge removal operations but only as a wet spot on the tank's outer wall. Leaks are monitored with video cameras in the tank annulus; however, only 25 percent of the annulus surface is visible because of the location of the cameras and risers.

Tank 4 is the next Savannah River Site tank to be cleaned; it also contains sludge and cooling coils. The order in which waste is retrieved from the tanks depends on the feed blending requirements for the Defense Waste Processing Facility.

The Hanford Site has somewhat less operational experience with sludge removal than Savannah River. Hanford experience is as follows:

• Tank C-106 had a significant amount of sludge removed (187,000 gallons) during the 1997-1998 campaign using sluicing techniques. Sludge removal was enhanced using modified sluicing and chemical dissolution (additional 32,500 gallons) in the 2003 campaign.

• Sludge removal using modified sluicing started in November 2005 and is in progress in Tank C-103 (75,000 gallons of sludge initially).

• Hanford gained experience removing hard deposits of saltcake in Tank S-112 by using water jets to break them up, facilitating dissolution.

• Waste retrieval using the vacuum retrieval system is deemed complete in Tanks C-202 and C-203 (both 208.2 m^3 [55,000-gallon] tanks).

• Waste retrieval using the vacuum retrieval system is under way for Tank C-201 (started in October 2005).

• Waste retrieval from Tank C-204 will begin in April 2006.

• Removal of 302 m^3 (80,000 gallons) of sludge from Tank C-108 has slated to start late in 2005 using the modified sluicing technique.

• Removal of 302 m^3 (80,000 gallons) of sludge from Tank C-101 using the Mobile Retrieval System is planned but a date has not been set.

Because of the number of tanks and waste types present at this site, the sludge removal experience gathered thus far is not necessarily representative of the type of waste retrieval challenges faced when cleaning the site's 18 tank farms. To date, all experience has been in a few tanks inside just two of these tank farms.

The Idaho site has retrieved waste from 7 of the 11 1135.6 m^3 (300,000-gallon) tanks and from all 4 of the 113.6 m^3 (30,000-gallon) tanks (see Table F.1 in Appendix F for details). The other tanks are still being used to hold sodium-bearing waste awaiting immobilization. The site has successfully deployed the same retrieval techniques (pumping and water washing) for the tanks thus far and plans to use the same technology for all of the remaining four tanks.

Retrieval Effectiveness

The most important measure of retrieval effectiveness is the amount of radioactive material left in the tank for on-site disposal (i.e., increasing retrieval effectiveness leaves decreasing amounts of radioactive material). While bulk retrieval may remove most of the radionuclides, the effectiveness of bulk retrieval technologies is essentially irrelevant to determining the amount of radionuclides left in a tank. As a consequence, this section focuses on the effectiveness of residual waste retrieval technologies. The effectiveness of DOE's retrieval efforts that have been completed at Hanford, the Savannah River Site, and Idaho National Laboratory is summarized in Table F-1 in Appendix F.

Based on the information in Table F-1, it is evident that waste retrieval has been very effective for certain tanks. For example, the heels in Savannah River Site Tanks 17 and 20 and all of the Idaho National Laboratory tanks cleaned to date each contain less than 3,000 Ci ($1. \times 1.11 \times 10^{14}$ Bq). Indicators of more effective retrieval appear to be (1) the absence of solid materials such as sludges and (2) the use of chemical cleaning on deposits where the chemicals would be expected to be effective. Experience with cleaning tanks with oxalic acid indicates that it is not universally effective in reducing the amount of radionuclides remaining in the tank. For example, it appears to be effective on many sludges but not on zeolite deposits.

Indicators of less effective retrieval appear to be the presence of recalcitrant solid materials, especially zeolite deposits containing relatively high concentrations of cesium-137. It is too soon to tell how the general and site-specific issues discussed above will affect retrieval effectiveness because retrieval from tanks having such materials is just beginning.

In summary, each tank is essentially unique in terms of waste type, tank design, and history. DOE's bulk waste removal strategy is sensible and has thus far been successful. DOE is learning from experience, improving its understanding of the unique conditions in each tank as it progresses with tank cleanup, and developing and adapting technology to accommodate these differences. DOE is to be commended for exchanging technical information among sites to allow lessons learned to be incorporated in the program. Activities such as the yearly Hanford and Savannah River Site technical exchange meetings, the workshop on lessons learned from the closure of Tanks 17 and 20 at the Savannah River Site in 1998, frequent conference calls, and activities previously undertaken by the Tanks Focus Area are helpful. The Savannah River Site is also planning a technical workshop in March 2006 to identify technologies, including commercially available ones that could be implemented for the heel removal and tank cleaning activities for Tanks 5, 6, and others, with emphasis on techniques that minimize secondary waste volume generation.

The Tanks Focus Area was a particularly helpful approach for centralizing information and lessons learned for all DOE tank sites. The Tanks Focus Area was a user-driven, needs-based program within DOE's Environmental Management Office of Science and Technology.[7] This program performed research, development, and deployment activities on tank waste characterization, monitoring, safe waste storage, retrieval, closure, pretreatment, and immobilization. The Tanks Focus Area's reports and the Innovative Technologies Summary Reports (the latter was another initiative of DOE's Office of Science and Technology) are still available on the Internet and provide a valuable although somewhat dated centralized information database on DOE's tank waste sites, including lessons learned.[8] (See further discussion in Chapter IX.)

Tank Waste Retrieval Operations at Other DOE Sites

DOE has two additional sites at which radioactive waste has been stored in underground tanks: (1) the West Valley Demonstration Project (WVDP), New York, and (2) Oak Ridge National Laboratory, Tennessee. Table III-1 summarizes the main features of the tanks at these two sites, which are discussed briefly below. More detailed descriptions of the tank cleanup efforts at these sites can be found in Appendix G. Some of the information is based on data published in 2002.[9] Since then, most of the waste at the two sites has been retrieved and the remaining out-of-service tanks at Oak Ridge have been closed. DOE reports that retrieval of high-level waste from the West Valley Demonstration Project tanks was completed in 2002 and the residual sodium-bearing wastewater was retrieved from the two largest underground tanks in 2003. The wastes at these sites have some characteristics in common with wastes at the Savannah River Site and Hanford, although the tanks in New York and Tennessee are not as large as the largest tanks at the Hanford and Savannah River Sites (see Table III-1).

West Valley Demonstration Project

Between 1966 and 1972, the West Valley Demonstration Project plant generated approximately 2,317 m^3 (612,000 gallons) of high-level waste. This waste consisted of 2,271 m^3 (600,000 gallons) of alkaline plutonium-uranium extraction (PUREX) sludge and supernatant liquid and approximately 45 m^3 (12,000 gallons) of acidic thorium extraction (THOREX) waste. The site had two tanks for these waste streams, and two additional tanks were used for secondary waste streams during waste processing. The retrieved wastes were immobilized in a vitrification facility, which operated between 1996 and 2002. The vitrified high-level waste canisters are intended for disposal in a geologic repository.

Oak Ridge National Laboratory

The liquid low-level radioactive waste system at Oak Ridge National Laboratory (ORNL) dates back to the mid-1940s when ORNL was constructed as part of the Manhattan Project. ORNL was a pilot site for Hanford chemical processes and had approximately 390 m^3 (103,000 gallons) of sludge stored in a system consisting of nine large Gunite[10] tanks and numerous smaller metal tank systems (Lewis et al., 2002a). In addition, the tanks contained small quantities of dried waste that had the consistency of chalk. The Gunite tank sludge (solids) contained approximately 85,000 Ci (3.1×10^{15} Bq). This radioactivity came from uranium, plutonium, thorium, and other long-lived isotopes, as well as from the high concentrations of cesium-137 and strontium-90, which have relatively short half-lives. The tanks also contained organic materials in trace amounts and other heavy

[7]DOE's Office of Science and Technology along with the Tanks Focus Area were disbanded in 2002 when the Office of Environmental Management was reorganized.

[8]Tanks Focus Area reports are available at *http://www.tanks.org* and the Innovative Technologies Summary reports available at *http://apps.em.doe.gov/OST/mainpubs.asp*.

[9]DOE provided data on West Valley and Oak Ridge in addition to the data taken from the cited references. Not all of these data agree, as reflected in differences between the numbers in Table III-1 and the text. The committee did not research or examine them further because these sites are not the focus of this study.

[10]Gunite is a mixture of Portland cement, sand, and water sprayed over a wire mesh and reinforcing rod frame.

TABLE III-1 Tank Waste Retrieval Programs at the West Valley Demonstration Project, New York and Oak Ridge National Laboratory, Tennessee

	West Valley Demonstration Project	Oak Ridge National Laboratory
Number of tanks to close	4	67 (does not include tanks currently in active service)
Tank types and material	2 carbon steel tanks (8D-1 and 8D-2) and 2 stainless steel tanks (8D-3 and 8D-4)	Various configurations of tanks. Includes 16 Gunite tanks; the remainder are stainless steel tanks (horizontal and vertical orientations that are direct buried, located in vaults, or in building basements)
Tank sizes	57 to 2,839 m^3 (15,000 to 750,000 gallons)	0.76 to 644 m^3 (200 to 170,000 gallons)
Tank diameters	21 m (70 feet) carbon steel tanks 4 m (12 feet) stainless steel tanks	1 to 15 m (3 to 50 feet)
Tank conditions	No leakers	No leakers; however, many inactive tanks collected in leakage prior to closure
Waste types	Alkaline (neutralized PUREX waste) in 8D-1 and 8D-2; acidic (THOREX waste) in 8D-4; decontamination process solutions in 8D-3	Low-level waste liquids, low-level and transuranic sludges
Waste volumes	2,317 m^3 (612,000 gallons)	1,893 m^3 (500,000 gallons)
Radionuclide contents	300,000 Ci not including short-lived decay products.	5,500 Ci
Internal obstructions	Significant (complex bottom structural grid work and some in-tank hardware)	In-tank hardware, including structural supports, piping, and instrumentation
Status of retrieval	More than 97% of long-lived radionuclides removed.[a]	Interim CERCLA closure complete for 67 inactive tanks (1995 through 2005). Retrieval and disposal of sludge from active (i.e., in-service) Melton Valley storage tanks to be completed as part of Transuranic waste program[b,c,d,e]
Bulk waste retrieval methods	Mobilization pumps, sluicing, and pumping	Sluicing (using nozzles, lances, and borehole miner) or mixing (using fluidic pulse jet, pulsed air, pulsating mixer pump, and Flygt mixers) and pumping[b]
Residual waste retrieval methods	Flushing, acid washing, robotics, and sluicing	Mixing (Flygt mixers and fluidic pulse jet), sluicing (borehole miner, high-pressure nozzles, and lances), and confined sluicing
Closure schedule	Not yet finalized; closure expected to take up to 20 years	67 Federal Facility Agreement Category C and D tanks (inactive tanks) closed through 2005; remaining active service tanks will be scheduled for closing as storage mission is completed
Maximum tank age (years at closure)	More than 50	More than 50
Site- and tank-specific considerations and uncertainties	Corrosion; "bathtub ring" along the walls of 8D-2; 8D-1 contained spent zeolites (3 m^3 of solids);[d,e] water in vaults	Waste not classified as high-level; in-tank chunks of Gunite; various types, sizes, and configurations of tanks

NOTES: CERCLA = Comprehensive Environmental Response, Compensation, and Liability Act. DOE provided the data in this table and the committee did not research or examine them further because these sites are not the focus of this study.

[a] Waste retrieval status as of February 2, 2006, from DOE-WV, 2006.
[b] Waste retrieval status information from DOE-ORNL, 2005.
[c] Elmore and Henderson, 2002a.
[d] Bamberger et al., 2001.
[e] Hamel and Damerow, 2001.
SOURCE: Adapted from Elmore and Henderson, 2002a.

metals. Groundwater had leaked into the tanks, adding approximately 1300 m^3 (345,000 gallons) of water. This water accumulated on top of the sludge in an aqueous supernatant layer. The supernate and the tank walls contained an additional estimated 15,000 Ci (5.6×10^{14} Bq). In tank cleanup operations, ORNL constituted a test bed for a number of retrieval technologies considered for deployment at other sites.

Comparing Table III-1 with Table II-5, which lists information about the tank wastes at the Savannah River Site, the Hanford Site, and Idaho National Laboratory, shows the similarities and the differences among the tank types and types of waste at Oak Ridge and West Valley. The experience with Silo 3 calcine retrieval at the Fernald Environmental Management Project discussed in Appendix G also bear some similarities to Idaho National Laboratory's calcine retrieval, although the radioactive contents of calcine are very different. Given the differences in the physical and chemical properties of the waste and tank designs, not all the experience gathered at these other sites is relevant to the three sites addressed in this report. However, sharing lessons learned and communication among the sites may still be helpful as DOE continues its cleanup mission.

Cost of Bulk and Residual Waste Retrieval Using Physical Techniques

The cost of waste retrieval includes costs for bulk waste retrieval and costs for residual waste retrieval. Despite the relatively limited number of waste retrievals that have been completed, the cost of waste retrieval has decreased substantially as a result of learning from experience. The best example of this occurred at Hanford (see Figure III-4) where the combined cost of the first retrieval effort on a large tank (C-106) was about \$140 million, whereas the comparable costs of more recent retrievals ranges from \$20 million to \$40 million. The projected cost for the next group of tanks (208.2 m^3 [55,000-gallon]) is around a few million dollars each and estimated to be about \$15 million for a large tank.

The first cost decrease in Figure III-4 (from Tank C-106 to Tanks S-112 and S-102) is attributable to switching from "past-practice" sluicing, which used very large volumes of water, to "modified sluicing," which uses water more efficiently to target, mobilize, and collect the sludge. The second cost decrease (to Tanks C-201 through C-204) is attributable to moving from permanently installed equipment and pipelines to portable modules that are reusable (to some extent),

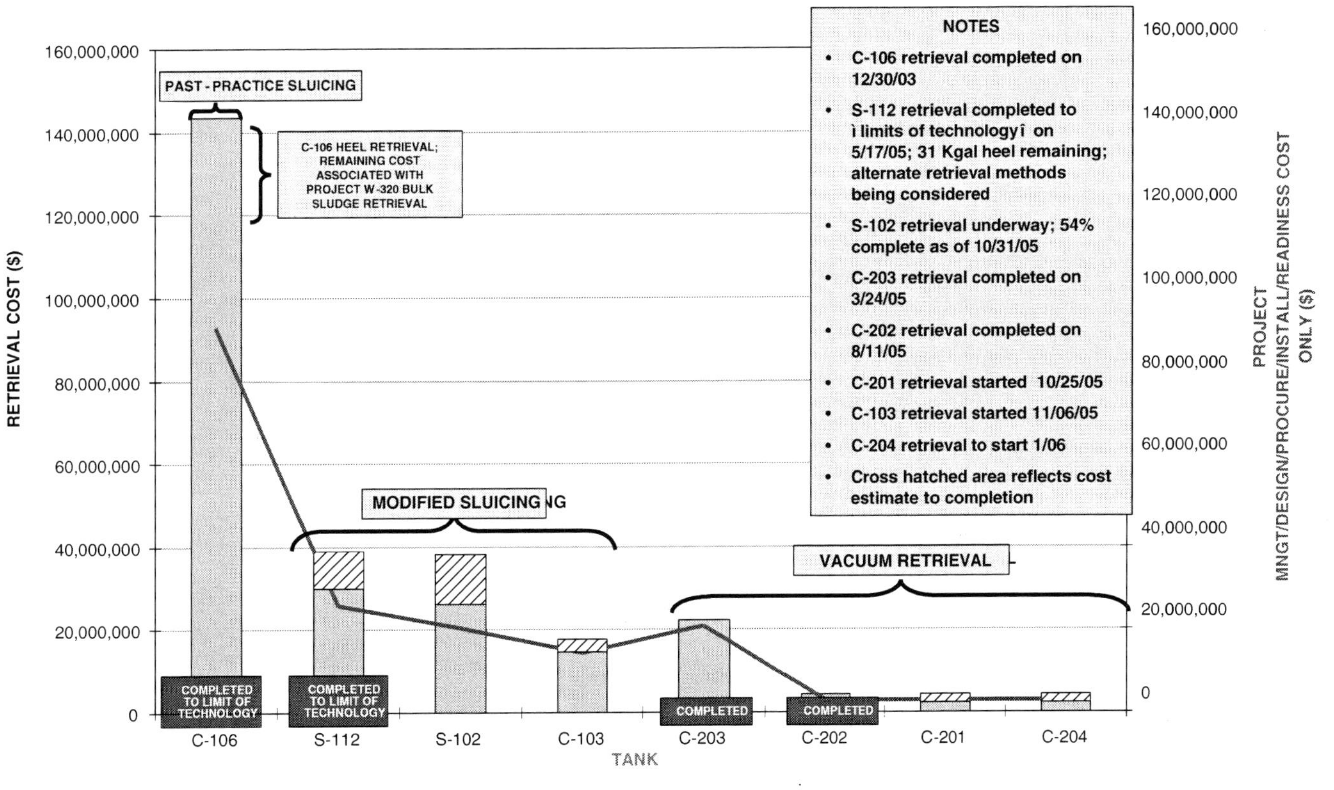

FIGURE III-4 Waste retrieval costs and status at Hanford as of November 17, 2005. SOURCE: Quintero, 2005a.

and the use of temporary above-ground transfer lines that substantially reduce infrastructure costs. The third (projected) cost decrease appears to be driven by a transition from existing hydraulic technologies (e.g., water jets) to the reusable Mobile Retrieval System (see Appendix G), which further reduces water usage by employing a vacuum device combined with a mechanical vehicle to push waste. It should be noted that in all cases, most of the cost is attributable to moving equipment to a tank and installing it, not to retrieval operations per se. The estimated cost for retrieving waste from the larger 100-series tanks using the Mobile Retrieval System is about \$20 million per tank. However, the cost savings realized so far do not necessarily reflect substantive cost savings in waste removal operations in the future, given the limited number of tanks that have been completely cleaned to date.

Retrieval at the Savannah River Site began with modified sluicing for bulk waste retrieval. As a consequence, combined retrieval costs are comparable to those at Hanford and range from \$23 to \$34 million per tank excluding costs for chemical cleaning. As previously mentioned, the Savannah River Site has just begun retrieving waste from Tank 5 using a submersible mixer pump. This technology is expected to reduce the retrieval cost to a range between \$6 and \$14 million. The cost reduction is attributable to the pump's added effectiveness compared to the long-shaft mixer pumps, so that the number of pumps required in each tank is reduced from four to two; also to the use of the submersible pumps in multiple tanks, which reduces the cost per tank. The spread in cost range is attributable to the need for superstructure above the tanks: tanks with small amounts of sludge do not require a superstructure, whereas tanks with large amounts of sludge would require such a superstructure (DOE- SRS, 2005d).

The design safety analysis at the Savannah River Site requires an evaluation for hydrogen gas release from tanks containing sludge when mixing-removal operations are planned. Safety analyses assume that the quantity of sludge in the tank dictates the amount of hydrogen gas retained in the sludge (same number of grams of hydrogen per inch of depth). Mixing the sludge releases gas that is otherwise held up in the sludge. There is a limit to the allowable rate of hydrogen released. The submersible mixer pumps mix everything above them and are assumed not to mix sludge 30-45 cm (12 to 18 inches) below. To reduce the gas release rate, submersible mixer pumps are suspended closer to the top of the sludge and work their way down. Suspending the submersible mixer pumps at different heights requires a superstructure. In the case of Tank 5, because of its gas release calculation, all of the sludge could be mixed simultaneously. Therefore, the Tank 5 submersible mixer pumps could be lowered to the tank bottom. In some of the other tanks (e.g., Tank 4), the gas release calculation from sludge mixing may dictate that submersible mixer pumps cannot be lowered to the tank bottom, and therefore some superstructure will be required to support them at different levels of mixing.

At the Idaho National Laboratory, waste has been retrieved from seven 1136 m^3 (300,000-gallon) tanks and four 114 m^3 (30,000-gallon) tanks from 2002 to 2005 at a total (development plus operations) of \$35 million. This yields an average cost of \$7 million per tank, although this number is likely to decrease slightly because some of the development costs will be allocated to the four remaining tanks where retrieval has not yet been completed. Although many of the Idaho tanks have cooling coils on their bottoms, the relatively low retrieval costs can be attributed to the favorable physical characteristics of the waste (i.e., absence of hard sludges, solid particles easily washed from walls and suspended in the tank liquids).

Cost of Additional Residual Waste Retrieval

One important factor in determining whether radionuclides have been removed to the maximum extent practical is the cost of retrieving additional residual waste from the tanks when the limit of the baseline technology has been reached. Savannah River Site staff has developed cost estimates for additional waste retrieval from Tanks 18 and 19 using three different approaches: (1) additional hydraulic sluicing using a closed loop similar to that in the Mobile Retrieval System now being deployed at Hanford, (2) an in-tank vehicle similar to the vacuum device now being deployed at Hanford as part of the Mobile Retrieval System, and (3) chemical cleaning with oxalic acid. The additional cost for such cleaning is estimated to range from \$10 to \$15 million per tank if a tank has been taken "out of service." This estimated cost for chemical cleaning can be compared to the estimated cost for chemical cleaning of Tank 16 (\$250,000) and a 1999 estimate of about \$1 million.

The significant increase in the recent estimate for Tanks 18 and 19 is due to the fact that these two tanks have been taken out of service, which means substantial efforts are required to reverse preparations made for closure (e.g., reconnecting support equipment and utilities) and additional safety and operational constraints. These same assumptions were used for the other two alternatives. Presumably, the cost of applying chemical cleaning to a tank that is still in service would be closer to \$1 million than to \$10 million given the additional safety requirements that were introduced since oxalic acid was applied to Tank 16 (Hill, 2005).

At the Hanford Site, costs for additional waste retrieval from Tank C-106 have been estimated to be \$1.24M to \$2.97M per m^3 (\$35,000 to \$84,000 per cubic foot), or a total of \$5.7M to \$13.5M, depending on the retrieval technology if it is assumed that approximately 4.5 m^3 (160 cubic feet) of waste would be removed (Sams, 2004). This assumption was made for purposes of analysis only and is by no means assured if the additional technology were to be actually deployed.

At the Idaho National Laboratory, development and deployment of a new technology for retrieval of additional residual waste is estimated to cost slightly more than development and deployment of the existing tank cleaning system. The cost of the new technology (mechanical, chemical, or arm based) was estimated by applying an escalation rate of 10 percent to the tank waste retrieval cost to date, yielding \$38.5 million per tank. Idaho National Laboratory believes the actual cost would likely be higher because the new technology may not be able to use much of the existing infrastructure at the site. A large fraction of the cost is independent of the technology used to retrieve waste because it involves the preparation of plans and other documentation, testing, readiness reviews, equipment mobilization, installation, demobilization, and leak detection and monitoring during waste retrieval.

Worker Dose Estimates During Retrieval

Cleaning up large underground tanks is a hazardous operation even for tanks that do not contain radioactive waste (Cole, 1992). The additional cumulative worker doses resulting from further retrieval efforts is estimated to be 7,150 millirem (mrem or 71.5 millisieverts, mSv) for 11 1135.6 m^3 (300,000-gallon) tanks[11] at the Idaho National Laboratory and ranges from 4,500 to 7,500 mrem (45 to 75 mSv) per tank at the Savannah River Site (DOE-ID, 2005a; DOE-SRS, 2005a). The larger values at the Savannah River Site are presumably a result of the substantial effort that would be required to prepare a "ready-for-closure" tank for retrieval operations. Site staff are still analyzing the information on worker dose risks gathered during waste retrieval (e.g., through radiation surveys and job task analysis). Hanford does not have worker dose estimates for waste retrieval. Information on worker doses during retrieval from Tank C-106 is not readily available because waste retrieval operations comprised many activities spread over a 7 to 10 years and the site did not keep track of cumulative worker dose at that time. It is possible to reconstruct the cumulative worker dose for the entire period, but the information has not yet been compiled (Quintero, 2006).

Estimated Dose to the Public from Closed Tanks

The reduction in dose to the public from retrieving additional residual waste has been estimated by DOE for both the Savannah River Site and Idaho National Laboratory tank waste determinations (DOE-SRS, 2005a; DOE-ID, 2005a) and for Tank C-106 at Hanford (Sams, 2004). The estimates are based on the extent to which the dose rate to a hypothetical member of the public would be reduced by removing or reducing the radionuclide inventory in the tank and multiplying this by an assumed 50 years of exposure to yield the dose avoided by an individual. If 100 percent of the residual waste were to be removed by additional retrieval, a situation that is not actually possible because of limitations on the efficiency of separations, the avoided all-pathways dose to an adult residing in proximity to the F Area Tank Farm at the Savannah River Site is 2 mrem for Tank 18 and 0.45 mrem (4.5 μSv) for Tank 19 (DOE-SRS, 2005a).[12] The Idaho Section 3116 waste determination for its tank farm estimates that the remaining radioactivity in the tank farm poses a potential radiation all-pathways dose to a member of the public on the order of 0.5 mrem (5 μSv) per year, with approximately half of the dose due to tank heels. Removing all tank residuals would achieve a dose reduction of approximately 0.25 mrem (2.5 μSv) per year, or 12.5 mrem (125 μSv) over 50 years, regardless of cost (DOE-ID, 2005a).

The Hanford Site estimates that removing an additional 4.5 m^3 (160 cubic feet) of residual waste from Tank C-106 would reduce the all-pathways doses from the current residual volume (10.5 m^3 [370 cubic feet]) from 2.5×10^{-3} to 1.39×10^{-3} mrem per year (2.5×10^{-5} to 1.39×10^{-5} mSv per year) calculated at the fence line of Area C (Sams, 2004, p. 2-37). Therefore, the avoided dose over 50 years would be 0.05 mrem (or 1.11×10^{-3} mrem [1.11×10^{-5} mSv] per year over 50 years) for Tank C-106. Hanford estimates that this further waste reduction entails an incremental lifetime cancer risk reduction of 8×10^{-9} at a cost of \$5.7 million to \$13.5 million. The avoided dose over 50 years at each site is significantly different because of differing natural settings (e.g., different hydrogeology) and assumptions used in the underlying performance assessments (see Chapter VI).

Adequacy of In-Tank Waste Characterization for Waste Retrieval

Waste retrieval requires knowledge of physical data about the waste (e.g., density, particle size, viscosity, fraction of solids) to facilitate pumping and to avoid plugging the pipelines. Such knowledge is gathered relatively easily by grab sampling. DOE's approach has been thus far to begin waste retrieval with the knowledge available on the waste and adapt retrieval technologies to the conditions encountered in the tanks.

One case that may prove to be more complicated than anticipated is retrieval of calcine waste at the Idaho National

[11]Tank cleaning worker exposure is estimated at 650 mrem (6.5 mSv) per tank for 23 workers for a total exposure of about 7.15 rem (71.5 mSv) for 11 1135.6 m^3 (300,000-gallon) tanks (DOE-ID, 2005a, p. 58). All doses cited are effective dose equivalent.

[12]The assumption is that members of the public construct a dwelling near, but outside of, the F Area Tank Farm on the Savannah River Site. The location of the residential dwelling is assumed to be downgradient near Fourmile Branch just downstream of the seepline for the entire 10,000-year period of analysis. The resident is assumed to use Fourmile Branch for recreational purposes and sustenance (Gilbreath, 2005).

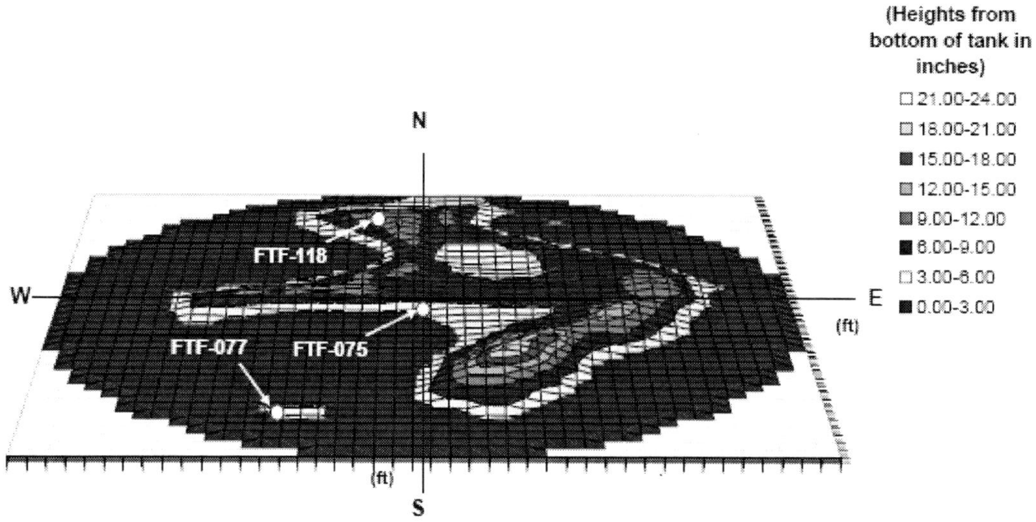

FIGURE III-5 Residual material configuration and sample locations for Tank 19 at the Savannah River Site. SOURCE: DOE-SRS, 2005a, p. 66.

Laboratory. (Retrieval of calcine is discussed at greater length in Appendix G.) DOE plans to retrieve this granular powder from the stainless steel silos (bin sets) by reversing the process by which it was emplaced: vacuuming it from the bins through the risers at the tops. DOE conducted a successful test with simulated waste: caked calcine was readily dislodged by physical contact with the vacuum head, and the calcine could be transported pneumatically. However, whether the actual calcine will behave like the simulant after residing in the bins for several decades is uncertain. Retrieval of a different calcine from a silo at DOE's Fernald site proved challenging because of compaction. A previous National Research Council (NRC, 1999b, p. 22) report states that, given the little characterization information about the Idaho calcine (see Chapter II),

> it would be difficult to conclude that there would be no problem with pneumatic retrieval. Indeed the committee believes that there will be problems but that they can probably be handled. However, this eventually might require mechanical operations to aid particle flow and more elaborate retrieval methods (e.g., a manipulator arm) than simple pneumatic transfer.

The committee agrees with this assessment. Previous studies (NRC, 1999b; CRESP, 2005) have concluded that the calcine in the bins can be safely stored for hundreds of years because of the absence of water in the bins, the stability of the calcine, and the dry environment. As a consequence, addressing the disposition of the wastes contained therein has justifiably been accorded a lower priority in the face of more pressing issues at the Idaho National Laboratory and other DOE sites. The committee identifies the calcine retrieval and bin disposition as issues for further examination.

Adequacy of Retrieved Waste Characterization to Support Operations and Disposal of Retrieved Wastes and Heels

Characterization of heels is necessary to determine their chemical and radionuclide composition (actinide content, alpha, gamma, and beta emitters) to calculate the "source term" in the performance assessment of the tank once closed. Tank heels left in the tanks must also be characterized to demonstrate that the waste has been adequately removed and that the residues can be left in the tanks. Heel characterization is performed by visual inspection through cameras, sampling, special analysis,[13] and predictions based on knowledge of wastes that were sent to the tank (process knowledge). Figure III-5 shows an estimate of the residual waste at the bottom of Tank 19 at Savannah River after residual waste removal based on a visual inspection.

Obtaining, packaging, shipping, and analysis of each sample resulted in personnel radiation exposure. Tank 19 sample dose rates were as high as 8,000 mrem (80 mSv) per hour dose to the hands and 4,000 mrem (40 mSv) per hour dose to the whole-body. The total whole-body worker dose received from the three solid samples was approximately 200 mrem (2 mSv). Tank 18 sample dose rates were as high as 150 mrem (1.5 mSv) per hour dose to the whole body. The total whole-body worker dose received from the six solid samples and the one liquid sample was approximately 120 mrem (1.2 mSv) (DOE-SRS, 2005a, p. 63). Detailed discussion of tank heel sampling can be found in the performance objective demonstration document for Tanks 18 and

[13]For example, using sampled radionuclide concentrations to estimate other radionuclide concentrations based on rates of radioactive decay or fission-yield ratios.

19 (Buice et al., 2005). The heel sampling campaign at the Savannah River Site showed that the heel contains less radioactive material than the bulk of the waste because it was thoroughly rinsed by waste slurring and tank washing. Those rinses removed some of the soluble radionuclides, which decreased the radioactive inventory and leachability of the heel compared to the bulk waste.[14] The time that water would be in contact with the waste in post-closure degradation scenarios is, however, much greater than the duration of tank cleaning operations; thus, even low leach rates for long-lived radionuclides, such as technetium-99 and neptunium-237, can result in groundwater contamination. Both the reduced release rates and the duration of contact with water are important factors in determining residual risks to a member of the public and hence how clean is clean enough.

Evaluation of How Clean Is Clean Enough?

Up to this point the sites have used similar information to decide when to cease retrieving waste. All three sites use the amount of solids removed from the tanks per cycle of mixing and pumping as the main criterion for determining that waste has been removed "to the maximum extent practical." The solids removal rate is usually high at the beginning of waste retrieval, regardless of the technique used, but as retrieval proceeds and there are fewer solids to recover, it decreases and eventually becomes constant at a low value. At some point, considerations such as the amount of secondary waste generated versus the amount of solids removed, cost, worker dose, and tank space limitations justify a decision to stop waste removal.

At the Savannah River Site, 46 cycles of sluicing and pumping were applied to remove the zeolite from Tank 19 and six cycles for Tank 18. Once the removal rate decreased significantly and went below 200 gallons of solids per cycle and 5,000 gallons of solids per cycle for Tank 19 and Tank 18, respectively, waste removal was deemed complete. The Draft Section 3116 Waste Determination reads (DOE-SRS, 2005a, p. 91):

> a number of obstacles prevented further residual waste removal. The obstacles fell into three categories: (1) fast-settling zeolite resins, (2) general tank access limitations and obstructions, and (3) mounds of insolubles. In addition to those obstacles, FFA [Federal Facility Agreement] requirements [i.e., tank closure milestones] for closure of Tank 19 and Tank 18, as well as tank space constraints were additional factors that prevented further residual waste removal."

However, the committee observes that if one retrieval method shows diminishing effectiveness, other methods may still prove effective. Simply repeating a procedure 46 times does not demonstrate that all practical efforts have been made. Also, constraints imposed by closure schedules should not be an important consideration in selecting retrieval methods if scheduling interferes with effective waste removal.

At the Hanford Site, before DOE can make a waste determination concerning classification of the residual waste, it must meet the waste retrieval requirements specified in the Tri-Party Agreement (TPA). The TPA directs DOE to "remove as much tank waste as technically possible, with tank waste residues not to exceed . . ." 30 cubic feet (.85 m^3) in the small 200-series tanks and 360 cubic feet in the large 100-series tanks. Then, to make a determination that the waste is not high-level waste under DOE Order 435.1, DOE must, among other things, demonstrate that the residues remaining in the tank "have been processed, or will be processed, to remove key radionuclides to the maximum extent that is technically and economically practical."

To meet this criterion, DOE must remove *waste* from the tanks to the maximum extent that is technically and economically practical because radionuclides generally cannot be removed from the tanks without removing other wastes as well. The metric that Hanford uses to decide when to cease retrieving residual waste is the volume of waste being retrieved versus the amount of new waste being created.

At Idaho National Laboratory, the waste retrieved from the tank was monitored for its radioactivity until it reached an asymptotic value (see Figure III-6). Visual examination (comparison of waste levels in the bottom of the tank to the thickness of known benchmarks) and waste sampling are also used at the three sites to assess the progress of waste retrieval.

In the two recent tank waste determinations at the Savannah River Site and Idaho, DOE presented (although in a somewhat qualitative way) its trade-offs among some of the key technical factors that lead it to conclude that enough waste has been retrieved from Tanks 18 and 19 at the Savannah River Site and will be retrieved from the entire tank farm in Idaho. DOE also made similar trade-offs in the documents recently submitted to the U.S. Nuclear Regulatory Commission (USNRC) showing how waste removal from Tank C-106 at Hanford has reached the technology limit, as required by the Tri-Party Agreement (Sams, 2004). In essence, DOE separately compares the estimated cost (dollars) and worker dose (millirem) from additional retrieval for each tank to the estimated dose to a hypothetical member of the public over 50 years (millirem) that would be avoided by the additional retrieval. This comparison (a ratio) is used as the primary justification in the Savannah River tank waste determination for waste having already been retrieved to the maximum extent practical (Gilbreath, 2005).

As stated previously, the 50-year avoided all-pathways dose to a member of the public residing in the proximity of the tanks ranges from 0.45 to 2 mrem (4.5 to 20 μSv) for the

[14]The sludge was sampled (for some constituents) before residual waste removal and the heel was sampled after, which establishes that rinsing lowers the radionuclide concentrations.

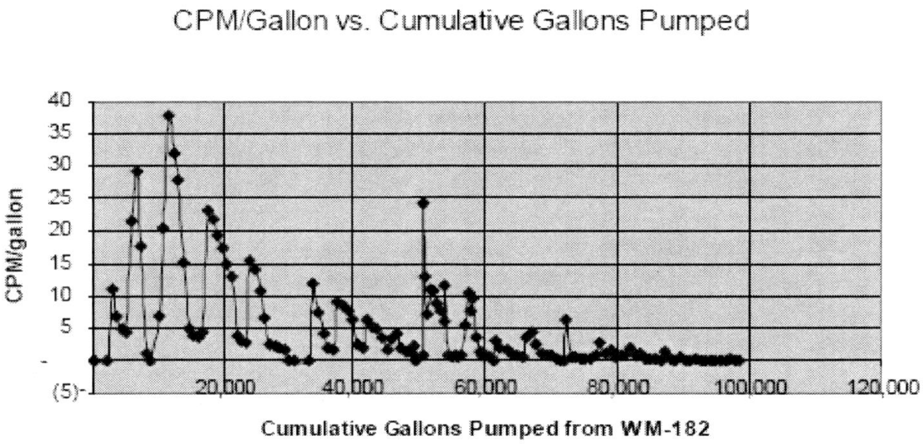

FIGURE III-6 Radioactivity (counts per minute per gallon) measured during cleaning operations at the Idaho National Laboratory. SOURCE: DOE-ID; 2005a, figure 13, p. 53.

Savannah River Site, and 12.5 to 23 mrem (125 to 230 μSv) for Idaho National Laboratory,[15] and is 0.05 mrem (5×10^{-4} mSv) for the Hanford Site. The cost per avoided unit of dose ranges from about $5 million per mrem to more than $30 million per millirem at the Savannah River Site, from $1.7 to $3.08 million per millirem at Idaho National Laboratory, and from $100 to $240 million per millirem to further remove waste from Tank C-106 at Hanford. The Idaho waste determination reports that the average worker dose for cleaning and closing the 11 1135.6 m^3 (300,000-gallon) tanks is expected to total about 7,100 mrem (71 mSv) for all occupational exposure. The ratio of increased worker dose to avoided 50-year dose to the public ranges from about 2,000 mrem (20 mSv) to more than 16,000 mrem (160 mSv) at the Savannah River Site and between 309 and 568 mrem (3.1 to 5.7 mSv) at Idaho. Both ratios assume that all of the residual waste is recovered. Similar information is not available for the Hanford Site because worker doses received during cleanup of Tank C-106 are still not available.

On this basis, DOE determined that the risk reduction does not justify further waste removal from Savannah River Site Tanks 18 and 19. The committee notes that the risk-benefit assessment used as a rationale for the waste determination at the Savannah River Site did not discuss the assumptions about the "dollar value" of a dose or provide any framework to compare the doses and costs calculated for additional waste retrieval to doses and costs in different scenarios (e.g., doses to the public after only the waste that is easy to retrieve is removed). The Idaho waste determination reaches the conclusion that further waste retrieval is not necessary in a different way (DOE-ID, 2005a, p. 58):

> [W]ith typical average doses to the public from natural sources and medical treatment in the range of 300-400 mrem per year, it is not judged practical or cost effective to reduce the estimated dose from [the tanks] by such a small amount.

The Hanford Site compares the additional waste retrieval costs for Tank C-106 to human health risk reduction (expressed as incremental lifetime cancer risk) based on residual waste volume in the tank (Sams, 2004). The cost ratios for the Savannah River Site were based on what DOE considers conservative assumptions but could be considerably lowered with foresight — for example, by undertaking additional residual waste retrieval before preparing the tank for closure. Additional discussion of doses to the public from the tank heels and assumptions used in the tanks performance assessments is provided in Chapter VI.

For completeness, at Oak Ridge National Laboratory the criteria for stopping waste retrieval were to leave (1) no visible sludge; and (2) residual dirty water in bottom of tank roughly equivalent to the water that would be added during decontamination of retrieval equipment during retraction. At the West Valley Demonstration Project site, flushing effectiveness was monitored using sampling, radiological dose rate changes, and visual inspection. The heels in the tanks were then characterized as described above to determine whether they meet the Class C criteria. Three important operational criteria that determined whether residuals were removed to the maximum extent practical were (1) the chances for melter or major equipment failure during waste retrieval; (2) loss of qualification of vitrified high-level waste form; and (3) significant increases of glass volume.

[15]The Idaho waste determination states that the avoided dose from the tank heels by a member of the public is 12.5 mrem over 50 years. However, the conclusion that more tank cleaning is not needed is based on the cumulative all-pathways doses, which are estimated to 0.46 mrem per year, hence 23 mrem over 50 years (DOE-ID, 2005a).

The committee observes that the question, How clean is clean enough? has both policy and technical dimensions because it involves trade-offs among risks to workers, the public, and the environment and cost. In theory, some combination of residual waste retrieval technologies could continue to be applied until the radionuclides in the waste residuals are reduced to negligible levels. In practice, the possibility of removing radionuclides to this extent is counterbalanced by diminishing returns; reduced leachability of residual radionuclides,[16] limits on funds; the dose to workers that occurs during retrieval efforts; the potential for or reality of leaks to the environment during retrieval; and the generation of increasing amounts of secondary waste. There are also nontechnical factors to be considered such as stakeholder values (see discussion in Chapter VIII). Data on the costs and worker doses associated with each tank cleaning option were scant. Input to decisions about the maximum extent practical and ALARA would be improved if DOE kept careful records of costs and worker doses as progress is made on these first tank cleanings and closures.

Part of the answer to the question, How clean is clean enough? is provided by performance assessments to evaluate whether waste left in the tanks meets the performance objectives set forth in the legislation, DOE orders, and federal facility agreements governing waste removal (see Chapter VI). As discussed in Chapters VI and VIII, results in the performance assessment are based on assumptions and scenarios that take place over long periods of time and involve great uncertainties. Assumptions about the duration of institutional controls are a particular case in point. If DOE's assumptions about institutional controls are not correct, members of the public could reside near the closed tanks or low-activity disposal site within a few hundred years, and doses to the public could be sufficiently large enough that the cost and worker risk associated with additional retrieval might be justified. The issue of how long institutional controls are assumed to keep people away from the closed tanks is a matter of policy that cannot be resolved scientifically.

Another part of the answer to the question, How clean is clean enough? is provided by DOE and its contractors through a discussion of technology limitations. Waste retrieval from DOE tanks is a one-of-a-kind and first-of-a-kind endeavor. There is no other activity that can serve as a benchmark for estimating radiation exposure to workers, technology possibilities and limitations, cost estimates, time lines, and so forth. The regulators and stakeholders have to rely on DOE for such inputs to the decision about whether waste has been retrieved to the maximum extent practical. Deciding the appropriate balance among these competing factors is the essence of deciding whether radionuclides have been removed to the maximum extent practical and, thus, that the tank is "clean enough" to proceed to closure.

Therefore, the debate often revolves around what "retrieval to the maximum extent practical" means and whether further decreasing the estimated risks to the public by retrieving additional waste from the tanks justifies additional worker exposure and retrieval costs. Because of the variability of the physical, chemical, and radiological characteristics of the residual waste in each tank, such decisions must be made on a tank-by-tank basis.

FINDINGS AND RECOMMENDATIONS

Finding III-1: Retrieval of waste from each tank is essentially unique in terms of waste types, tank design, and history. DOE is learning from experience, improving its understanding of the unique conditions in each tank as it progresses with tank cleanup, and developing and adapting technology to accommodate these differences.

Recommendation III-1: Depending on the particular requirements of individual tanks, DOE should select and use the most effective sequence of waste retrieval tools from the available suite of tools to ensure that waste is removed to the maximum extent practical. When the limit of a given technology is reached, DOE should evaluate the need to use other waste retrieval tools.

Finding III-2a: Sequential application of available waste retrieval technologies, such as sluicing of bulk waste and the use of pressurized water jets, in-tank vehicles, vacuum devices, or chemical cleaning, has been very effective in retrieving waste from some tanks at all three sites, with residual waste inventories amounting to a few thousand curies or less in a large tank.

Finding III-2b: Available waste retrieval technologies have been less effective on other tanks, primarily those that contain sludges that are thick and difficult to mobilize and solid deposits such as zeolites that can be dislodged but are difficult to remove.

Finding III-2c: DOE will face additional challenges as it continues its tank retrieval efforts. The primary challenges are a forest of vertical cooling coils that will impede the access and maneuverability of retrieval devices in most of the tanks at the Savannah River Site and the potential for additional waste leakage to the environment from tanks at Hanford as water is introduced to retrieve wastes. A combination of such technologies as leak detection, waste mobilization, and waste removal devices that use little water could address the Hanford challenges and are now being deployed. Solutions to the challenge at the Savannah River Site must await the results of ongoing retrieval efforts to better characterize the nature and extent of the problems posed by the coils.

[16]As noted previously, the leachability of residual waste is reduced because the easily solubilized materials has been removed by washing and slurrying.

Recommendation III-2: During tank cleanup and closure operations, DOE should continue to adapt and develop effective technologies for waste retrieval with emphasis on tanks with obstructions and recalcitrant waste. (See also Recommendation IX-2 in Chapter IX.)

Finding III-3: DOE has estimated the cost, public dose avoided, and worker dose increase from deploying residual waste retrieval technologies to determine that radionuclides have been recovered to the maximum extent practical. Long-term risks are based on the results provided by the performance assessment and the assumptions in it, particularly those about the duration of institutional controls. If DOE's assumptions about institutional controls are not correct and members of the public reside near the closed tanks within a few hundred years, doses to the public could be sufficiently large that the cost and worker risk associated with additional retrieval might be justified. The issue of how long institutional controls are assumed to keep people away from closed tanks is a matter of policy that cannot be resolved scientifically. (See Chapter VI, Recommendation VI-2, which is based on Finding III-3 and Finding VI-2.)

Finding III-4: DOE has pneumatic technology for retrieving calcine from the bins at the Idaho National Laboratory and has successfully demonstrated this technology on simulated waste, including caked waste and waste exposed to humid conditions that could cause agglomeration. However, the exact characteristics of the waste in the bins is not precisely known and DOE may face challenges (such as very hard calcine deposits) when it proceeds with retrieval. Because the waste is in a very stable storage situation that could remain for decades if not centuries, DOE has deferred retrieval of this waste or even decisions regarding it in favor of addressing higher priorities at DOE sites.

Recommendation III-4: DOE should continue modest efforts to anticipate calcine characteristics and develop appropriate retrieval technologies for these situations.

Finding III-5: DOE is to be commended for exchanging technical information among sites to allow lessons learned to be incorporated in the program.

Recommendation III-5: Activities such as the yearly Hanford and Savannah River Site technical exchange meetings, technical workshops on tank waste retrieval and tank closure, frequent conference calls, and activities previously undertaken by the Tanks Focus Area are helpful and should be continued.

IV

Processing and Treatment of Retrieved Tank Waste

The purpose of the waste processing described in this chapter is to separate the radioactive constituents in wastes that have been retrieved from Department of Energy (DOE) tanks, as discussed in Chapter III, from the much larger amount of nonradioactive constituents.[1] This separation process has long been a necessary part of DOE's strategy to reduce the volume of high-level waste that must eventually be disposed of in a deep geologic repository. DOE plans to immobilize and dispose the large-volume, low-activity waste stream in near-surface facilities at the DOE sites.

Section 3116 of the 2005 National Defense Authorization Act (NDAA) provides the current legal underpinning of DOE's radionuclide separation strategy for the Savannah River Site. It states that the term high-level radioactive waste (HLW) does not include radioactive waste that "... has had highly radioactive radionuclides removed to the maximum extent practical." This provision recognizes that it is not possible to remove all of the radioactive constituents from the retrieved waste stream, just as it is not possible to retrieve all wastes from the tanks.

This chapter assesses DOE's plans for processing retrieved tank wastes to produce waste streams that are suitable for on-site disposal. In particular the committee considered the following:

- DOE's knowledge of the physical, chemical, and radiological characteristics of the waste in the tanks (i.e., Is this knowledge adequate to support DOE's plans for radionuclide separation?);
- Actions that DOE should consider to ensure that processing plans comply with the performance objectives for land disposal facilities and other requirements (i.e., does the processing reduce to the maximum extent practical the amounts of radionuclides that must be dealt with by land disposal?); and
- Existing technology alternatives and technology gaps (i.e., Are there existing technologies that could be more robust than the current baseline technology, and is there a need for a new technology to overcome the uncertainties with the current baseline technology?).

OVERALL APPROACH

Chapter II describes DOE's basic strategy for processing retrieved tank wastes at the Hanford and the Savannah River Sites. Although the Hanford waste is more diverse than that at the Savannah River Site, they are generally similar in terms of their origins and their physical, chemical, and radiological properties. As discussed in Chapter II and further in this chapter, insoluble sludges in tank waste at both sites contain most of the long-lived radionuclides (and half or more of the total radioactivity). Sludges comprise only about 10 percent of the waste volume and would be very difficult to process for significant additional radionuclide separations.[2] Technical, cost, and risk considerations led site engineers to agree early that the only realistic sludge option was conversion to a stable solid (e.g., vitrification) suitable for shipment to and disposal in a geologic repository.

The soluble salt wastes amount to about 9 to 10 times the volume of sludge; the technical feasibility of separating cesium from salt wastes was demonstrated in the early 1970s (see Sidebar II-1). The option of *not* separating the radionuclides, but instead disposing of all the salt waste in a geologic repository, is precluded by cost, risk of shipping such large amounts of highly radioactive waste, and the practical limit on the quantity of waste to be disposed of in the repository. A requisite for such separations, however, is that the separation process be designed and operated to be highly

[1]Most of the volume of DOE tank waste is made up of nonradioactive constituents that originated mainly from cladding and matrix materials in the spent fuels that were reprocessed, chemicals used in reprocessing, and chemicals used to maintain the waste system.

[2]For example, an initial step would be to redissolve the sludges in acid, which in itself would greatly increase the waste volumes.

effective. Otherwise substantial amounts of radionuclides will remain in the waste and its suitability for determination to be non-HLW under Section 3116 may be questioned.

Processing to remove radionuclides separates tank waste into two streams (see Figure II-1). One is a relatively small volume stream in which radioactive species have been concentrated (high-level waste stream). This stream is further processed and immobilized on-site (e.g., into canisters of vitrified waste) for shipment and disposal in a deep geologic repository, such as the one proposed by DOE at Yucca Mountain, Nevada. The second, much larger, stream from which "radionuclides have been removed to the maximum extent practical" (low-activity waste stream) is intended to be disposed as non-high-level waste at the DOE site where it originated. The low-activity waste stream would include most of the nonradioactive waste constituents and little of the radioactivity.

Wastes at the Idaho National Laboratory are substantially different from those at Hanford and SRS (see Chapter II). At the present time, DOE envisions that all Idaho National Laboratory waste, a much smaller volume than Hanford or Savannah River Site waste, will be removed from that site. The limited waste processing planned for Idaho National Laboratory is discussed later in this chapter.

Waste Processing at the Savannah River Site

The alkaline tank wastes at the Savannah River Site are in three phases. As noted in Chapter II, water-soluble saltcake and supernatant salt solution (supernate) comprise about 90 percent of the waste volume and contain about 50 percent of the wastes' total radioactivity, notably from cesium-137 (see Figure II-4). The water-insoluble sludge phase comprises only about 10 percent of waste volume and the remaining 50 percent of the radioactivity. Concentrated in the sludge phase are the longer-lived actinide radionuclides such as plutonium-239, and some fission products, including strontium-90.

Beginning in the early 1980s, DOE used this natural separation of the Savannah River Site tank wastes to formulate plans for their permanent disposal. Once retrieved from the waste tanks, the sludge (mainly oxides and hydroxides of iron, aluminum, manganese, other metals, and insoluble radionuclides) was to be converted into a stable glass form (vitrified) for disposal in a high-level waste repository. The water-soluble waste (mainly sodium nitrate, nitrite, and hydroxide) was to be processed to remove radionuclides, which would be vitrified along with the sludge. The low-activity fraction of the waste (sometimes referred to as the "decontaminated salt solution") would be stabilized with grout and disposed on-site. This strategy has not changed. The Savannah River Site Defense Waste Processing Facility began vitrifying sludges in 1996. DOE claims that its uninterrupted operation is a key component of the Savannah River Site tank closure program.

The site contractor intended to remove cesium from its salt wastes by an in-tank precipitation (ITP) process, which was first tested in one of the Savannah River Site waste tanks in 1983. The process was never put into operation, and DOE abandoned it in 1998 because of technical and safety issues. Possible reasons for the failure and alternative processes to replace the in-tank precipitation process were reviewed in detail by the National Research Council (NRC, 2000a). Current plans for salt processing largely follow recommendations of that National Research Council report—in particular, that ". . . SRS should consider tailoring the processing operations to tank waste contents, with the goal of reducing processing time and costs and freeing up tank space" (NRC, 2000a, p. 85).

DOE now proposes to process its salt wastes by utilizing three different processes[3] that will be available at different times and have different throughput capacities and radionuclide removal capabilities (see Figure IV-1). Two low-capacity processes are expected to be available sooner and are referred to as "interim" processes by DOE (Phase 1, Figure IV-1): (1) the deliquification, dissolution, and adjustment (DDA) process, which could begin immediately upon permitting by the State of South Carolina, and (2) the actinide removal process and modular caustic-side solvent extraction unit (ARP/MCU), which is expected to begin operations in 2007, when the facility is ready. Federal requirements for salt waste processing have been satisfied by consultation with the U.S. Nuclear Regulatory Commission (USNRC) and the recent approval of the salt waste determination by the Secretary of Energy (DOE-SRS, 2006) in accordance with Section 3116 of the 2005 NDAA.

The timelines shown in Figure IV-1 were not definite at the time this report was completed. The high-capacity chemical processing facility, called the Salt Waste Processing Facility (SWPF) was scheduled to be operational in 2009 but is now delayed until 2011.

DOE's current plan for retrieval and processing is shown in Figures IV-2 and IV-3 (DOE-SRS, 2005b). The first process, the DDA process (Step 1, Figure VI-1), selectively retrieves salt waste that has relatively low concentrations of cesium. The process begins with removing the overlying supernate, draining most of the interstitial liquid from the saltcake (DOE estimates that this step will be about 70 percent effective), and storing the liquid for later processing. Process water or dilute salt solution is then added to the tank to dissolve the remaining saltcake, and insoluble constituents are allowed to settle out.[4]

[3]DOE refers to this as a two-phase, three-step approach. Although this wording suggests that all the wastes undergo each process, that is not DOE's plan.

[4]As noted previously, the insoluble component contains actinides and strontium. The radionuclide removal effectiveness of this settling process is not known, although DOE describes it as removing a significant portion of the radionuclides.

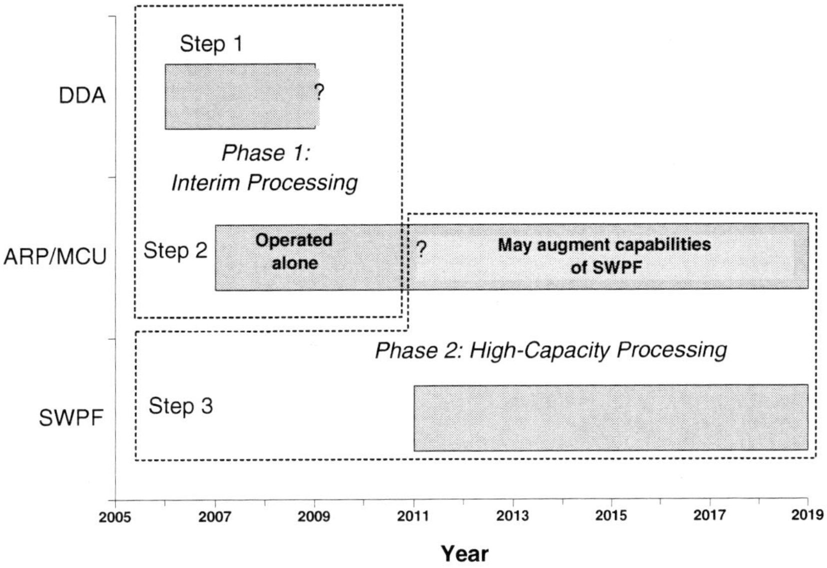

FIGURE IV-1 Time line for salt waste processing at the Savannah River Site as described by DOE (two-phase, three-step approach). NOTES: ARP = actinide removal process; DDA = deliquification, dissolution, and adjustment; MCU = modular caustic-side solvent extraction unit; SWPF = Salt Waste Processing Facility.

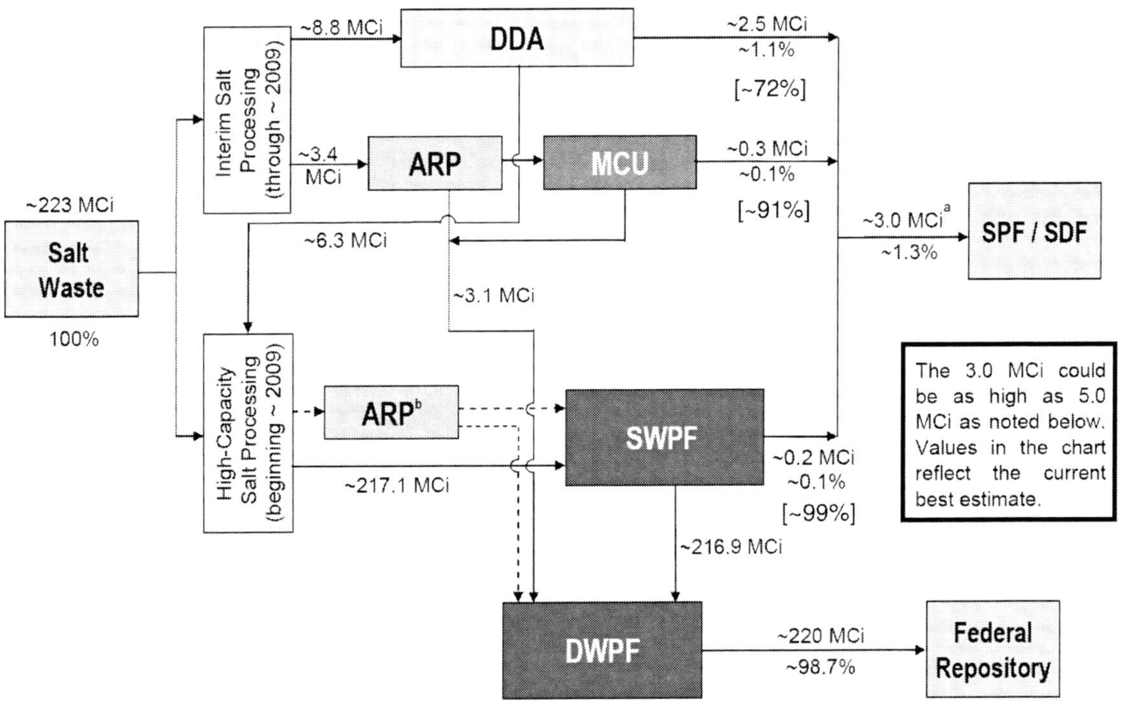

[a] Due to the uncertainty associated with the current characterization of the saltcake waste, the actual curie content of this material may be as high as 5 MCi and the percentages would change accordingly. Curie numbers include daughter products of Cs-137 and Sr-90.

[b] ARP Facilities will have the capability to supplement the actinide removal capacity of SWPF if required.

FIGURE IV-2 Radioactivity flows in DOE's salt waste processing plans. Percentages in brackets are the radioactivity separation efficiencies of the processes. NOTES: ARP = actinide removal process; DDA = deliquification, dissolution, and adjustment; MCU = modular caustic-side solvent extraction unit; SWPF = Salt Waste Processing Facility; SDF = Saltstone Disposal Facility; SPF = Saltstone Production Facility. 1 MCi = 3.7×10^{16} Bq. SOURCE: Adapted from DOE-SRS, 2005b.

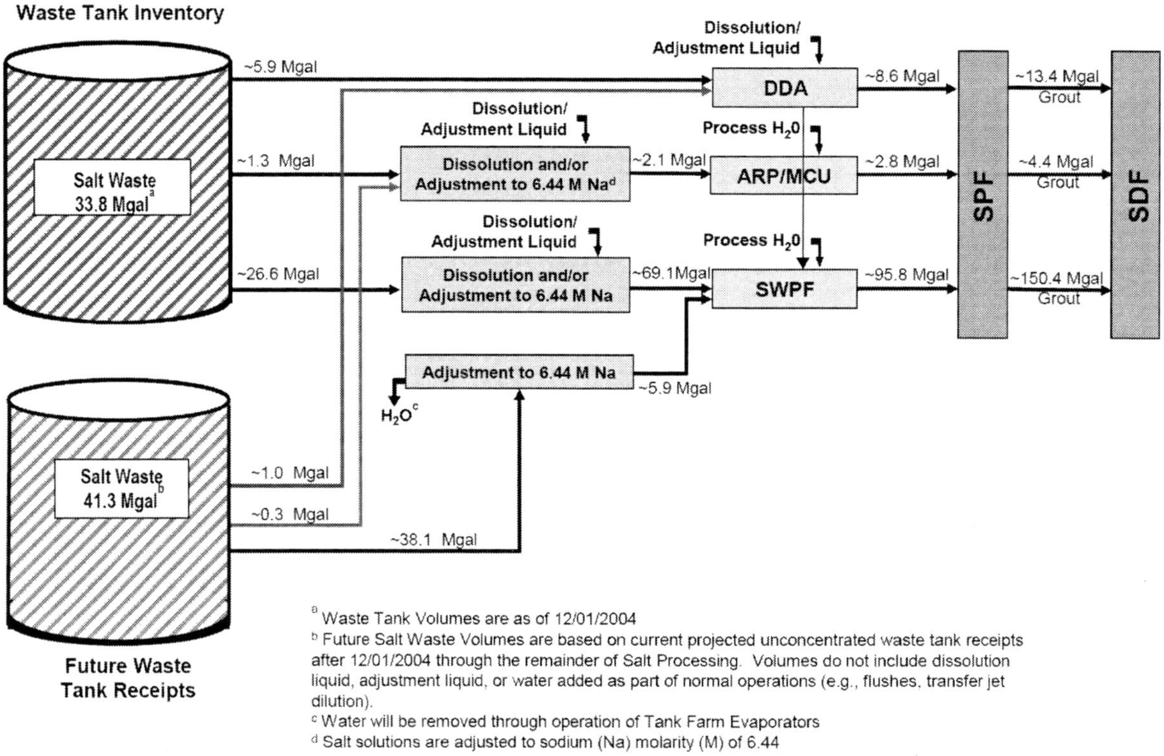

FIGURE IV-3 Volume flows in DOE's salt waste processing plans. NOTE: 1 Mgal (million gallons) = approximately 3,800 m^3. ARP = actinide removal process; DDA = deliquification, dissolution, and adjustment; MCU = modular caustic-side solvent extraction unit; SWPF = Salt Waste Processing Facility; SDF = Saltstone Disposal Facility; SPF = Saltstone Production Facility. SOURCE: Adapted from DOE-SRS, 2005b.

Finally, after adjusting the sodium concentration in dissolved salt waste, DOE plans to send this waste to the Saltstone Production Facility (SPF) for immobilization in the Saltstone Vaults. Thus, the DDA process uses physical rather than chemical means to accomplish cesium separation. The physical separation method used in DDA cannot achieve degrees of separation similar to those in the chemical processes used in the other facilities, described below.

In the second process (Step 2, Figure IV-1), DOE plans to apply two chemical processes in sequence: an actinide removal process (ARP) and a modular caustic-side solvent extraction unit (MCU). In the ARP, monosodium titanate is added to a tank to sorb strontium and the actinides. The monosodium titanate is recovered by filtration. The MCU will use a solvent extraction process to recover cesium from the salt waste. The ARP/MCU processing facility will have a smaller throughput than the other two processes and is designed to operate on its own and later, in conjunction with the high-capacity Salt Waste Processing Facility, if needed. The recovered products from both ARP and MCU will be sent to the Defense Waste Processing Facility to be incorporated into glass logs, and the processed salt waste will be incorporated into saltstone. ARP/MCU is scheduled to begin operation in 2007 and end operation when the Salt Waste Processing Facility comes on-line, but it could run until the end of the waste processing campaign (projected to be 2019) (DOE-SRS, 2006).

The high-capacity processing will apply the actinide and cesium removal processes of the ARP/MCU in the Salt Waste Processing Facility, which is designed for a much greater salt waste throughput and higher removal efficiency (more stages of separation). The ARP may continue to be used for additional recovery of strontium and actinides from selected wastes. Recovered products will be sent to the Defense Waste Processing Facility to be incorporated into glass logs, and the processed salt waste will be incorporated into saltstone. This larger-volume treatment phase is scheduled to operate from 2011 until the salt waste processing is completed.

The approach shown in Figure IV-1, using different processes at different times, was conceived for two reasons: (1) to allow DOE to continue tank remediation and operation of the Defense Waste Processing Facility during the time required to construct and permit the Salt Waste Processing

TABLE IV-1 Projected Efficacy of Salt Waste Treatment Facilities

	DDA	ARP/MCU	SWPF
Date expected to be in operation	2005	2007	2011
Date expected to cease operations	2009	2011	2019
Volume to be processed, million gallons	6.9	2.1	75
Volume to be sent to saltstone, million gallons	8.6	2.8	95.8
Radioactivity to be removed, MCi	8.8	3.4	217.1
Radioactivity to be sent to saltstone, MCi[a]	2.5	0.3	0.2
Projected radioactivity removed, %	71.6	91.1	99.9
Projected share of radioactivity in the saltstone, %	83.3	10	6.7

[a] DOE indicated that because of the uncertainty associated with the current characterization of the saltcake waste, the total actual radioactivity of the material going to saltstone may be as high as 5 MCi (185 PBq). Other uncertainties associated with values of radioactivity and the time lines have not been determined. Values in curies include contributions from the daughter products of cesium-137 and strontium-90. 1 MCi = 3.7×10^{16} Bq. 1 million gallons = 3,785 m^3.

Facility; and (2) to free up tank space to support site operations and batch preparation for the Salt Waste Processing Facility (Hintze, 2005; Spears, 2005).

DOE plans to address the "tank space crisis" in the near term by implementing the interim salt waste DDA process. The DDA process would alleviate some of the space problem. However, during the short time it will be in operation, DDA would process less than 10 percent of the salt waste and would leave behind at least five times as much radioactivity in the saltstone compared to the ARP/MCU and the Salt Waste Processing Facility that will treat the other 90 percent of the salt waste. Because mixing the low-activity waste into a waste-form grout (saltstone) is an essentially irreversible action, the decision to send the DDA waste stream directly to saltstone permanently commits a substantial amount of radioactivity to the site. In other words, although the DDA process would free up tank space, this space is attained at the cost of a large increase in radioactivity left on-site, compared to processing the waste through the planned chemical processing facilities (ARP/MCU and SWPF). Even though these higher levels of radioactivity, primarily from cesium, may not cause projected doses from the Saltstone Vaults to exceed dose limits, the limited separation achieved with DDA raises the question: Does this process remove radionuclides to the maximum extent practical?

Table IV-1 compares the efficacies of salt waste treatment processes. The table shows that DDA is significantly less effective than ARP/MCU and the Salt Waste Processing Facility in removing radioactivity from salt waste. DOE indicated that up to 5 MCi (1.85×10^{17} Bq) of radioactivity could be sent to saltstone depending on the actual radionuclide concentrations in the saltcake and the efficiency of the DDA process. The contribution of radioactivity sent to saltstone from DDA alone could be as high as 4.5 MCi (1.67×10^{17} Bq), which represents almost 90 percent of the total amount of radioactivity sent to saltstone from all three salt waste processes. The potential for such an increase concerns the committee, especially when coupled with the committee's observations about the limitations of DOES performance assessments, discussed in Chapter VI.

The committee is further concerned about the implications of further delays in startup of the high-capacity process to remove radionuclides from salt wastes. The committee noted in its interim report that the original schedule to bring the facilities on-line (ARP/MCU by 2007 and the high-capacity Salt Waste Processing Facility by 2009) and operating to specifications (i.e., processing waste at the expected throughput and meeting the waste acceptance criteria) was ambitious. In fact, DOE recently announced a 26-month delay in initial operation of the SWPF as a result of a change in seismic design specifications. Given the constraints under which DOE states it must operate (e.g., restricting use of noncompliant tanks), this would seem to leave DOE with the undesirable choices that it sought to avoid: extend the use of interim processing leading to increased amounts of radionuclides being disposed in saltstone, or reduce or cease DWPF operations in the face of a tank space crisis. In its salt waste determination (DOE-SRS, 2006), DOE stated its intention to put no more radioactive material in the saltstone than described in the draft salt waste determination (DOE-SRS, 2005b). The committee is unable to offer further insights on this issue because DOE was still formulating its plans as this report was finalized.

Based on DOE's current experience with the Salt Waste Processing Facility and prior experience with developing and initiating operations at major waste processing facilities,[5] the committee judges that it would be prudent for DOE to plan for the possibility that salt waste will not be removed from the tanks at the planned pace. If realized, this possibility would make available tank space even more scarce. In

[5]Several General Accounting Office (now the Government Accountability Office) reports have commented on the challenges of bringing on-line and operating large-scale waste processing facilities (GAO, 1997a, 1997b, 1999, 2003, 2004).

other words, DOE needs a contingency plan for tank space. More generally, the committee cautions that in a schedule-driven system there is the danger that wastes could be sent through the process that is currently available rather than one that is most suited to removing radionuclides to the maximum extent practical from each waste stream. The committee recognizes, of course, that other considerations (e.g., safety, risk, cost) are involved in such decisions. In its interim report, the committee offered some suggestions to address this tank space crisis (see Appendix E).

Low-Level Waste Immobilization and Disposal at the Savannah River Site

At the Savannah River Site, the low-activity salt solution resulting from the separation processes described above is mixed into a waste-form grout known as saltstone. Depending on the specific constituents of the salt solution, the grout is formulated using appropriate proportions of portland cement, fly ash, and ground granulated blast-furnace slag, water, and chemical admixtures. The grout is pumped into concrete vaults, where it solidifies. Saltstone has a low oxidation-reduction potential (Eh) to stabilize key radionuclides (such as technetium-99) in less soluble forms to reduce the likelihood of their leaching out or migrating in the groundwater (Rosenberger et al., 2005).[6] This design is most effective in waste-form grouts like saltstone in which radionuclides are mixed relatively uniformly. The concrete vault has a concrete roof and will eventually have an engineered cap covering the entire installation. The engineered cap, together with the roof, walls, and floor of the vault, directs water away from the saltstone to minimize the leaching of radionuclides or toxic heavy metals from the saltstone into the groundwater. As of 2005, DOE had poured 25,000 m^3 (880,000 cubic feet) of saltstone containing only 225 Ci (8.3×10^{12} Bq) of radioactivity.

Waste Processing at Hanford

With the notable exception of the campaign to extract cesium-137 and strontium-90 from its tank wastes in the 1970s (see Sidebar II-1), Hanford has had relatively less experience in waste processing than the Savannah River Site and its tank wastes are more heterogeneous. As noted earlier, the Hanford approach to radionuclide removal is conceptually the same as that at the Savannah River Site (Figure IV-4). DOE plans to process the retrieved waste to concentrate most of the radioactivity in a high-activity waste stream and leave most of the nonradioactive chemicals and relatively small amounts of radionuclides in a relatively low-activity waste. This reduces the volume of high-activity waste to be vitrified and sent to a geological repository.

The current plan at Hanford is to produce up to 14,500 canisters (15,700 m^3) of vitrified waste (DOE, 2002) containing approximately 184 MCi (6.81×10^{18} Bq) of radioactivity to be sent to a repository, and around 270,000 m^3 (9.5 million cubic feet) of stabilized low-activity waste containing about 7.1 MCi (2.63×10^{17} Bq) of radioactivity for disposal on-site (DOE-RL, 2004a). However, the planned vitrification facility for low-activity waste (part of the Waste Treatment and Immobilization Plant discussed below) does not have the capacity to process all of the low-activity waste by the 2028 completion date agreed to in the federal facility agreement for Hanford (Hanford FFACO, 2003). To meet that milestone, DOE is planning "supplemental treatment" as an alternative to Waste Treatment and Immobilization Plant. The DOE accelerated cleanup effort has proposed cost and schedule savings by sending more than half and up to two thirds of the low-activity waste to "supplemental treatment" (Figure VI-4).

Waste Retrieval and Staging

Waste will be retrieved from the single-shell tanks (SSTs) and transferred to the double-shell tanks (DSTs) where it will be blended and sampled to ensure compliance with Waste Treatment and Immobilization Plant (also called the WTP) feed specifications. The Waste Treatment and Immobilization Plant, which is now under construction, will include processes for separating retrieved tank wastes into high and low activity fractions as well as vitrification facilities to be discussed below. The high activity fraction will be vitrified and shipped off-site to a geologic disposal facility (i.e., Yucca Mountain, if it is licensed and constructed). The Waste Treatment and Immobilization Plant will also vitrify about half of the low activity fraction, which will be disposed of on-site. The Waste Treatment and Immobilization Plant recently encountered schedule and cost overrun problems. To help ensure the project's eventual success, the Secretary of Energy has initiated a detailed review of the plant's chemical process flow sheet and likely throughput.

Waste Treatment and Immobilization Plant Pretreatment

The objective of pretreatment is to separate the waste into a low-radioactivity fraction that contains the bulk of the chemical waste for on-site disposal as low-activity waste and a highly radioactive fraction containing the bulk of the radioactivity and minimal chemical mass for off-site disposal in a federal geologic repository as HLW. The treatment approach depicted in Figure IV-5 results in the on-site disposal of approximately 90 percent of the waste mass and the off-site disposal of about 10 the balance of the total. Off-site disposal of the HLW coupled with off-site disposal of radioactive materials resulting from prior radioactive isotope removal

[6] The function of chemically reducing grout is discussed more extensively in Chapter V.

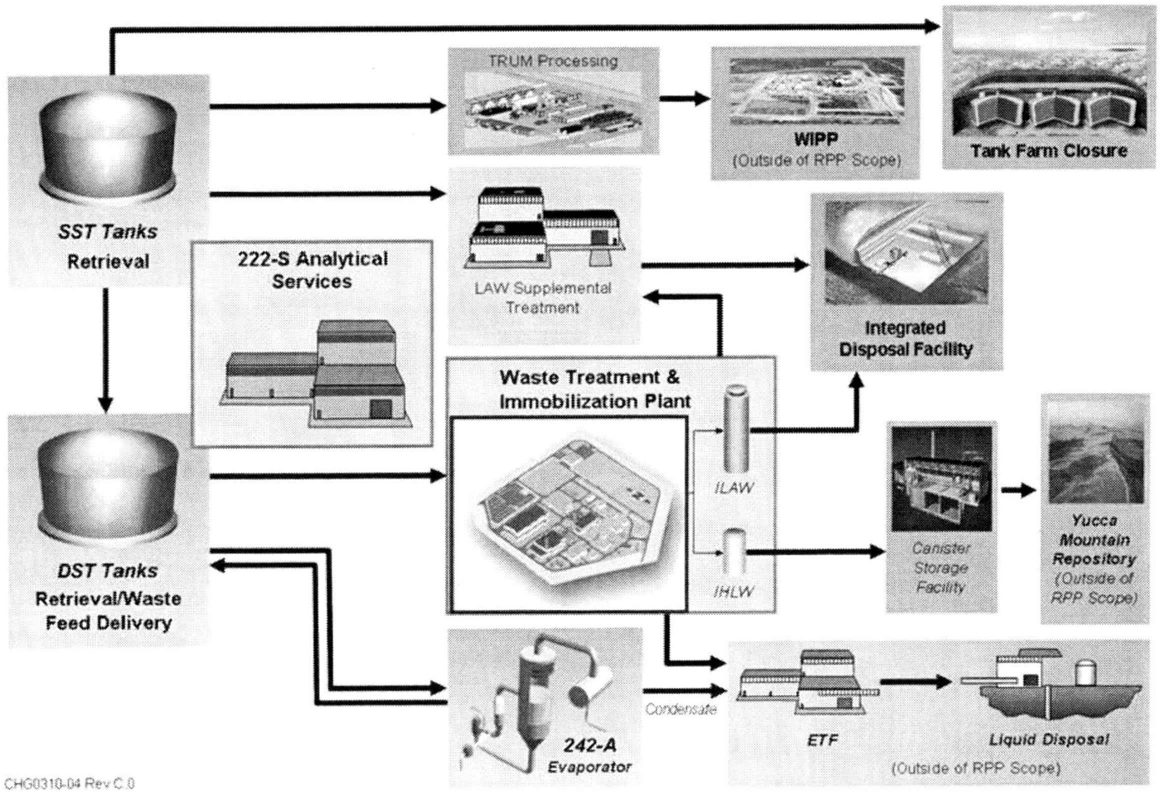

FIGURE IV-4 Hanford tank waste treatment and disposal plan. SOURCE: Schepens, 2005.
NOTES: DST = double-shell tank; ETF = efficient treatment facility; IHLW = immobilized high-level waste; ILAW = immobilized low-level waste; LAW = low-activity waste; TRU = transuranic material; WIPP = Waste Isolation Pilot Plant

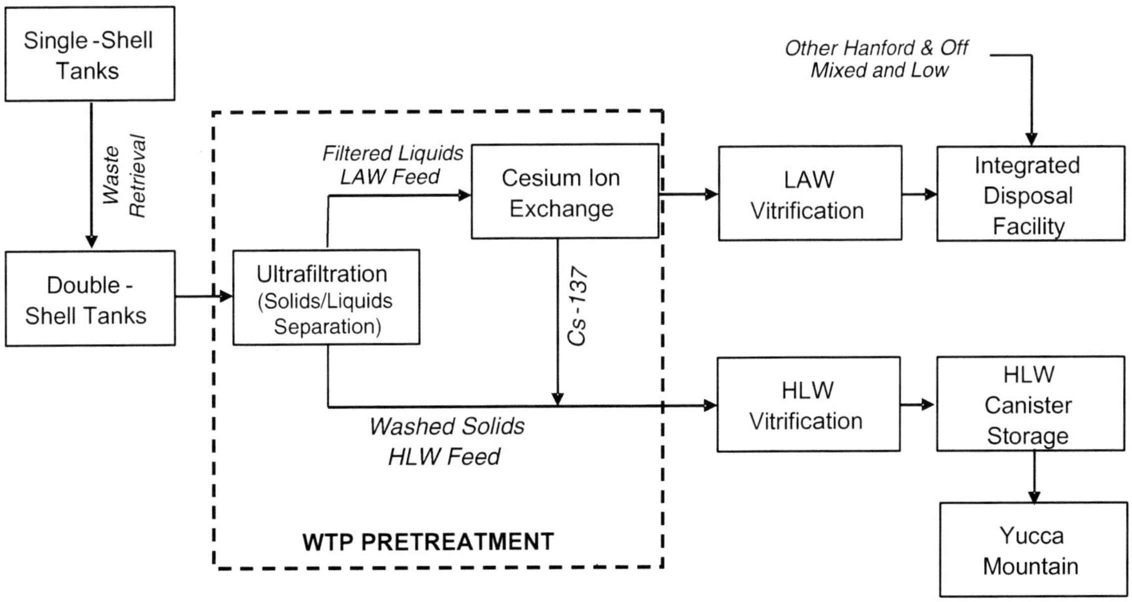

FIGURE IV-5 Waste Treatment and Immobilization Plant pretreatment steps.

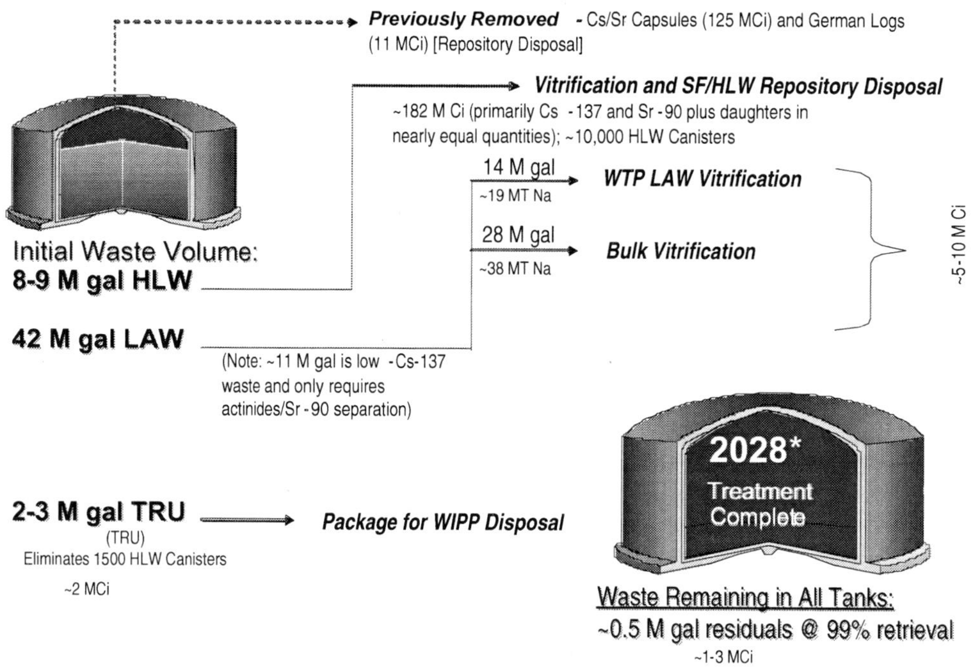

FIGURE IV-6 Multiple treatment and disposition pathways at Hanford. In its presentations to the committee, DOE calculations show 40 to 60 percent of the low-activity waste undergoing bulk vitrification. SOURCE: Mann, 2005; Schepens, 2005. NOTES: 1 MCi = 3.7×10^{16} Bq; 1 M gal = 3,785 m^3; HLW = high-level waste; LAW = low-activity waste; TRU = transuranic; WIPP = Waste Isolation Pilot Plant; WTP = Waste Treatment and Immobilization Plant.

campaigns (e.g., the cesium and strontium capsules), would result in approximately 97 percent of the radioactive inventory that is (or was) in the tanks being disposed off-site (See Figures IV-5 and IV-6).

DOE, in consultation with the USNRC, reached provisional agreement that the criterion to "remove radionuclides to the maximum extent technically and economically practical" can be met by different methods for different tanks, including a solid-liquid separation process or precipitation-filtration to remove most of the strontium-90 and actinides, and ion exchange (for some wastes) to remove cesium-137. The pretreatment processes implement the agreement between DOE and the Commission (Paperiello, 1997).

Dividing the waste into high-level and low-activity fractions occurs primarily by solid-liquid separation (ultrafiltration). The solids removed contain nearly all of the actinides, nearly all of the strontium-90, and approximately 25 percent of the cesium-137. The solids are washed to remove bulk chemicals (e.g., sodium and aluminum) and reduce the chromium content. The washed sludge solids are mixed with glass formers and fed to the high-level waste melters. The filtered liquid waste stream contains the cesium-137 and other soluble radionuclides. Cesium-137 is removed by ion exchange except as explained below for low cesium wastes. The pretreated liquids are mixed with glass formers and the chemicals washed from the sludge and then are fed to the low-activity waste melters in the Waste Treatment and Immobilization Plant. Some low-activity wastes from the plant may also be sent to supplemental treatment, as indicated in Figure IV-5.

Solids-liquids separations are a key pretreatment step to remove the actinides and strontium-90 for subsequent processing as high-level waste. In the 1970s, chemical complexants were used in Hanford's B-Plant in processes to remove greater than 40 percent of heat generating radionuclides (cesium and strontium) then in the tank waste and to provide source material for the cesium and strontium capsules (see Sidebar II-1). Wastes from that processing, which contained the complexants, are stored in two tanks. Interactions of the complexants with actinides and strontium in those tanks resulted in these radionuclides becoming soluble—moving from the sludge phase into the supernate. Consequently, a precipitation process will be used in the Waste Treatment and Immobilization Plant for the liquids in those two tanks to make them compatible with the pretreatment process described above.

Some tanks contain low concentrations of cesium-137 (Low-Cesium Waste) either because the cesium was removed or because the waste did not result from reprocessing. DOE developed a waste management plan and analysis (Petersen, 1996) indicating that cesium-137 ion exchange was not economically practical at cesium-137 concentrations less than

0.05 Ci (1.85×10^9 Bq) per liter (in a 7 molar sodium solution). That report was reviewed by the USNRC and was a basis for the letter to DOE referred to above (Paperiello, 1997). DOE is currently updating those analyses. For cesium waste feeds containing substantially less than 0.05 Ci per liter, no cesium removal is anticipated. For feeds approaching 0.05 curies per liter, cesium removal techniques such as selective dissolution or even a simple ion exchange may be used, primarily to keep worker exposures as low as reasonably achievable (ALARA). According to Figure IV-6 a total of 5-10 MCi (1.85×10^{17} to 3.7×10^{17} Bq) will be disposed on site at Hanford.[7]

Waste Treatment and Immobilization Plant High-Level Waste Vitrification

The separated solids and the cesium-137 removed from the low-activity waste by ion exchange constitute the waste feed stream to high-level waste vitrification. High-level waste vitrification is the immobilization method of choice nationally and internationally; it also produces a waste form that meets planned repository waste acceptance criteria.

Bulk Vitrification of Low-Activity Waste

The Waste Treatment and Immobilization Plant Low-Activity Waste Vitrification Facility may have the capacity to vitrify as little as one-third (see Figure IV-6) of the low-activity waste by the Tri-Party Agreement milestone completion date of 2028. Unless additional low-activity waste processing capacity is provided, treatment operations could extend to approximately 2050. The initial plan was to build a second Waste Treatment and Immobilization Plant Low-Activity Waste Vitrification Facility soon after the first facility started up. DOE, its regulators (the Washington State Department of Ecology and the Environmental Protection Agency), and a group of internal and external experts participated in a 2002 study to determine the feasibility of supplemental treatment technologies to help meet the Tri-Party Agreement milestone for treatment without the high capital cost of a second facility.

Three supplemental technologies were selected for lab tests with surrogate wastes: a cement-based material called "cast stone," steam reforming,[8] and bulk vitrification. Cast stone was eliminated because of waste form performance (i.e., it could not meet the Washington State Department of Ecology's groundwater requirements). Steam reforming looked promising but had significant life-cycle cost uncertainties related to scale-up from the available vendor's unit to that required for Hanford's supplemental treatment, and the probable need to consolidate the granular product into a more monolithic waste form (Choho and Gasper, 2002). DOE decided to conduct steam-reforming tests at Idaho on sodium-bearing waste and tank waste surrogates at Hanford. At Hanford, DOE is conducting full-scale bulk vitrification tests on surrogate wastes and actual tank wastes (up to 300,000 gallons of waste from tank S-109, a low-cesium waste tank) (see Figure IV-7).

DOE has a 2006 Tri-Party Agreement milestone to select the technology it will use to supplement the Waste Treatment and Immobilization Plant Low-Activity Waste Vitrification Facility capacity. Although DOE has not yet selected a supplemental treatment technology, initial technology "downselections" have favored bulk vitrification (Choho and Gasper, 2002).

The committee toured the pilot-scale, nonradioactive, bulk vitrification facility at Hanford; heard presentations on this technology from DOE and contractors; and reviewed available literature. The committee observed that bulk vitrification is a much different approach to processing than the Savannah River Site saltstone. In particular, it is a high-temperature process (greater than 1000°C), which will change the chemical nature of the waste (e.g., decompose nitrates) and produce a substantial off-gas stream from the decomposition. Some fraction of the radionuclides (cesium and technetium) may migrate under the thermal gradient in

FIGURE IV-7 Bulk vitrification processing is performed in a refractory-lined commercially available container. SOURCE: AMEC Earth & Environmental, Inc.

[7]Although Hanford plans are still developing, the waste flows described in documents provided to the committee would route more cesium-137 to low-activity waste disposal than what is planned for saltstone at the Savannah River Site.

[8]In a typical steam-reforming process, superheated steam along with the material to be treated and co-reactants are introduced into a fluidized bed reactor where water evaporates, organic materials are destroyed, and waste constituents are converted to a granular, leach-resistant solid (NRC, 2005b: p. 37; see also DOE-EM, 2005 and DOE-ID, 2002).

the molten material or be volatilized. At this time there are no data to evaluate these possible effects.

In addition, the Defense Nuclear Facilities Safety Board (DNFSB) has raised concerns about the proposed design of the full-scale test facility, which had no containment beyond the equipment itself. Although intended to process only low-curie feed, it is unusual to operate such a facility without successive barriers to confine radioactive materials. The DNFSB has noted that these and other cost-saving measures could compromise safety (DNFSB, 2005).

Waste Processing at the Idaho National Laboratory

According to information available to the committee, the Idaho National Laboratory has no plan for on-site disposal of waste from reprocessing nuclear fuels that would fall within the scope of this study. Although final decisions have not been made, the current plan is for retrieved calcine to be sent to a repository without radionuclide or bulk-chemical separation after simple packaging or after being immobilized in an inert matrix. Sodium bearing waste is a highly acidic waste that contains trivial amounts of solids and no saltcake. It has been retrieved from some tanks, has been consolidated in a smaller number of tanks, and will be conditioned for disposal using steam reforming. Steam reforming is a commercially available technology that has been used for a variety of radioactive wastes including those from the nuclear power industry. A previous National Research Council study recommended steam reforming the sodium-bearing waste (NRC, 2005c). DOE considers sodium-bearing waste to be defense transuranic waste, which can be disposed of in the Waste Isolation Pilot Plant in New Mexico, although the State of New Mexico has not agreed with DOE's position to date.

FINDINGS AND RECOMMENDATIONS

The following are the committee's findings and recommendations with respect to waste processing to remove radionuclides from retrieved tank waste for on-site disposal of portions of that waste.

Finding IV-1a: The Salt Waste Processing Facility at the Savannah River Site and the Waste Treatment and Immobilization Plant at Hanford are essential components of tank waste removal, processing, and closure systems at these sites. They are one-of-a-kind facilities that present technical risks in their design, construction, and operation.

Finding IV-1b: Both the Savannah River and Hanford sites presented the committee with an enormous amount of waste characterization data based on actual sampling, process histories, and model calculations.[9] While such characterization data can never be fully complete given the heterogeneities of DOE tank wastes and the one-of-a-kind nature of their planned processing, the committee judges DOE's knowledge of the physical, chemical, and radiological characteristics of these sites' tank wastes to be adequate for DOE to proceed with its plans for processing retrieved wastes as described in this chapter.

Recommendation IV-1: For the purpose of waste processing and the design of processing facilities, DOE should continue to characterize its waste, but this should be done after waste retrieval and mixing, when truly representative samples can be taken. Even then, the contents and their concentrations need only be known sufficiently for reliable and efficient processing to take place and to provide the radionuclide inventory adequate for subsequent performance assessments. Some processing methodologies may have more stringent quality control requirements than others. In these cases the amount of characterization required may be increased or more adaptable processes could be sought.

Finding IV-2a: Each site is pursuing different technologies for immobilizing its processed non-high-level waste[10]—saltstone at the Savannah River Site, steam reforming at the Idaho National Laboratory, and vitrification at Hanford.

Finding IV-2b: The Hanford bulk vitrification process is less well developed technically than either the Savannah River Site saltstone or the Idaho National Laboratory steam reforming. Bulk vitrification operates at high temperatures, which may volatilize much of the waste and increase technical and safety risks.

Recommendation IV-2: Before issuing a record of decision on supplemental treatment at Hanford, DOE should carefully and transparently review bulk vitrification versus the Savannah River Site saltstone and the Idaho National Laboratory steam reforming. This review should be conducted by a panel of technical experts independent of DOE.

Finding IV-3: The Savannah River Site is facing serious challenges due to limited available tank space and the need for additional tank space to maintain operation of the Defense Waste Processing Facility and meet tank closure commitments. The Salt Waste Processing Facility relies on more efficient technologies to remove radionuclides from the Savannah River Site tanks than the deliquification, dissolution, and adjustment (DDA) process. However, it cannot

[9]See Tables II-1, II-2, and II-3 in Chapter II for summaries of these data.

[10]This finding refers to the low-activity fractions of tank waste that the Savannah River Site and Hanford will dispose on-site and the sodium bearing waste at the Idaho National Laboratory that DOE considers to be transuranic waste.

be brought into operation before 2011. The committee is concerned that the schedule for tank closure and the tank space crisis may increase the need to use DDA and possibly extend its operations, which could lead to disposal of additional radioactive material on-site in saltstone.

Recommendation IV-3a: To reduce the quantities of radionuclides to be disposed of on-site, DOE should develop alternates or enhancements to the deliquification, dissolution, and adjustment treatment process to solve its tank space problems.

Recommendation IV-3b: DOE and its regulators, with public stakeholder involvement, should objectively balance costs and risks (near and long term) of schedule delays in Savannah River Site salt processing against those of sending increased quantities of radionuclides to on-site disposal in order to preserve tank closure schedules.

V

Tank Grouting and Closure

Cementitious materials are used worldwide to immobilize low- and intermediate-level radioactive wastes (IAEA, 1999, 2000, 2004).[1] Historically, grout has been one of the most commonly used materials for solidifying and stabilizing these wastes, and its technology is at a mature stage of development. Grout stabilization of Resource Conservation and Recovery Act (RCRA) heavy metals (e.g., chromium, lead, mercury) is standard technology for producing waste forms that meet U.S. Environmental Protection Agency (EPA) requirements (NRC, 1999c). The committee agrees in principle with the Department of Energy's (DOE's) choice of these Portland-cement based grouts for immobilizing tank waste residues by filling emptied tanks with grout, recognizing that this is an essentially irreversible action. However, there are numerous caveats that arise from DOE's unusual applications of these materials that are outside the construction industry's experience. The committee discusses these issues in the first half of this chapter.

The committee recommended in its interim report that DOE "decouple" tank waste removal and tank closure actions where there are indications that significant amounts of radioactive material are present in the tank after cleanout operations have ended. The committee also recommended that DOE work with the State of South Carolina to revise closure milestones, if necessary (see Appendix E, Recommendation 1). In this chapter, the committee reiterates this finding and recommendation and extends them to the Hanford and Idaho sites.

Decoupling does not imply delaying a site's tank closure program. Decoupling means that for a given tank, once the planned waste removal program has been completed, there is an objective evaluation of the result. Only after this evaluation is a decision made to proceed with the essentially irreversible step of tank grouting or to execute additional waste removal operations. Additional removal operations would likely employ new approaches according to lessons learned from the previous operations. Decoupling is discussed in the second part of this chapter.

USE OF GROUT FOR TANK CLOSURES

By immobilizing the waste and acting as a barrier around it, grouting can reduce the likelihood that the waste will cause harm (see Chapter VI). However, grouting does nothing to reduce the hazard of the waste itself and can be viewed as "treating the symptom rather than the disease." Grouting sludge heels or other tank wastes on-site is not a substitute for removing radioactive materials to the maximum extent practical, as discussed elsewhere in this report, and disposing of them in a geologic repository.

Freshly prepared grout can be mixed intimately with waste to be stabilized (e.g., salt waste at the Savannah River Site) and the mixture pumped into its final containment where it solidifies (such as the Savannah River Site's saltstone vaults; see Chapter IV). Alternatively, especially for stabilizing low-level solid wastes, the grout is often poured in and around objects in a drum or larger container, and the containers themselves can be embedded in grout. As will be discussed in this section, DOE's plans to grout tank waste residues are a hybrid of both practices.

Experience with concrete in the construction and oil industries is extensive, and the materials are relatively inexpensive. Like any materials used in engineering and construction, these materials have well-defined operating

[1]Portland cement is a mixture of silicates and aluminates of calcium obtained by firing (usually) a mixture of limestone and clay in a rotary kiln. The solid material (clinker) formed on cooling this molten mass is ground with a small quantity of gypsum. Portland cement reacts with water to form a solid that is stable in water. "Concrete" is a solid product that results from mixing cement, water, aggregate (sand and gravel or crushed stone), and admixtures that may affect its chemical or physical properties. "Grout" is a mixture of cement and water with or without aggregate, proportioned to produce a pourable consistency. Waste materials (sludge, salt) generally behave as additives and become chemically or physically incorporated into the solid material.

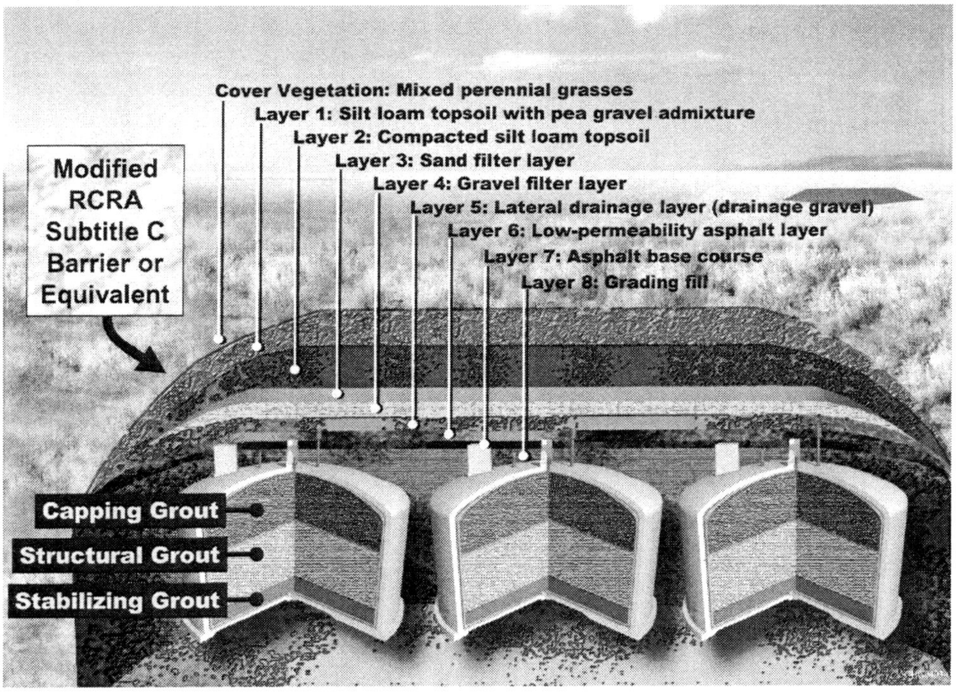

FIGURE V-1 Engineered barrier system to close tank at the Hanford site. A similar plan of engineered grout layers plus a cap system is adopted at the Savannah River Site and the Idaho National Laboratory. SOURCE: Sams, 2005.

envelopes in which they can be used with confidence, but outside these limits their performance may be uncertain. The use of grout for the long periods involved in the disposal of radioactive waste (more than 100 years) is outside the general operating envelope for cementitious materials in industrial applications. There are no good precedents for cementitious materials to maintain very low permeability to water and other properties necessary to retain radionuclides for the very long times required by DOE.

After a tank has been emptied, it must be filled with a solid material to prevent potential collapse of the roof and walls due to the weight of the overburden and the lateral pressure from the surrounding soil. Such collapse would not occur immediately after emptying the tanks but would be the result of corrosion and aging of the tank structure. The potential collapse of the structures could cause a subsidence of the ground surface (final tank farm closure grading), affecting surface water drainage. Filling the tank with solid material limits such a collapse.

DOE plans to close emptied tanks by placing one or more layers of engineered grout in them to provide the structural support described above, encapsulate and stabilize the tank heel, and act as a physical barrier that inhibits the flow of water through the residual waste. Some tanks would have a high-strength layer of grout that would serve as an intruder barrier. Engineered covers to retard infiltration to the tanks after closure are also under consideration at the three sites (see Figure V-1).

Engineered Grouts for Tank Closure

Specially formulated grouts have been developed to backfill the tanks after the waste retrieval is deemed complete. These grouts are being used for tank closures at the Savannah River Site and are planned for tank closures at the Idaho National Laboratory, as discussed in the sections that follow. A layered system of different types of grout is part of the engineered barrier system designed to reduce groundwater infiltration into the radioactive sludge layer.

The main requirements of the engineered grouts used to immobilize radioactive waste follow:

- They must be suitable for pumping into the tanks, typically through long pipes or "tremies" for placement in a tank without segregation throughout;
- They must provide near- and long-term high pH and chemically reducing capabilities to maintain the radionuclides and toxic heavy metals, such as technetium and neptunium, in their least mobile chemical forms (i.e., low-oxidation state or reduced form) (Buice et al., 2005); and
- They must minimize the flow of water through the material (and the consequent leaching of radionuclides and metals from the grout).

The cementitious materials to be used to fill the tanks are a mix of Portland cement, ground granulated blast furnace slag, and fly ash. Portland cement enables the grout to set (solidify) and gain strength in a reasonable amount of time. It also gives the grout a high pH (approximately 12, which is highly alkaline). Slag (a byproduct of the steel industry) gives the grout a low reduction-oxidation potential, or Eh (i.e., a "reducing" grout). Fly ash (a by-product of coal-fired power plants) helps minimize thermal cracking by limiting the heat generated by the grout during the curing process. The high pH and low Eh should reduce the solubility and mobility of many radionuclides if they are well mixed with the cement matrix or if contacted by water that has been sufficiently altered by the cement matrix.

The specific proportions of cementitious materials in the grout are modified to optimize its ability to immobilize the waste, based on an analysis of the waste. At the Savannah River Site, DOE also intends for the grout to serve as a barrier to inadvertent intrusion by burrowing animals or humans drilling or excavating, because it would be clearly different from the native soil.

Uncertainties in DOE Tank Grouting

The committee agrees with DOE's selection of grout for tank closures because of the extensive experience base and relatively low cost for DOE's near-term (approximately 20 year), large-scale needs to immobilize waste. A previous review of materials for stabilizing waste (NRC, 1999c) did not identify any promising material that might be superior to grout for DOE's tank closures. The committee does not foresee the development of better alternatives and neither DOE nor the committee judges it necessary to explore alternative tank fill materials. However, as noted previously, the use of grout for tank closure is unique both in the basic construction challenges it presents and in DOE's use of the material to encapsulate tank residues for very long periods of time.

In reviewing DOE's plans for tank grouting, which are detailed in the following section, the committee developed two sets of concerns (see Sidebars V-1 and V-2). These concerns highlight and summarize lacunae in present knowledge, that DOE must address—many on a tank-by-tank basis—to ensure effective radionuclide immobilization. Simply pumping grout into mostly emptied tanks may not fulfill DOE's responsibilities to its regulators, public stakeholders, or Congress under Section 3116 of the 2005 National Defense Authorization Act (NDAA).

Savannah River Site staff is doing extensive work in developing grout formulations for tank wastes and estimating how these grouts might perform, working to address some of the concerns discussed in Sidebars V-1 and V-2. Recent studies improve grout production and batching, grout flow, and measurement of the effective diffusion coefficient of technetium-99 in reducing bulk fill grout. An ongoing cooperative program with the Khlopin Radium Institute in Russia is addressing the modeling of technetium-99 stabilization in grout and possible improvements. An evaluation of alternative materials and admixtures for achieving zero-bleed (i.e., no water separation from grout mix), self-leveling grouts is also going on. The Savannah River Site is also continuously updating its knowledge base on radionuclide leaching from grout as new data are generated, and it is continuing research to combine design features of the reducing grout and bulk fill grout (Langton, 2005).

Although most of the information gathered by the committee was provided by Savannah River Site research personnel, Hanford and the Idaho National Laboratory have also done work in this area and benefit from the knowledge gathered at the Savannah River Site (Quigley, 2005; Sams, 2005). In a more fundamental approach, researchers at the National Institute of Standards and Technology (NIST) have developed a model that contributes significantly to understanding and predicting changes in the microstructure and transport properties of grout materials over long times (Garboczi et al., 2004).

Tank Grouting at the Savannah River Site

Tanks 17 and 20 at Savannah River Site were emptied (see Chapter III), and they have been backfilled with three layers of cement grout and closed. Plans are being finalized to grout and close Tanks 18 and 19. The grout materials are designed to reduce the mobility of any radionuclides and toxic heavy metals remaining in the tanks after cleaning and to lend structural stability to the tanks themselves.

In Tanks 17 and 20, the bottom layer, which is in contact with the radioactive residual sludge, is an engineered grout called "smart grout." The middle layer, the thickest of the three, is a low-strength grout (bulk fill), and the top layer is a harder grout intended to serve as a barrier against inadvertent intruders. The smart grout was formulated to generate less heat of hydration than ordinary portland cement grout and was placed in a series of lifts to allow time for some of the heat of hydration to dissipate to minimize cracking. The plan for Tanks 18 and 19, and currently for future tanks, is to have two layers of cement grout: a thick layer similar to the smart grout and a top layer of higher-strength grout to act as a barrier to intruders (DOE-SRS, 2005a).

DOE recognizes that there is effectively no mixing of grout with the insoluble waste heel. Having resisted attempts to remove them (see Chapter III), waste heels are likely to be in inaccessible locations and practically immovable. In addition, there are physical limitations on where the grout can be discharged into a tank (tremie[2] placement) and differences in density and viscosity between the cementious material and the tank heel (DOE-SRS, 2005a; USNRC, 2005). In Tanks 17 and 20 a series of tremie placements was made around the circumference of the tank to lay down the first grout layer and contain the tank residues rather than displace them

[2]A tremie is a pipe used to convey and deposit grout or concrete rather than simply pouring the material from a height.

SIDEBAR V-1
Construction Challenges Pertaining to Grouting of Tanks

The ability of grouted waste and grout-filled waste tanks to provide the long-term radionuclide immobilization that DOE is anticipating, as described throughout this report, depends greatly on success in meeting several challenges, many of which are well known by the construction industry. Each tank will present variations on these challenges based on its limitations for physical access; internal obstructions; and the amount, location, and properties of residual waste deposits. The adequacy of DOE's tank closures will depend on careful consideration of each of the following on a tank-by-tank basis:

- **Pumping**. Because of the limited access to the tank and the large volume of grout that must be placed, it is likely that the fill material (concrete or grout) would be placed by pumping; hence it must remain pumpable during the entire time of placement.
- **Flow characteristics**. The grout must flow to the walls of the tank from the point(s) of placement while retaining its integrity—that is, it cannot segregate into its constituent ingredients.
- **Degree of mixing with or encapsulation of waste**. Based on the results of mockups and the few tanks that have been grouted, the heel material does not mix with the grout to an appreciable degree. Also, since the grout remains on top of the heel, encapsulation of the waste is incomplete. However, placement techniques may influence the final distribution of heel and grout material, resulting in better encapsulation of the heel.
- **Inhomogeneity of grout**. Because the waste liquid remains on top of the heel in some tanks, there may be some mixing or sorption of the liquid into the grout. In some cases, dry cementitious materials are to be placed pneumatically on top of the liquid waste to stabilize the free liquids. The resulting inhomogeneities must be evaluated to determine whether they affect the overall performance of the grouted tank.
- **Effectiveness of grout (filler) in immobilizing waste**. If the walls and pipe surfaces of a tank cannot be adequately cleaned, some radioactive waste will remain above the elevation of the grouted bottom layer. Thus, the low-strength "filler" would require some ability to immobilize the waste. If the low-strength filler does not have the required capability, it may be necessary in some cases to use higher-quality grout to fill the entire tank.
- **Heat generation**. The hydration reactions of all cementitious materials evolve heat. Because grout does not conduct heat well, the temperature within the grout can rise significantly and lead to cracking. Heat generation must be controlled by proportioning the grout appropriately using materials that generate little heat and then allowing the heat to dissipate in a way that avoids thermal cracking (see cold joints, below). For the use of grout to be acceptable, either the grouting must not result in thermal cracking, or cracking must not result in significant adverse effects on the performance of the grouted tanks. In either case, testing and analysis are needed to verify DOE's expectations
- **Long-term monitoring**. As discussed in detail in Chapter VII, Sect. 3116 of the NDAA requires a post-closure monitoring program. One desirable component of such a program is to monitor the performance of the waste form to ensure it is performing as expected and to provide early detection of radionuclide release. To accomplish this, it would be helpful if grout construction could be designed to allow the desired monitoring to occur, as is recommended in Chapter VII.
- **Cold joints**. One means of managing the heat generated is to use a series of grout placements (i.e., lifts) rather than place all the grout at once, allowing the heat to dissipate between each lift. In normal construction, specific measures must be taken to ensure that the concrete behaves as a monolith across such "cold joints" between the lifts. Because of the limited access into the tank and the hostile environment, it may not be possible to take such measures. Thus, the presence of cold joints must be taken into account in assessing the performance of the grouted tanks.

Alternative grouting formulations and techniques need to be tested in mockups (as is done routinely in construction projects), which allows the contractor to gain experience, and verify the properties of the grout as placed. In addition to the construction industry, the oil industry has developed a great deal of expertise in grouting of areas with difficult or limited access. The U.S. Army Corps of Engineers also has personnel who have participated in the development of related technology whose expertise could be brought to bear on these problems.

toward the walls. Documents indicate that there were small areas of incomplete grout coverage at the intersections of grout deposited by different tremie placements (USNRC, 1997a).

The smart grout covered the fixed, insoluble waste particles (the solid heel, containing primarily actinides and strontium) and displaced the liquids. The liquids, which contain technetium, other soluble radionuclides, and some suspended insoluble particles, were largely displaced to the top of the grout. They were absorbed by a second layer of dry grout to provide further immobilization of the waste. After placing more smart grout on top of the dry grout, an improved version of a "controlled low-strength material" was then added above this to fill much of the tank, inhibiting water flow (a hydraulic barrier) and preventing collapse. Finally, a third layer of a higher-strength grout material was used to fill the voids around the risers and to act as an intruder barrier.

DOE's estimates of grout behavior over time do not assume that the waste is mixed in the grout, but they do

SIDEBAR V-2
Unusual Requirements of DOE Grout Applications

DOE has relied heavily on knowledge of the characteristics of grout and its long-term behavior gained by the construction industry. While this approach is appropriate, it is not necessarily sufficient for the purposes of stabilizing radioactive wastes. The following are some topics on which experience from construction applications alone provides an inadequate knowledge base for DOE's applications. Additional information, most likely from a research and development program, is needed to provide the necessary understanding of the behavior of the fill material (grout or concrete) over the long term so that appropriate grout formulations can be selected and performance assessments can be based on valid assumptions.

- **Compatibility of grout with liquid waste.** In most cases, some liquid waste remains on top of the heel. As the grout flows into place, it either mixes with or absorbs water from this liquid. The radionuclides, toxic heavy metals, and other chemical constituents of this liquid may locally affect such characteristics of the grout as its ability to set or its durability.[a]
- **Exposure to radiation.** Experience in the nuclear industry has established that properties of cementitious materials can be affected by high levels of radiation (e.g., in reactor shielding). Lower levels of radiation may or may not affect the properties of the grout material.[b] Radiation levels from tank wastes are much lower than in reactor applications and decrease with time. However, the effects of persistent radiation from the tank waste on grout performance have not been evaluated. It is possible that the radiation levels are not sufficient to cause deterioration, or that even deteriorated grout is satisfactory in this application, but this has to be established.
- **Deterioration of the tank floor and sides.** It is anticipated that the carbon steel tank floors and sides will eventually corrode away. The concrete slabs and vaults in which the tanks sit were not designed as long-term containment. Performance assessments must continue to account for the effects of the eventual loss of these barriers.
- **Reducing capabilities of grout.** The ability of the grout to stabilize radionuclides and toxic heavy metals rests on its high pH and low Eh. High pH is important in the construction industry because it helps protect embedded steel against corrosion. Thus, there is some understanding of the mechanisms of loss of pH over time. However, Eh has no particular relevance in construction; thus, much less is known about its persistence over time.
- **Extremely long service lives.** In the construction industry, a typical service life is on the order of 50 to 100 years, and regular maintenance is necessary to achieve it. While examples of ancient concrete still survive today, they are exceptional and have little relation to modern construction materials and techniques or to service conditions that will be encountered in the tanks. DOE seeks to place grout that retains its properties in some form for 500, 1,000, or even 10,000 years. Often the strategies for durability in the construction industry involve postponement or slowing of deterioration rather than prevention. These strategies have to be reconsidered for the extended service lives required in the tanks.
- **High groundwater table.** In a few cases, the elevation of the groundwater table is above that of the tank floor. Coupled with the likely deterioration described above, this could result in the tank heel coming in contact with groundwater that has not been substantially altered by the chemically tailored grout atop the heel. The performance assessment of these tanks must include this condition, and DOE may want to consider more thorough cleaning or other means to reduce the risks associated with these tanks.

---

[a] A similar issue has been raised in the context of low-activity waste disposal given the potential for interaction between the chemicals in the waste and the grout. While chemical compatibility could be a problem for saltstone, the committee has not examined the issue and has no evidence that it is a problem.

[b] Although the doses required are high (see, e.g., Utsunomiya et al., 2003), the principal concern would be that penetrating beta rays could cause solid state radiolysis in hydrated phases, such as those present in grout and zeolites.

assume that the grout continues to be an intact hydraulic barrier for 500 years and maintains its alkalinity and reducing capability for 10,000 years. Despite the considerable amount of work performed by DOE contractors, the committee received little quantitative (experimental or other) information to support the 500-year and 10,000-year assumptions.

Langton and coauthors describe the different needs and challenges for waste tanks at each DOE site, tank fill materials placement requirements, leaching and durability properties, and technology needs to demonstrate tank fill physical and leaching properties (Langton et al., 2001). The committee is aware of a qualitative analysis of the tank waste grout from 1992 (Lokken et al., 1992), but of only one recent experimental study, which is on the leaching characteristics of grout with respect to technetium-99 (Harbour et al., 2004). DOE's *Performance Objectives Demonstration Document for the Closure of Tank 19 and Tank 18* (Buice et al., 2005) contains a set of calculations concerning reducing capabilities of the grout and examines the 10,000-year assumption. These calculations are discussed in Chapter VI. The committee believes that the short- and long-term performance of tank fill materials warrant further research to bridge a knowledge gap (see Chapter IX).

Tank Grouting at Hanford

Hanford has not yet finalized its closure plan for single-shell tanks. An environmental impact statement has been initiated to assess the final closure configuration of the single-shell tanks after retrieval. For planning purposes, Hanford is currently assuming a landfill closure configuration that utilizes grout as the fill material based on research and field experience gained at the Savannah River Site. Gravel, concrete, and other materials have been considered as fill materials, but their performance is inferior to—or their handling is more complex than—that of grout. According to presentations to the committee, the Hanford tanks would be filled in three layers of flowable grout:

- Layer 1: a 30 to 90 cm (1 to 3 feet) layer of free-flowing grout that will cover waste residuals and debris on the tank bottom and support subsequent fills.
- Layer 2: grout that will enhance stability of the tank structure and fill the majority of the tank.
- Layer 3: high-compressive-strength grout placed in the remaining void space to discourage intrusion.

Savannah River National Laboratory staff performed scaled testing (lab, bench, and large scale) to develop grouts with properties suitable for Layer 1 in the Hanford tank and waste environments, which differ from conditions at the Savannah River Site (Harbour et al., 2004). This study found that waste particles at the bottom of the tanks would be only partially encapsulated by the grout. However, the grout would be able to penetrate many of the interstitial regions. The study also found that the stabilizing layer should provide a reducing environment, thus decreasing the mobility of contaminants of concern. The layer would also provide a physical barrier to slow the release of these contaminants in the environment. Both multipoint and single-point tremie placements have been evaluated to accommodate various riser configurations, but additional tests are needed (Langton et al., 2003).

Hanford is also planning a tank closure demonstration on one of the smaller C-200 series tanks after completion of waste retrieval. The purpose of the demonstration is to verify tank stabilization by core sampling of the grout layer. This work will also include characterizing contaminated soil outside the tank and stabilizing it by impermeable barrier installation; characterizing and stabilizing one diversion box and direct buried pipelines by in situ grouting; and characterizing and isolating in-trench pipelines. DOE also plans to continue grout formulations studies for Hanford-specific applications. Given the early stage of its tank closure plan, Hanford has the opportunity to benefit from continuing dialogue on tank closure with the other DOE sites (including Oak Ridge and West Valley).

Tank Grouting at the Idaho National Laboratory

Idaho National Laboratory staff demonstrated a method of placing grout onto a tank floor to permit retrieval of additional slurry from the tank using a variable-depth steam jet. Five sequential placements of the grout pushed liquid toward the jet intake, allowing removal of additional liquid from the large-diameter tanks (see Figure V-2).

The sequential placement technique was developed when the site did a 1999 mockup test. A mockup of a tank was constructed at an Idaho Falls industrial facility. This was a full-scale horizontal slice of the bottom of a tank, only a few feet in height, but 50 feet in diameter, with cooling coil structures and simulated residual solids. Grout mixtures were tested to validate assumptions of flowability in both the tank and the surrounding vault areas, and the ability of the grout to move the in-tank solids toward the steam jet (INEEL, 1999).

The main assumptions about the tank fill material used in the Idaho National Laboratory tank closure performance assessment are that the outer vault grout fails at 100 years, tank and tank grout fail at 500 years, and piping fails at 500 years. The main difference from the other two sites' plans for tank grouting is that Idaho's tank closure plan does not include a layer of high-compressive-strength filler material to serve as an intruder barrier. Idaho National Laboratory staff assumes that intruders who might attempt to drill in the tank farm area would expect to encounter basalt flows; therefore, the presence of a high-compressive-strength grout on the top of the tanks would not necessarily prevent drilling.

The grout formulation has not yet been finalized at the Idaho site. Appendix C of the Draft 3116 waste determination (DOE-ID, 2005a, Appendix C) reads:

> The grout planned for use at Idaho is expected to exhibit strongly reducing conditions, as in Hanford and Savannah River tank closure plans. However, current Tank Farm Facility analysis concludes that reducing conditions in the grout are not necessary to demonstrate compliance with performance objectives.

Idaho's Department of Environmental Quality (IDEQ) has approved partial closure plans (i.e., not for the whole tank farm) for the 1136 m^3 (300,000-gallon) tanks WM-182, 183, 184, 185, and 186. DOE has submitted, but IDEQ has not yet approved, partial closure plans for Tanks WM-180 and 181 (300,000-gallon tanks) and WM-103, 104, 105, and 106 (30,000-gallon tanks). DOE has not submitted closure plans for the remaining 300,000-gallon tanks or other portions of the tank farm because those tanks store waste that DOE plans to treat by steam reforming and ship for disposal off-site.

The site is collaborating closely with the Savannah River and Hanford Sites on grout formulation and placement methods. Idaho National Laboratory is also working with other DOE sites, the Pacific Northwest National Laboratory,

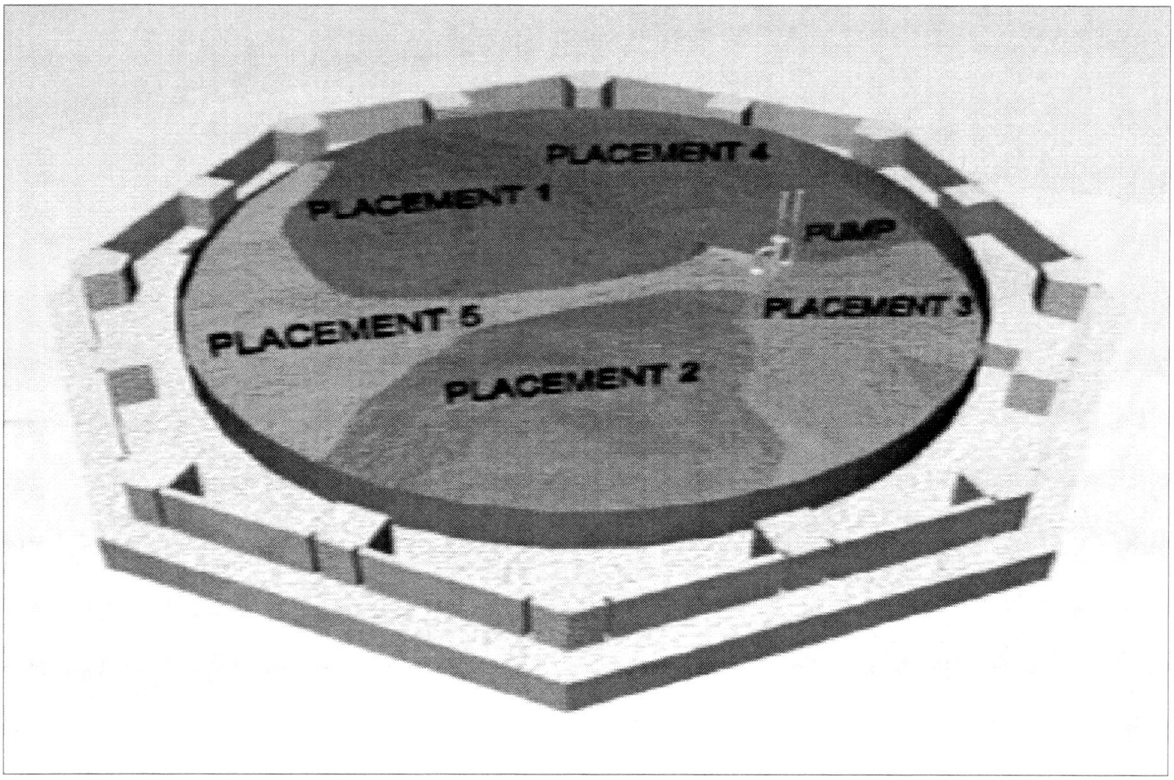

FIGURE V-2 Plan for sequential placement of grout in tanks at the Idaho National Laboratory. SOURCE: Lockie et al., 2005.

and the United Kingdom Atomic Energy Authority on issues related to tank closure.

DECOUPLING WASTE REMOVAL FROM TANK CLOSURE

In its interim report about the Savannah River Site, the committee found that tank closure milestones make tank waste removal and tank grouting schedules appear "coupled" (i.e., one following the other as soon as possible) for some tanks (see Appendix E, Finding 1b). For example, in the case of the draft Section 3116 waste determination for Tanks 18 and 19, the *milestone* for closing these two tanks was one of the top three criteria for determining that "waste has been removed to the maximum extent practical" (DOE-SRS, 2005a). In that report the committee recommended that retrieval and closure not necessarily be closely coupled, especially for tanks containing significant amounts of residual radionclides. Subsequent to the interim report, DOE and the State of South Carolina reiterated their preference for closing tanks soon after retrieval is completed (see the section on objections to decoupling, below). The committee remains concerned that DOE is defining what is practical, in at least some cases, by what is required to meet a milestone or by the letter of the law (radiation doses at a far-future time), rather than making decisions based on sound science and engineering judgment.

According to information reviewed by the committee, the volume of sludge residues left in a tank after waste retrieval is completed may vary by two orders of magnitude (a hundredfold). This is not necessarily bad; it simply reflects the inherent uncertainty in expected tank cleaning results at this early point in DOE's program. Reducing waste volume in a million-gallon tank down to 100,000 gallons is 90 percent removal; reducing waste volumes to 10,000 gallons would be 99 percent removal. Table VI-3 shows some very optimistic assumptions in DOE's environmental impact statement for tank closure at the Savannah River Site. Many or most tanks were assumed to contain only 100 gallons after retrieval was completed. The accompanying discussion in Chapter VI suggests that residues of 5,000 to 10,000 gallons or more are more consistent with experience with the methods used in the most recent residual waste removal campaigns.

The committee concurs with DOE that achieving near-term risk reduction by removing 90 to 99 percent of the waste volume can and should be accomplished as soon as possible. However, the committee does not agree with what appears to be a milestone-driven rush to grout a tank essentially permanently and irrevocably even if much more radioactive material remains than expected. For a problematic tank, decoupling waste removal from grout closure—that is, allowing opportunity for objective assessment of the results and reassessment of the path forward—is essential.

Advantages of Decoupling

There are several advantages to decoupling tank closure from tank cleanup. The first advantage is that for tanks that prove difficult to clean (either because the tank configuration obstructs access or because the waste is recalcitrant), options can be kept open in the near term (5 to 10 years) to remove additional waste and/or to use improved immobilizing material to fill the tank. Filling a tank with grout is essentially an irreversible action. The second advantage of decoupling tank closure is to allow periodic reassessment of technology developments and alternatives to reduce long-term risks presented by the tank heels.

A third advantage in delaying closure of these tanks is that it allows time to gather operational experience for tanks containing cooling coils and other waste retrieval challenges (see Chapter III). DOE obtained reasonable results in retrieving waste from Tanks 17 and 20, leaving behind very little residual waste. Tanks 18 and 19, which have undergone waste removal, are estimated to have an order of magnitude more radioactivity than Tanks 17 and 20, but the greater challenges lay ahead. DOE started its tank waste removal and closure campaign with Type IV tanks, which are simpler to work with because of the absence of cooling coils. This approach makes sense with respect to retrieval technology, because it allows DOE to learn from the simpler tanks before tackling the more complex ones. Tanks with coils may present an additional challenge because they are likely to have more solids encrusted on the interior surfaces and those solids will be difficult to reach. This is because (a) there is more surface area to which waste material can adhere and (b) there are more obstructions that make retrieval more difficult.

DOE has developed operational experience with in-tank activities such as sampling, slurrying, pumping, removing waste heels with water jets (sluicing), and operating other remotely controlled equipment. In some cases, DOE may need more time than is allowed by the Savannah River Site Federal Facility Agreement closure milestone to apply what it has learned, test, identify any new challenges, and evaluate new technologies to maximize the removal of waste and stabilize residual waste in the more difficult tanks (see also Chapter IX).

The fourth advantage of delaying closure of these tanks is that it would allow for a focused research and development program to enhance tank waste removal, improve waste immobilization, and improve tank stabilization as recommended in Chapter IX. A previous National Research Council report also recommended further research in waste retrieval and immobilization prior to tank closure (NRC, 2001a). As noted in the committee's interim report, the long-term performance of tank fill materials appears not to have been established adequately; the committee discusses uncertainties in the long-term performance of these materials in Chapter VI. To lend confidence to the assumptions used in the performance assessment, a delay in tank closure would give DOE more time to evaluate grout formulation and techniques and to conduct studies of projected long-term performance by laboratory and field testing of tank fill materials (see Chapter IX).

DOE itself recognizes the potential benefits of decoupling tank cleanup from tank closure. A series of reports requested by the Tanks Focus Area and developed by Pacific Northwest National Laboratory describes the concept and applicability of placing a tank and its residual contents into a safe, stable, and minimum maintenance condition pending final closure options, what is defined as "tank lay-up" (Elmore and Henderson, 2001a, 2001b, 2002a). In these documents, tank lay-up is viewed as a potential necessity to bridge the time gap between tank cleanup and final closure, because sometimes the decision to close a tank is not made for many years after the tanks have been emptied (e.g., see the West Valley discussion in Chapter III and Appendix G); in these reports, tank lay-up is assumed to last for up to 20 years. Tank lay-up activities are discussed at five DOE sites (Hanford, Idaho National Laboratory, Oak Ridge National Laboratory, Savannah River Site, and West Valley Demonstration Project). The reports clearly discuss how lay-up depends on the number and physical condition of the tanks; expected lay-up period; uncertainty in closure requirements; perceived risks associated with waste heels; and the regulatory environment. The more recent of the two reports (Elmore and Henderson, 2002a; pp. 2-3) states:

> Tank lay-up activities are expected to reduce the perceived risks associated with the tanks: Likewise, subsequent hazard/accident analyses on a tank-by-tank basis could result in the following:
>
> - Lowering the hazard classification for certain facilities, which could impact conduct of operations, hazardous waste management, emergency preparedness, and training
> - Reduction in the number of safety-class, safety-significant, and defense-in-depth structures, systems, and components, which could reduce the number of required engineered and administrative controls
> - Reduction in the number of technical safety requirements (e.g., safety limits, limiting control settings, limiting conditions for operation)
> - Reduction in monitoring or surveillance frequencies (e.g., liquid/solids levels, waste temperatures, vapor space pressures, leak detection probing, corrosion prevention)
> - Reduction in tank reporting requirements
> - Reduction of maintenance on the tanks and supporting and interfacing systems (e.g., vapor space filtration, liquid level devices, temperature probes, light-duty utility arm [LDUA], core sampling system)

- Reduction in the interface requirements associated with nontank facilities and systems
- Reduction in configuration management requirements, procedure maintenance, number and depth of assessments, required personnel training, hazardous materials and radiation protection requirements, and other requirements to be determined on a Site and tank basis.

The most recent summary report on tank lay-up activities also recommends that DOE share lessons learned on tank closure activities among its sites (Elmore and Henderson, 2002b).

The committee does not advocate decoupling the removal and closure schedule based only on the future possibility of discovering better technologies for cleanup and closure without identifiable current prospects. Rather, the committee encourages developing or adapting specific technologies that are at least in the applied research stage and researching a narrow set of questions that, if answered, could enhance tank heel removal and closure effectiveness. The committee selected a time frame (5 to 10 years) that is in reasonable accord with the overall schedule for tank farm closure and would not extend tank closure indefinitely into the future.

Concerns with Decoupling

The recommendation in the interim report was not well received by DOE and the State of South Carolina (see Appendix E). In interviews with reporters, South Carolina Department of Health and Environmental Control representatives reiterated their commitment to the schedule for closing tanks and disagreed with the committee's conclusion that delaying filling of tanks with grout would be beneficial from the perspective of risk. These representatives argued that unless previously agreed to milestones for tank closure continue to be met, progress will stall. This concern could be addressed if separate milestones were established for tank waste retrieval and for closure.

The committee notes that to delay grouting of specific tanks may not delay the final closure milestone for the entire tank farm, which will take several years.[3] If new technologies become available in the near future (i.e., 5 to 10 years), it may be possible to clean up and close tanks faster (possibly leaving less waste behind), thus meeting the final milestone for closing the tank farms.

Even if the decoupling did result in some delay, the federal facility compliance agreements could be modified, as they have on many other occasions, provided that the action improves the outcome. In 2002 the General Accounting Office (GAO, now the Government Accountability Office) issued a report on the implications of DOE's compliance agreements in waste cleanup (GAO, 2002). The GAO found that compliance agreements have not been a barrier to previous DOE management improvement initiatives. Regulators generally supported these initiatives, saying that they support efforts to implement faster, less costly ways to reduce environmental risks at the sites, as long as DOE's approach did not reduce funding for individual sites (GAO, 2002).

The second objection raised against delaying tank closure is that a tank could collapse due to lateral pressure from the surrounding soil, or from the weight of the overburden. In its interim report, the committee recommended that DOE consider the risks from postponing tank closure compared to the risk reductions that could be achieved if the postponement improves heel removal. A qualitative assessment by DOE of the issues associated with aged and abandoned underground structures and vessels includes the potential for roof and side wall collapse; filling with water from runoff (bathtub effect); and internal seepage, which can lead to overflowing, leaking, or leaching; and buoyancy (Langton et al., 2001). However, the committee is not advocating abandoning the empty tanks on-site and has seen no quantitative assessment of the risks of postponing tank grouting. According to DOE, the tanks are not in near-term danger of collapsing after bulk waste retrieval;[4] indeed, the structural support provided by the tank fill is not likely to be needed until DOE is ready for ultimate closure of the tank farm. In most cases, postponing closure of tanks that contain significant amounts of residual waste for several years would appear to have essentially no effect on near- or long-term risk, while leaving open the possibility of further risk reduction if more of the waste can be removed.

The third objection against delaying tank closure is that once equipment is in place for tank waste removal (e.g., the superstructure for in-tank operations), it is convenient to proceed to use the same equipment for closure, rather than moving it to another tank and reequipping the first tank when it is ready for closure. This may be a valid concern if DOE is using a superstructure that is difficult or costly to move; it is not clear how much of an inconvenience this would impose.

Therefore, the committee recommended in its interim report that DOE evaluate advantages and disadvantages for the entire waste management operation at a given site from both a risk and a cost perspective. If DOE can relax other constraints on tank waste removal, such as the tank space problem, delaying tank closure could free up funds planned for closure activities, and those funds could be devoted to

[3]At Hanford the closure schedule is 2024 for single-shell tanks and 2032 for double-shell tanks; at the Savannah River Site the closure milestones are 2022 for Type I, II, and IV, and 2028 for Type III tanks; at Idaho the tanks will be closed in six phases from 2005 to 2016; there are no milestones for closing the calcine bins.

[4]It is the committee's understanding that the geometry of the tanks is inherently stable (i.e., resistant to collapse). The emptied tanks, therefore, need not be filled until immediately prior to closure of the entire tank farm and placement of the engineered cap (if used).

enhancing waste removal, waste processing, and confidence in the near- and long-term performance of the waste immobilization and tank fill materials. Similarly, research and development require funds, but could, if successful, result in lower costs and increased safety overall (see Chapter IX).

FINDINGS AND RECOMMENDATIONS

Finding V-1: To protect future inhabitants on or near the present DOE sites, the primary objective of DOE's tank closure program is to remove reprocessing wastes from DOE sites and permanently isolate the radionuclides in a geologic repository, such as Yucca Mountain. Grouting and other technologies (e.g., Hanford's low-activity waste vitrification) to immobilize the wastes left on-site are secondary lines of defense for protecting future near- or on-site inhabitants.

Recommendation V-1: DOE should maintain its primary objective of removing radioactive tank wastes from DOE sites. Immobilization of wastes left on-site cannot be a substitute or justification for not removing tank wastes from the sites to the maximum extent practical (e.g., to meet schedule commitments).

Finding V-2: When a tank has a relatively simple configuration (i.e., without a network of cooling coils or other obstacles) and can be cleaned to an acceptable degree, it is reasonable to continue with tank closure soon after retrieval. However, when the residue in a given tank after cleaning still contains significant amounts of radioactive material, proceeding immediately to closure effectively precludes any further removal of residue from the tank. In its interim report, the committee recommended that DOE consider decoupling tank cleanup and closure activities.

Recommendation V-2: In cases where significant amounts of radioactive residues remain after tank cleaning, efforts should be directed to emptying and cleanup of other tanks while more effective retrieval techniques are sought. The committee judges that this approach would result in improved risk reduction. This decoupling need not delay the scheduled closure of the overall tank farm.

Finding V-3: Some of DOE's performance assessments for residual wastes in storage tanks incorporate assumptions about the ability of the grout to retain its structural integrity and chemical properties over centuries and even millennia without a firm basis in either empirical data or fundamental scientific principles. In the near term, decisions about the formulation of grouts for tank fill are being made on the basis of experience in very different applications and, in some cases, on data from short-term tests on saltstone. The committee has not seen any reports of long-term testing or more fundamental research directed at the unique aspects of DOE applications, particularly the binding capacity of grouts and changes in various properties over the extended times contemplated by DOE.

Recommendation V-3: The committee recommends that DOE initiate a focused research and development program over a 5- to 10-year period, and longer where necessary, to improve fundamental understanding of the long-term performance of tank fill material and tailoring grout formulations to different tanks or group of tanks. The program should involve collaboration among government laboratories, universities, and industry.

Further details, findings, and recommendations on research and development can be found in Chapter IX.

VI

Performance Assessment

The statement of task charges the committee with evaluating: ". . . (2) any actions additional to those contained in current plans that the Department should consider to ensure that its plans to manage its radioactive waste streams will comply with the performance objectives of Part 61 of Title 10, Code of Federal Regulations; . . . (4) existing technology alternatives to the current management plan for the waste streams mentioned above and, for each such alternative, an assessment of the cost, consequences for worker safety, and long-term consequences for environmental and human health. . . ." Because of the short time available for the completion of this report and the fact that some information from the Department of Energy (DOE) was not available, it was not possible to analyze cost, worker safety, or long-term human and environmental health consequences of alternatives to the current waste management plan. However, the committee did evaluate some of DOE's performance assessments, which are meant to demonstrate compliance with the performance objectives and evaluate the long-term consequences of DOE's plans.[1]

A performance assessment is a quantitative evaluation of the anticipated behavior of a disposal facility that projects the extent of contaminant migration from the facility and the potential impacts of releases on human health and the environment. Such a systematic examination of the engineered and natural environment can also give analysts a qualitative sense of the likelihood of different outcomes by allowing them to examine reasonable and bounding scenarios. DOE uses performance assessments to establish waste concentration limits, waste form requirements, and facility design requirements that are needed to protect long-term public health and safety and the environment (Mann et al., 2001). DOE also uses a facility's performance assessment as its most prominent tool to demonstrate compliance with performance objectives, i.e., that dose limits[2] will not be exceeded. The performance assessment that is used to seek approval for disposal plans may be somewhat different from other performance assessments for the same facility (see Sidebar VI-1), but performance assessment does not end with approval (or disapproval) of disposal plans.

Performance assessment and monitoring are connected before and after closure. Performance assessments use data from monitoring during operations and cleanup as input values for model parameters and to identify anomalies that may require revision of conceptual models. Monitoring programs use performance assessments to assist the selection and location of data acquisition systems. In considering this relationship, the committee benefited from the proceedings of a recent workshop on performance monitoring sponsored by DOE (DOE, 2005). Because there are ongoing requirements for performance assessment, the committee views a performance assessment as an evolutionary document that must be updated and changed as new information becomes available and as changes occur at the site (changes in both the physical situation and the site activities—e.g., operations, closure, post-closure monitoring and surveillance).

Performance assessments (see Sidebar VI-2) are based on conceptual and numerical models that simplify reality to predict contaminant releases, migration, and consequent exposures of people and biota under a set of assumptions and scenarios. As noted in an earlier National Research Council report, "[p]roperly done, risk assessment is a powerful tool for systematically organizing the information and understanding the behavior and impacts of radioactive waste at a particular location" (NRC, 2005b). The committee regards performance assessment as a valuable tool for examining the

[1]The committee did not undertake a detailed review of every input parameter, feature of the model, or method used. The U.S. Nuclear Regulatory Commission is conducting just such a detailed review. The committee focused on the methods, key assumptions, and results.

[2]DOE calls the numerical limits for inadvertent intruders "performance measures," but for simplicity the committee refers to them as dose limits.

SIDEBAR VI-1
Conservative or Realistic Assessment

Where this report mentions conservative or realistic assessments, the committee uses these terms in the context described below. A realistic performance assessment implies that the conceptual models on which the calculations are founded, and the values assigned to the model parameters, are structured to provide a "best estimate" of system performance, given the available data and current conceptual understanding. It is most appropriate to adopt this approach when using a performance assessment for planning or design approaches, such as improving facility design or establishing waste form requirements, or when confirming the performance of a disposal site.

In demonstrating compliance in a regulatory context, it is often appropriate to adopt approaches that produce conservative estimates that overstate the magnitude of the impact. In this way, uncertainties in the description of certain physical or chemical processes, and in the estimates of model parameters, can be given less weight in evaluating the results of the performance assessment.

SIDEBAR VI-2
The Performance Assessment Process

During the last three decades, a generally agreed process has developed for carrying out the technical steps in a performance assessment (see Appendix I for a more thorough description of the performance assessment process), although different methods and models are used at each site and even for different facilities within a site. In many cases, the nontechnical steps, from identification of the consequences of concern to the approach used in characterizing risks, are just as important as the technical steps and indeed are inextricable parts of these steps (see, e.g., NRC, 1994, 2005b; PCCRARM 1997a, 1997b). The methodology for carrying out the nontechnical steps is not as standardized, and these factors are discussed in Chapter VIII. Even within the technical realm, there are debates about the best way to implement some parts of the technical methodology. However, the requisite components of the methodology are not really at issue.

For example, some practitioners prefer to examine uncertainty and variability using Monte Carlo methods, which run computer models many times using probability- and frequency-based distributions of values for input parameters to yield a probability distribution of results. Other practitioners, including analysts that conducted all of DOE's tank waste performance assessments, use deterministic methods and carry out sensitivity studies (with no embedded probability weighting) to illustrate the effects of uncertainty and variability. All, however, agree that uncertainty and variability must be accounted for and presented in an assessment.

A previous National Research Council report, *Risk and Decisions about Disposition of Transuranic and High-Level Radioactive Waste* (NRC, 2005b), states that "The key feature of the risk analysis process described in this [report] is that the data, modeling, and any other calculations in estimating risk must be structured to inform a specific and well-defined decision," and "analytical detail and complexity should be limited to the minimum necessary to distinguish the best option or options." In this approach, complexity is added only as it provides needed greater fidelity to the behavior of the real system being modeled.

waste form, the disposal facility, the disposal environment, and the likely future interactions among the waste, people, and surrounding environment. A good performance assessment can help identify which factors are most important in ensuring safe disposal of the waste, and even what actions may increase or reduce risk. The committee does not, however, believe that a performance assessment can accurately determine the concentrations and quantities of long-lived contaminants that will leach from a disposal facility far into the future. Accepting the absolute numerical results of a performance assessment entails accepting that the assumptions and scenarios adequately represent physical reality. Because the absolute numerical results, including the numerical predictions of doses from radionuclides, are approximations and subject to unquantifiable uncertainties, the committee views performance assessments as only a part of the demonstration of compliance. Performance assessment is a tool to support decision making; it is not a definitive statement of future conditions at a site.

DOE'S PERFORMANCE ASSESSMENTS

The committee has examined the performance assessments that were available before 2006 for tank heels (at all three sites) and the separated low-activity waste from the tanks (at the Savannah River Site and Hanford). DOE has documented its code selection process (see, e.g., Mann et al., 1999), and DOE's performance assessments use models that are recognized and accepted by the performance assessment community and specialists who model contaminant transport through different media. The committee has made no judgment whether DOE's current tank closure plans now meet the performance objectives, which is a regulatory decision. Instead, the committee focused on technical aspects of several of DOE's performance assessments. Although the committee did not have sufficient time or resources to confirm the results of the models used, it did examine the methods, key assumptions, and results to evaluate what confidence should be placed in the performance assessments.

In preparing its interim report, which focused on the Savannah River Site, the committee identified several concerns about the modeling approach DOE was using to demonstrate compliance with the performance objectives in 10 CFR 61, some of which are discussed below. At each committee meeting at the three sites, the committee asked DOE to provide presentations making the case why its proposed approach for managing and disposing of tank wastes is safe and acceptable.

Concurrent with the committee's study, the U.S. Nuclear Regulatory Commission (USNRC) provided DOE with a long list of both general and detailed requests for additional information to support its review of DOE's draft salt waste determination (DOE-SRS, 2005b) at the Savannah River Site. DOE and the USNRC also held supplemental discussions in public meetings to clarify technical points and reach a mutual understanding of their respective modeling philosophies for performance assessment.

From these interactions, DOE clearly learned what information and what kinds of transparency independent analysts need for them to evaluate DOE's proposed actions. The clarity, quality, and completeness of DOE's analyses, presentation of the information, and reasoning have improved dramatically since the committee's interim report was finalized. DOE has developed new documents called "performance objectives demonstration documents" (PODDs) that describe the objectives, conceptual models, assumptions, and reasoning that supports decisions far better than the committee had seen in prior DOE performance assessments. The PODDs also include sensitivity studies, which allow reviewers to evaluate the consequences of modeling assumptions and input parameter values.

The PODDs address committee concerns, such as how the projected performance changes with grout durability, which is examined through a set of scenarios that constitute deterministic sensitivity studies. Some other concerns, such as questions about how long the pH and chemically reducing properties of grout need to persist in order to give acceptable performance, have not been addressed fully—the sensitivity studies provided in the September 30, 2005 Savannah River Site draft waste determination for closure of two tanks and its associated PODD (Buice et al., 2005; DOE-SRS, 2005a) are not entirely sufficient for this purpose. The committee's concerns about the point of compliance (the location at which compliance with performance objectives must be demonstrated) that DOE has selected and DOE's assumptions in intruder scenarios remain. An additional concern about the combined inventory estimated to remain in the tanks at Savannah River Site has arisen since the committee issued its interim report. These concerns are described in the discussion of the Savannah River Site results, later in this chapter.

DOE's plans to grout and close the tanks have raised a number of concerns among environmental groups (Makhijani et al., 1986; NRDC, 2003, 2004a, 2004b, 2005; Makhijani, 2004; Makhijani and Boyd, 2004; Perks, 2004; Smith, 2004). The main concern is that, given that the tank heels and grout do not mix, grout is not viewed as a form of waste immobilization, but rather as a layer on top of the tank heels. Even incorporating waste into grout will not prevent leaching of contaminants into groundwater. The only means to prevent leaching over the long term is to keep water out. Environmental public interest groups warned that the waste would eventually leach into the groundwater near the Columbia River in Washington, the Snake River Aquifer in Idaho, and the water table near the Savannah River (e.g., NRDC, 2003). Smith provides detailed concerns about the long-term performance of grout related to uncertainties in the evolution of leaching properties, degradation mechanisms, hydraulic properties, chemical properties, patterns of waste and grout distribution in the tanks, and temperature effects on grout and waste with time (Smith, 2004). Makhijani and others raise a number of further concerns related assumptions about long-term stewardship, long-term predictions of environmental changes, and human behavior at the Savannah River Site (Makhijani et al., 1986; Makhijani, 2004; Makhijani and Boyd, 2004).

In its 2004 performance assessment for the Saltstone Vaults and in earlier assessments for closed tanks at the Savannah River Site (DOE-SRS 1997a,1997b), DOE assumed that the grout would maintain its physical integrity as a hydraulic barrier for 500 years. The committee was particularly concerned that: (a) there was scant scientific support for the 500-year assumption, and (b) DOE's approach of treating modeling elements, such as all of the grout in a tank, as uniform or homogeneous ignores phenomena that are dominated by heterogeneities, such as fracture flow. In response to the USNRC's request for additional information and supplemental discussions, DOE carried out sensitivity studies that included an "early failure" scenario in which the grout maintains a low hydraulic conductivity for only 100 years (with chemical reducing conditions remaining the same as in the original analysis). This scenario indicated little change in the quantities of radionuclides released (an 8 percent change in the projected dose) and, therefore, showed that assumptions about the physical integrity of the grout are not critical to the results of the performance assessment. In all subsequent calculations for tank closures, DOE did not take credit for the concrete tank vault and carbon steel tank, and it is assumed that there are cracks in the basemat starting at 500 years after closure (Buice et al., 2005). (See Chapter V for a detailed discussion of tank grouting.)

DOE made similar assumptions about the pH and chemically reducing capability of the grout in the tanks; i.e., that reducing conditions would be maintained for 10,000 years. DOE's response to the USNRC's request noted above was to carry out calculations in which concentric cracks form in the grout in the tank, thus creating preferred flow paths and locally diminishing the reducing conditions (Buice et al., 2005). DOE's calculations indicate that most of the grout maintains its reducing capacity. However, the bulk of the grout is not particularly relevant, given that the waste does not substantially mix with the grout. Grout that comes in contact with water flowing through fractures may not maintain its reducing capacity indefinitely; hence leached radionuclides may not remain in a less mobile, chemically reduced valence state. Lukens and others (2005) have shown that the technetium species in grout are rapidly oxidized by oxygen if exposed to it, such as they could be along the cracks. According to DOE (Buice et al., 2005),

> Additional calculations were conducted to determine the rate at which the bottom seven inches (lowest modeling node) of reductant was consumed: with no cracks, 96% of the grout remained reduced after 10,000 years, whereas with three cracks, 83% of the grout remained reduced after 10,000 years.

It is not clear whether such modeling can predict the grout's reducing capabilities over thousands of years, or whether the impact of the quicker decline in grout reducing capability is important to the performance assessment results. While these assumptions seem reasonable when described by Savannah River Site scientists, there is no experience or other means to directly substantiate that such long-term assumptions are valid. In the construction industry the chemical reducing capabilities of a material are irrelevant, and "long term" means 75 to 100 years (with an active maintenance program).

Long-term performance of tank fill material is site dependent due to soil conditions, climate, and near-field chemistry (i.e., interactions among the chemicals in the tank contaminants and effects of any residue from previous treatments). For example, because the sludge does not mix completely with the grout but remains interlayered with or at least partially encapsulated by it, the pH and reducing capability of water migrating through the grout are important factors in its ability to minimize the mobility of radionuclides and toxic heavy metals from the sludge into the groundwater.

The first Savannah River Site tanks selected for closure do not contain some important complicating factors that affect tank cleaning, such as extensive vertical cooling coils (although they do contain zeolite). However, most of the tanks to be closed in the future contain cooling coils, which may become pathways for water infiltration to the residual waste if the grout inside or surrounding the coils shrinks significantly. Moreover, Tanks 9 through 12 are mostly submerged in the water table at all times. For these tanks, there is an additional possibility of water ingress from the sides or the bottoms.[3] In these cases the radioactive waste residuals may not have the full protection of the layers of grout to reduce influx, and groundwater inflows may not be buffered by the overlying grout to the same degree (see Chapter V, Finding and Recommendation V-3).

Performance Objectives and Exposure Scenarios

As noted in Chapter I, Section 3116 of the 2005 NDAA requires removal of "highly radioactive radionuclides to the maximum extent practical" and compliance with the performance objectives in 10 CFR 61 for wastes disposed on-site. The performance objectives (and their supporting guidance documents) lay out a set of numerical dose limits for people who could be exposed (workers, members of the public, and inadvertent intruders); a requirement that releases to the environment and doses during operations be made as low as reasonably achievable (ALARA); and requirements concerning site stability (see Appendix C). Thus, Congress laid out an approach for waste management and disposal that requires minimization of the inventory of radioactive material that is to be disposed at the sites (within the limits of what is practical), meeting or surpassing a set of numerical dose criteria, and ensuring that contamination of the environment is as low as reasonably achievable.

In addition to the assumed scenarios for release and transport of contaminants from the disposal location through the environment, DOE's performance assessments rely on assumed exposure scenarios. For a member of the general population, the scenario typically involves exposures from drinking water, consuming fish and vegetables, bathing, and other activities that could bring people in contact with contaminants carried through groundwater. Although air and soil pathways must be considered, they typically are not as important for underground disposal facilities such as those considered here.

The point of compliance is a critical element of a performance assessment used in support of a compliance decision. A point of compliance is a location some distance away from the disposal facility boundary. Inside the point of compliance is the facility operations area, which is subject to institutional controls for at least 100 years after closure. After site closure, anyone inside the boundary marked by these points of compliance (i.e., the controlled area) is considered an intruder. DOE must ensure that people in the controlled area (even after the period of institutional control) would not

[3]DOE has not yet established closure plans for these tanks, so the committee has not commented on them in detail.

receive environmental exposures that exceed the limits in DOE and USNRC guidance concerning inadvertent intruders. Beyond the point of compliance, DOE must ensure that members of the public would not receive environmental exposures that exceed the dose limits in DOE Order 435.1 and 10 CFR 61.

The point of compliance most commonly is set 100 m (109 yards) from the disposal facility boundary based on DOE Order 435.1 (DOE, 2001a) and USNRC guidance (USNRC, 1997b). Other locations based on plans for future land use and physical and institutional controls that may be in place are allowed under both commercial low-level waste disposal regulations and DOE Order 435.1 to prevent decision makers from missing important risks.[4] In all cases, institutional controls are assumed to prevent inadvertent intrusion for 100 years after closure of the disposal facility.

In principle, the time of compliance, too, can restrict consideration of risks in a way that neglects potentially important aspects of the disposal decisions, but the committee found no cases where DOE's analyses cut off important results because of the time of compliance. The time of compliance required for low-level waste facilities under DOE Order 435.1 is 1,000 years, but many of DOE's analyses extend up to 10,000 years to account for dose peaks that arrive after the mandated time of compliance.

Various intruder scenarios are considered in the DOE assessments: a construction scenario, a recreation scenario, and an agriculture scenario. In some cases, it is assumed that an intruder drills directly through a waste vault, tank, or transfer line. Some cases assume that drinking water and irrigation water are drawn from an aquifer that underlies the disposal facility.

Finally, for low-level waste disposal facilities, DOE Order 435.1 states (DOE, 2001a):

> [t]he performance assessment and composite analysis shall be maintained to evaluate changes that could affect the performance, design, and operating bases for the facility . . . maintenance shall include the conduct of research, field studies, and monitoring needed to address uncertainties or gaps in existing data. . . . Additional iterations of the performance assessment and composite analysis shall be conducted as necessary during the post-closure period.

The manual explicitly notes that review and revision are required when changes "alter the conclusions or the conceptual model(s) of the existing performance assessment or composite analysis." It also requires an annual "determination of the continued adequacy of the performance assessment and composite analysis [based on] the results of data collection and analysis from research, field studies, and monitoring."

This is perfectly in accord with the committee's view of a performance assessment as a living document. It is not clear in practice, however, that DOE's performance assessments and composite analyses are treated as living documents after disposal authorization is received. For example, the Savannah River Site's composite analysis for the E-Area Vaults Saltstone Disposal Facilities (WSRC, 1997) was last revised in 1999 (Cook et al., 1999) and does not reflect DOE's current understanding of inventories and expected performance of disposal facilities at the site. This means that the composite analysis has not been up to date as DOE, the State of South Carolina, and the USNRC have evaluated DOE's plans for disposal of tank waste on the site. DOE has in place a program for maintaining the adequacy of the performance assessments and composite analysis (WSRC, 2000). DOE informed the committee that the composite analysis is scheduled for revision in the near future. The need for current assessments of all the contaminants and facilities that contribute to risk at the site is discussed in the context of decision making in Chapter VIII.

The Savannah River Site Performance Assessment Results

DOE assumes that the Savannah River Site will remain under federal government ownership within its current boundaries in perpetuity. (See executive summary of DOE-SRS, 2005f for assumptions about future land use; see Appendix J of this report for a map of the site and the General Separations Area.) DOE further assumes that the site will continue to remain zoned for industrial, industrial support, and general uses. DOE intends to restrict the area around the facilities where tank waste will remain on-site from residential use for 10,000 years. In its performance assessment, DOE assumes that active institutional controls to prevent inadvertent human intrusion are effective for 100 years. In its performance assessments, DOE has considered the possibility that after 100 years, an inadvertent intruder could construct a residence near one of the disposal facilities (DOE-SRS, 2005b; Ross, 2005).

DOE uses different points of compliance for the different tank waste disposal facilities. Table VI-1 summarizes the points of compliance that DOE uses for members of the public and inadvertent intruders at each of the facilities considered at the Savannah River Site. For the closed tanks, the general population is defined as people residing outside the General Separations Area (GSA), which includes both the F and H Tank Farms, and the Z Area Saltstone Disposal Facility, where tank wastes will remain, in addition to the Defense Waste Processing Facility, the Old Burial Ground, the E Area low-level waste disposal facilities, and the canyons themselves (see Appendix J). DOE states in its performance assessment demonstration document for the tank closures, "[a] key assumption to the modeling analysis is

[4] "The point of compliance shall correspond to the point of highest projected dose or concentration beyond a 100 meter buffer zone surrounding the disposed waste. A larger or smaller buffer zone may be used if adequate justification is provided" (DOE, 2001a).

TABLE VI-1 Points of Compliance for Tank Wastes Planned for On-site Disposal at the Savannah River Site

	F Tank Farm	H Tank Farm	Saltstone Vaults
Members of the public Resident who constructs and lives in a dwelling at the point of compliance	"Just downstream" and on the opposite bank from a seepline where groundwater outcrops to a stream. The seepline is located 1.8 km downgradient of the tank farm. Exposure is via surface water.[a]	"Just downstream" and on the opposite bank of a seepline where groundwater outcrops to a stream. The seepline is located 1.2 km downgradient of the tank farm. Exposure is via surface water.[a]	100 m from the boundaries of the facility. Note that only doses arising from air and groundwater pathways have currently been considered at the point of compliance.
Inadvertent intruders Resident who constructs and lives in a dwelling and drills a well inside facility boundary	For soils,[b] inside the facility boundary. For groundwater, assumes a well drilled inside the facility boundary through a transfer line into the third aquifer below the surface.	For soils,[b] inside the facility boundary. For groundwater, assumes a well drilled inside the facility boundary through a transfer line into the third aquifer below the surface.	For soils, at the facility boundary. For groundwater, assumes a well drilled into the top aquifer (water table) at the facility boundary.

[a] Exposure pathways for member of public: (1) incidental ingestion of soil from shoreline deposits (during recreational activities), (2) direct radiation from seepline, (3) air inhalation (I-129 volatilization) at seepline, (4) dermal contact with Four Mile Branch, (5) drinking water from Four Mile Branch, (6) ingestion of fish from Four Mile Branch, (7) direct radiation from Four Mile Branch, (8) ingestion of milk from cows fed vegetation grown on soil irrigated with Four Mile Branch water, (9) ingestion of meat from cows fed vegetation grown on soil irrigated with Four Mile Branch water, and (10) ingestion of produce irrigated with Four Mile Branch water. Only 5, 6, and 10 are significant contributors to the peak dose.

[b] Contaminated soils considered in the intruder scenario.

that no unrestricted use of the land or groundwater for the GSA will be permitted as presented in the SRS End State Vision [6] and Savannah River Site (SRS) Long Range Comprehensive Plan. . . ." (Buice et al., 2005). The committee notes that the GSA is a rather large area (approximately 20 km^2 [about 8 square miles], judging from maps) and the points of compliance for the tank farms are rather distant from the boundaries of the facilities (1.8 km [1.1 mile]).

In contrast, the point of compliance for the Saltstone Disposal Facility is set 100 m from the disposal facility boundary. Concentrations of contaminants in the aquifers at the Savannah River Site (and therefore doses from the consumption of groundwater) decrease with distance from a leaking facility because of mixing, dilution, sorption, and decay of contaminants as they migrate away from the facility. As a result, the estimated radiation dose received by a member of the public at a point of compliance that is far from the disposal facility will be lower, and thus more likely to meet performance objectives. Other factors being equal, a larger area with contaminated groundwater has a greater likelihood than a smaller area does of an inadvertent intruder drilling a well and drawing contaminated water after active institutional controls cease to be enforced.

The selection of the point of compliance has both policy and technical dimensions. Good policy, however, requires that there be justification, including at least technical coherence, for the selection. DOE justifies its point of compliance for the tanks by turning to the land-use plans for the site, which envision government control and only industrial uses of the General Separations Area, in perpetuity. The South Carolina Department of Health and Environmental Control approved these points of compliance in closing Tanks 17 and 20, but recently raised questions about whether they are acceptable for future tank closures (SCDHEC, 2005a). Nonetheless, the points selected at the Savannah River Site tank farms afford the lower, intruder level of protection (i.e., higher dose limit) over a large area (approximately 20 square km^2), and even the intruder dose might be much higher than expected if some assumptions in the intruder scenarios, which are discussed below, do not prove to be correct (see Chapter VIII).

The agricultural intruder scenario for the Tanks 18 and 19 performance assessment assumes that an inadvertent intruder would draw household water from the Congaree aquifer, the deepest of three aquifers below the tank farms where estimated potential contaminant levels are lowest, rather than from the upper (water-table) or Barnwell-McBean aquifers where estimated potential contaminant levels are higher. (See Figure J-3 in Appendix J for a diagram showing these hydrologic units.) DOE supports this assumption by reasoning that an intruder would not choose to draw well water from the water-table or the Barnwell-McBean aquifers, which have a yield of 11 to 19 liters (3 to 5 gallons) per minute.[5] Instead, an intruder would know that a higher-yield aquifer lies below the Barnwell-McBean aquifer and would drill into that. DOE also assumes that the intruder constructs the household water well in a manner ensuring that cross-contamination of the aquifers does not occur during either

[5] Any well in the United States that yields less than six gallons per minute is generally considered a low-yield well. In performance assessments at other sites, DOE has considered exposure scenarios based on its interpretation of land uses by local Native American peoples. Alternative uses that could entail, for example, use of lower yield wells were not considered at the Savannah River Site.

TABLE VI-2 Dose Predictions at Different Locations from the Savannah River Site F Tank Farm

	Intruder		General Population
Source	Drinking Water Dose[a] from Well in Barnwell-McBean Aquifer 100 m from Tank Farm Boundary	Drinking Water Dose[a] from Well at Barnwell-McBean Aquifer Seepline	All Pathways Whole-Body Dose[b]
Tank 18			
Technetium-99	2.1 mrem/yr at 525 yr	0.14 mrem/yr at 665 yr	
Neptunium-237	23 mrem/yr at 4935 yr	1.3 mrem/yr at 6755 yr	0.04 mrem/yr at 6405 yr
Tank 19			
Technetium-99	5.8 mrem/yr at 595 yr	0.29 mrem/yr at 735 yr	0.009 mrem/yr at 735 yr
Neptunium-237	3.0 mrem/yr at 5355 yr	0.13 mrem/yr at 7245 yr	
F Tank Farm			
Technetium-99	57 mrem/yr at 525 yr	2.5 mrem/yr at 595 yr	0.07 mrem/yr
Neptunium-237	30 mrem/yr at 4795 yr	0.13 mrem/yr at 7245 yr	
Iodine-129	3.2 mrem/yr at 385 yr	0.26 mrem/yr at 455 yr	0.01 mrem/yr

NOTE: 1 mrem = 0.01 mSv.
[a] 50-year committed effective dose equivalent.
[b] DOE describes this dose as the 50-year committed dose equivalent to the whole body.

SOURCE: Buice et al., 2005.

well drilling or well use. Thus, the intruder's actions do not result in migration of contamination from the upper aquifer to the lower ones and do not increase the radiation dose received by the intruder as a result of aquifer cross-contamination.

While one might hope that an inadvertent resident would both construct his or her well properly and draw from the more productive, cleaner aquifer, it seems only prudent to examine what the consequences would be if these assumptions were incorrect. The contaminant concentrations that DOE predicts in the water-table aquifer are higher than those in the Congaree aquifer by factors ranging from about 60 to about 250, while those in the intermediate Barnwell-McBean aquifer are higher than those in the Congaree aquifer by factors ranging from about 250 to about 1,000. The more conservative approach of considering exposures from the more contaminated aquifers is more consistent with standard practice. Table VI-2 presents projected doses for people drawing drinking water from the Barnwell-McBean aquifer at two locations within the boundary defined by the chosen point of compliance, alongside the results for a member of the public residing across the Four Mile Branch. This scenario also assumes that drinking water is the only source of radiation exposure for the intruder. Although other scenarios take into account the possibility that radionuclides could be concentrated in animals and plants consumed by the intruder, this one does not (see Findings and Recommendation VI-2).

All of the doses are within the limits for intruders described in the guidance for implementing the performance objectives, although the higher concentrations of neptunium-237 and technetium-99 in the Barnwell-McBean Aquifer at 100 meters bear watching to ensure that they do not exceed the performance objectives. The committee was surprised to note that a large fraction (perhaps more than 75 percent) of the estimated neptunium-237 drinking water dose from a well in the Barnwell-McBean Aquifer 100 meters from the tank farm boundary is attributed to neptunium-237 migration from just one tank: Tank 18. The PODD for Tanks 18 and 19 (Buice et al., 2005) presents the estimated residual radionuclide inventory in tanks that have already been cleaned out (Tanks 17-20) and the projected residual inventory for F Tank Farm tanks that have not been cleaned out (Tanks 1-8, 25-28, 33-34, and 44-47). These inventories were used to calculate the concentrations and doses from the whole F Tank Farm, as presented in Table VI-2. The estimated remaining quantity of neptunium-237 in Tank 18 is 0.118 Ci (4.37×10^9 Bq) and the combined total estimate of neptunium-237 in all of the other tanks in the F Tank Farm is 0.0535 Ci (1.98×10^9 Bq) or approximately 45 percent of the inventory in Tank 18.

DOE explains the imbalance in the neptunium-237 inventory based on three factors (the following numbered points are quoted from Ross, 2006a):

1. The F-Area laboratory was used to analyze H Canyon Np-237 processing samples. The waste from the F-Area Laboratory was discarded to Tank 18 (and carried over to Tank 19 through evaporator processing) and, therefore, these tanks would have higher than average concentrations of Np-237.[6]

[6]DOE notes that the addition of the laboratory waste was not tracked in the waste characterization system (WCS) tables (DOE-SRS, 1999).

TABLE VI-3 Assumed Volume of Residual Waste Remaining in Closed HLW Tanks at the Savannah River Site.[a]

Tank Number	Area	Tank Type	Residual Material Volume (gallons)	Tank Number	Area	Tank Type	Residual Material Volume (gallons)
1	F	I	100	27	F	III	1,000
2	F	I	100	28	F	III	1,000
3	F	I	100	29	H	III	100
4	F	I	100	30	H	III	100
5	F	I	100	31	H	III	100
6	F	I	100	32	H	III	100
7	F	I	100	33	F	III	100
8	F	I	100	34	F	III	100
9	H	I	100	35	H	III	100
10	H	I	100	36	H	III	100
11	H	I	100	37	H	III	100
12	H	I	100	38	H	III	100
13	H	II	100	39	H	III	100
14	H	II	100	40	H	III	100
15	H	II	100	41	H	III	100
16	H	II	100	42	H	III	100
17[b]	F	IV	2,200	43	H	III	100
18	F	IV	1,000	44	F	III	1,000
19	F	IV	1,000	45	F	III	1,000
20[b]	F	IV	1,000	46	F	III	1,000
21	H	IV	100	47	F	III	1,000
22	H	IV	100	48	H	III	100
23	H	IV	1,000	49	H	III	100
24	H	IV	100	50	H	III	1,000
25	F	III	1,000	51	H	III	100
26	F	III	1,000				

[a] These volumes are an assumption for modeling purposes only and do not represent a commitment or goal for waste removal.
[b] Tank has been closed.

SOURCE: DOE-SRS, 2002.

2. The mass of material remaining in Tank 18 is 15,335 kg (16,357 kg using the upper 95% confidence density). For the purposes of estimating the future F-Tank Farm residual material radionuclide inventories, it was assumed that future waste removal activities would be successful in reducing residual material inventories to the [environmental impact statement (EIS)] predicted final inventories of 88.4 kg for high heat tanks and 884 kg for low heat tanks. As future tanks are closed, the mass of the residual materials will be adjusted as necessary for those tanks and the predicted radionuclide inventories at closure will be replaced with inventories based on analysis of residual materials.

3. Several conservative assumptions were made that also contributed to the elevated Np-237 inventory in Tank 18 (and Tank 19) . . . : (a) The upper 95% confidence sample results were used instead of the average. This increased the Np-237 inventory in Tank 18 by almost 16%. (b) The upper 95% confidence sample density was used instead of the average density. This increased the Np-237 inventory by another 7%. (c) The residual liquid Np-237 inventory was calculated based on a less-than-detection sample result for Np-237. (d) An additional 0.033 Ci [1.2×10^9 Bq] of Np-237 was calculated to be in tank wall corrosion products. This increased the Np-237 inventory by another 40%.

The committee examined some of the details of point two above, the prediction of final tank inventories quoted from the environmental impact statement. Table VI-3 shows the data from the *Savannah River Site High Level Waste Tank Closure Environmental Impact Statement* (DOE-SRS, 2002, Table C.3.1-2 in Appendix C).

The committee observes that these predictions are characterized in the tank closure environmental impact statement as assumptions, rather than commitments or even goals. Further, the assumed volumes of residual material in Tanks 17 and 20 (2,200 gallons [8.3 m^3] for Tank 17 and 1,000 gallons [3.8 m^3] for Tank 20), which are the only tanks that had been closed at the time the tank closure environmental impact statement was issued (2002), represent the upper bound of the assumed residuals among all of the tanks. Indeed, DOE assumed that Tanks 18 and 19 would each have 1,000 gallons residual. The current estimates in the performance objective demonstration document (Buice et al., 2005) are 4,300 gallons (16.3 m^3) in Tank 18 and 15,100 gallons (57.2 m^3) in Tank 19. There is zeolite in Tanks 18 and 19, which proved difficult to remove, so DOE proposes to leave it in place, but the performance objective demonstration document reports that Tanks 7, 25, and 27, all of which are in the F Tank Farm, also contain zeolite. Of the tanks that have not been cleaned out, some of them have large quantities of sludge; several of them do not. Access within Tanks 18 and 19 should be better than others because neither of the tanks has cooling coils in contrast to all of the tanks in the F Tank Farm yet to be cleaned, which do. For these reasons, Tanks 18 and 19 should be easier to clean than some of the other future tanks.

The committee views the assumptions about residual inventories as both optimistic and unsupported. It would be more prudent for DOE to base its assumptions on what is known about the actual tanks and their contents, and on what is known about the effectiveness of the tank waste removal technologies that DOE plans to employ in those tanks (see Chapter III). With this more informed estimate, DOE and others could then examine sensitivities to see what estimated consequences arise if assumptions in DOE's scenarios and models turn out to be inaccurate. Without such an approach, the committee does not have confidence in the results of the performance assessment for the tank farm (see Finding and Recommendation 3).

The Idaho National Laboratory Performance Assessment Results

In 2003, the Idaho National Laboratory released a performance assessment for the INTEC tank farm facility (TFF; DOE-ID, 2003b). The performance assessment is a comprehensive document that contains essentially all of the elements that the committee has identified as forming the blueprint for an acceptable performance assessment (see the flowchart in Appendix I). Widely accepted simulation codes were used to predict the release of radionuclides from the vaults and tanks, and to model flow and transport both in the vadose zone above the water table and in the saturated zone. Infiltration through the vadose zone was modeled using equivalent porous medium concepts, with estimates of hydraulic conductivity of the basalt layers linked to geologic descriptions of the individual flows. The hydrogeology model embedded within the performance assessment was calibrated in an attempt to match site conditions as they were best understood, data uncertainties were discussed, and a wide-ranging sensitivity study was presented. This performance assessment led the Idaho National Laboratory to conclude that closure plans for the tank farm could proceed safely and would be protective of the environment. The sensitivity study indicated that the potential for the doses to exceed the performance objectives was low and that a combination of worst-case assumptions would have to be realized to result in doses that exceeded the performance objectives for members of the public.

The performance assessment was reviewed by the USNRC, acting in an advisory role (USNRC, 2003), unlike its current role under Section 3116 of the 2005 NDAA. In addition, the USNRC developed its own performance assessment models for use in this review. The USNRC came to the conclusion that the Idaho National Laboratory had developed a reasonable source term estimate and had adequately modeled engineering system degradation, release, hydrology, and transport. Several recommendations were provided to increase confidence in the model predictions, such as including an effort to better estimate the radionuclide inventory in the sand pads beneath several tanks and to evaluate the sensitivity of the model results to the possibility of oxidizing conditions in the grout that would affect solid-solution distribution coefficients (K_ds).

The 2003 performance assessment used a conservative estimate of the radionuclide inventory in the tank heels, because the Idaho National Laboratory had yet to develop experience with residual waste removal. Tank cleaning that was carried out subsequent to the development of the performance assessment has shown that the radioactive inventories in the tank heels could be reduced to a value significantly less than a "best-case" condition that was modeled in the performance assessment as part of its sensitivity study. Updates of projected doses calculated on the basis of current estimates of the waste residuals that were developed from new operational experience are given in the *Draft Section 3116 Determination Idaho Nuclear Technology and Engineering Center Tank Farm Facility* (DOE-ID, 2005a). These calculations yield lower doses than the conservative estimates provided in the 2003 performance assessment (see Table VI-4).

Concerning the point of compliance at the Idaho tank farm facility for a member of the public, DOE states (DOE-ID, 2005a),

> The groundwater model analysis shows that the contamination plume center (where the highest concentrations enter the regional aquifer) would be 600 m (1,969 ft) southward in the downgradient direction from the center of the southernmost TFF tank. The contamination plume center is taken as the source of drinking water after the institutional control period of 100 years has expired.

Using this location appears to be a conservative approach.

A key element of the performance assessment calculations involves modeling of infiltration and radionuclide transport through the deep vadose zone beneath the Idaho National Laboratory. It is on the basis of this modeling that the maximum impacts on groundwater from radionuclide release are predicted to occur at a point on the water table some 600 meters to the south of the tank farm facility. The hydrostratigraphy beneath the tank farm facility is complex, as are the inferred flow patterns with multiple, discontinuous,

TABLE VI-4 The Idaho National Laboratory Tank Farm Closure Performance Assessment Results[a]

Performance Objectives (dose limit)	Performance Assessment Results (early estimate of inventory at closure)	Performance Assessment Results (current estimate of inventory at closure)[b,c]	Doses under worst-case K_ds, elevated infiltration, and conservative inventory
All-pathways dose to the public (not exceeding 25 mrem/yr)	1.86 mrem/yr[d]	0.46 mrem/yr	15.0 mrem/yr (at 342 years due to strontium-90)
Acute drilling scenario (less than 500 mrem)	232 mrem	152 mrem	
Acute construction scenario (less than 500 mrem)	0.80 mrem	0.23 mrem	
Chronic post-drilling scenario (less than 100 mrem/yr)	91.1 mrem/yr	25 mrem/yr	
Chronic post-construction scenario (less than 100 mrem/yr)	26.1 mrem/yr	3.15 mrem/yr	

NOTES: 1 mrem = 0.01 mSv.
[a] All doses are described by DOE as 50-year committed organ dose equivalents ($h_{t,50}$).
[b] The peak annual dose to the thyroid is approximately 6 mrem/yr compared to the 10 CFR 61.41 limit of 75 mrem/yr.
[c] The peak annual dose to any other organ is approximately 0.15 mrem/yr compared to the 10 CFR 61.41 limit of 25 mrem/yr.
[d] The groundwater pathway contributed 1.35 mrem/yr.

perched zones forming on lower-permeability units in the subsurface. The flow system may be undergoing transient readjustments to long-term changes in the infiltration regime at the facility. Numerical simulations of flow and transport can be exceedingly challenging when conducted over scales of hundreds of meters, particularly when conducted over scales of hundreds and thousands of years, especially within the vadose zone. Further complexity is introduced in the vadose zone by the presence of open fractures in the basalt flows. Predictions are likely to be non-unique (i.e., multiple solutions satisfy the boundary conditions and other requirements) and subject to considerable uncertainty.

Given the core role of this pathway element in the performance assessment, and the inherent complexity of the vadose zone, the committee sees merit in a detailed, independent evaluation of flow and radionuclide transport to the water table at the tank farm facility. The intent would not be simply to reproduce the computations in the performance assessment with a different software package, but to construct an independently derived conceptual model, numerical simulation, calibration, and prediction. This modeling task may benefit from the recent data analysis and insight gained in the model calculations associated with the ongoing soil remediation program at Idaho Nuclear Technology and Engineering Center and work at Box Canyon (see, e.g., Unger et al., 2004). This latter program has developed a groundwater flow model that extends from the ground surface to the Snake River Plain aquifer and a detailed geochemical model of the near-surface zone. A strategy of independent confirmation could provide considerable support to the performance assessment calculations that have been developed to date or raise important questions about the results.

Hanford Performance Assessment Results

DOE issued the *Preliminary Performance Assessment for Waste Management Area C at the Hanford Site, Washington* in 2003 (Mann and Connelly, 2003). DOE, the Washington Department of Ecology, and the contractor that generated the performance assessment, CH2M-Hill Hanford Group, Inc., convened a panel to review the performance assessment (Kosson et al., 2004). The panel criticized several aspects of the report, which reflected the preliminary nature of the analysis. Hanford Site personnel said that the review panel provided valuable feedback that will improve the quality of the revised performance assessment significantly. DOE is working toward a major update of the performance assessment based on the review, but the report had not been issued by the end of 2005. The committee concluded it was not worthwhile to review the preliminary tank farm performance assessment in depth because the work would be outdated immediately upon issuance of DOE's revised performance assessment, which the committee was told is imminent. The committee commends DOE and others involved for seeking peer review.

Just as the Savannah River Site has the Saltstone Disposal Facility, Hanford has a facility for disposal of immobi-

TABLE VI-5 Hanford Low-Activity Waste Disposal Facility (IDF) Performance Assessment Results[a]

Performance Objectives (dose limit)	Performance Assessment Results
All-pathways dose to the public (not exceeding 25 mrem/yr)	1.8 mrem/yr[b]
Acute drilling scenario (less than 500 mrem)	1.06 mrem
Chronic post-drilling scenario (less than 100 mrem/yr)	26.8 mrem/yr
Groundwater protection (less than 4 mrem from beta and gamma emitters)	0.7 mrem/yr

NOTES: 1 mrem = 0.01 mSv.
[a] All doses are 50-year committed effective dose equivalents.
[b] Peak occurs 2,400 years after closure.

lized low-activity waste from its waste tanks. DOE issued a performance assessment for the waste in 2001 (Mann et al., 2001). A revision or update was made in 2003 to account for: (1) a DOE decision not to separate technetium-99 from the low-activity waste streams, and (2) the likelihood that the waste would be disposed in an integrated disposal facility,[7] and (3) the different possible waste forms for the supplemental low-activity waste treatment.[8] The results of the update are presented in Table VI-5.

The doses to members of the public are controlled by only three radionuclides: iodine-129, technetium-99, and neptunium-237. For the first 5,000 years after facility closure, iodine-129 contributes approximately 90 percent of the dose in a scenario in which the member of the public is considered to be a farmer (the most restrictive of three scenarios examined for members of the public). Technetium-99 contributes about 10 percent of the dose in that same time frame. Further out in time, neptunium-237 increasingly contributes to the dose (44 percent at 10,000 years after closure). Important to the plume arrival times is the set of models and assumptions used to assess release from different waste forms. Low-activity waste from the Hanford tanks is considered either Category 1 waste (low-concentrations of radionuclides but the waste is not stabilized) or Category 3 waste (higher concentrations and stabilized or encased with grout). The "early" dose peak (at 2,400 years) results mostly from transport of iodine-129 and technetium-99 from, "only a relatively few Category 1 packages . . . (i.e., those packages with high technetium/iodine content)" (Mann, 2003). DOE notes that those packages could be disposed of as Category 3 packages, if necessary.

[7]The integrated disposal facility is a single disposal facility for immobilized low-activity waste from tank cleanup and other waste from the site.

[8]The different waste forms correspond to the different supplemental treatment options: glass from bulk vitrification, a granular solid from steam reforming, and grout from cast stone.

Peer Review of Performance Assessments

As noted above, DOE's assessments have improved as a result of the reviews carried out under the 2005 NDAA. The committee commends DOE for the improved quality of its presentation of the waste determinations and the supporting documentation (including performance assessments) and reasoning. The dramatic improvement reinforces the committee's judgment concerning the value of independent peer review.

The National Research Council has issued several reports on peer review (see, e.g., NRC, 1995, 2000b, 2002a), including reports specifically advising the DOE's Office of Environmental Management (NRC, 1996a, 1997b, 1998). Two critical elements that are worth noting in this context are independence and the level of effort. DOE and its contractors described having used only internal reviews for many or most of their performance assessments until recently. One contractor told the committee that it was an adjustment to write for outsiders rather than for themselves. That adjustment is not just about how information is presented, but also about supporting assumptions and reasoning for those who may not already think in the same way as the authors. The USNRC's reviews in its legislated role under Section 3116 of the 2005 NDAA provide such independence. To fulfill its potential, a peer review must be carried out at an appropriate level of effort. The scale of information (and the sheer number of documents) to be reviewed for the draft waste determinations requires a much greater effort than was provided in, for example, previous USNRC consultations. The level of effort must be matched to the task (see Finding and Recommendation VI-1).

FINDINGS AND RECOMMENDATIONS

Finding VI-1: Independent peer review, which is a cornerstone of good scientific and engineering practice, has helped DOE to improve the clarity, quality, and completeness of its waste determinations and performance assessments.

Recommendation VI-1: DOE should: (1) seek independent peer review of the data collection and analysis relevant to risk done to support the waste determinations and performance assessments before submitting draft waste determinations under Section 3116 of the 2005 NDAA; (2) arrange for independent reviews similar to those under Section 3116 for any waste determinations made under DOE Order 435.1; and (3) publish more of its data and analyses in peer-reviewed literature and accessible reports so that they can be reviewed by the technical community.

Finding VI-2a: Performance assessments are complex structures of models and assumptions. The objective of the assessment is to provide a conservative or realistic projection of contaminant release and movement and of human exposure over time to determine compliance with performance objectives. The performance assessment provides a reasonable method for predicting the ability of a site to meet the performance objectives of 10 CFR 61, provided that: (1) the assessment includes deterministic or probabilistic modeling coupled with uncertainty analysis; (2) the site characterization data enable analysts to model likely pathways for movement of potential contaminants from the site; (3) parameter values based on data and assumptions (e.g., those concerning the longevity of institutional controls) are conservative or realistic; and (4) the assessment is updated as site conditions change and as DOE's knowledge of site conditions and other factors improves.

Finding VI-2b: Considering the range of assumptions, the limitations of predictions regarding engineered systems with a finite design life in dynamic environments, the challenges in modeling contaminant transport through the environment (especially in the vadose zone), and unpredictable future human behavior, the performance assessment's numerical output is insufficient in itself to determine compliance with the performance objectives.

Finding VI-2c: Accepting the performance assessment results entails accepting the scenarios and assumptions that underlie those results.

Recommendation VI-2: DOE should continue to make clearer the basis for decision making and for assumptions in its performance assessments; it should ensure that the features of the models used are explicit and transparent, that the models are verifiable, and that assessments present an evaluation of uncertainty. One important example is that when DOE uses a nonstandard point of compliance, it should state clearly the potential exposures closer to the disposal facility in case assumptions about human behavior and institutions do not turn out to be true.

Finding VI-3: The committee views the assumptions about future residual inventories in the F Tank Farm as both optimistic and unsupported. Given these assumptions, the committee does not have confidence in the results of the performance assessment for the tank farm.

Recommendation VI-3: DOE should base its projections of future residual inventories on what is known about the actual tanks and their contents, and on what is known about the effectiveness of the tank waste removal technologies that DOE plans to employ in those tanks.

VII

Monitoring

The committee has been charged by Congress in Section 3146 of the National Defense Authorization Act (NDAA) of Fiscal Year 2005 to evaluate "the adequacy of the Department's plans for monitoring disposal sites and the surrounding environment to verify compliance with [the 10 CFR 61] performance objectives" and has been asked to recommend "the best means of monitoring any on-site disposal sites from the waste streams referred to above to include soil, groundwater, and surface water monitoring."[1] This chapter is one of the shortest in the report because the Department of Energy (DOE) has not yet, for the most part, developed plans for post-closure monitoring of the tanks, so the committee could not evaluate the adequacy of such plans. At the Idaho National Laboratory, long-term monitoring plans have been (and will continue to be) developed in compliance with records of decision for radioactive waste that spilled, leaked, or was injected into the ground at the Idaho Nuclear Technology Engineering Center, where the tank farm is located. These plans are discussed specifically, but in the rest of its evaluation, the committee has focused on the overall approach and structure used in DOE's current monitoring programs at the sites. The committee's findings and recommendations are consistent with this high-level review and do not focus on specific methods or on small details such as the placement of a particular monitoring well. Further, this chapter is devoted to technical issues, not to legal or regulatory issues, which are discussed in Chapter VIII.

As defined in 10 CFR 61, "monitoring means observing and making measurements to provide data to evaluate the performance and characteristics of the disposal site." DOE conducts monitoring in different stages, which can be related to the different stages of activity at the disposal sites: site characterization, operations, closure, observation or surveillance, and active institutional control (see, e.g., USNRC, 1997b). Each stage of monitoring may require different types of data. In addition, regulatory programs (under the Resource Conservation and Recovery Act [RCRA], the Comprehensive Environmental Restoration, Compensation, and Liability Act [CERCLA], and so forth) have specific data needs. Although monitoring at different stages and for different purposes has different goals and may try to answer different questions, it is desirable to coordinate all monitoring at a site. In addition to these different programmatic stages or time frames, monitoring takes place at different spatial scales, namely facility or disposal site monitoring and overall or site-wide monitoring (e.g., monitoring of the Saltstone Disposal Facility and site-wide monitoring for the Savannah River Site). Similarly, monitoring for conditions at these different spatial scales may have different objectives. For example, monitoring to confirm the expected performance behavior of a particular disposal facility would seek to encounter any contaminant plume from the facility, while monitoring at a site-wide level might gather data on the hydrologic budget at the site.

The committee has identified seven features of a good monitoring program. A good monitoring program: (1) is goal oriented; (2) has an integrated vision of monitoring for the overall site; (3) seeks relevant information in the right places; (4) observes the environment (both natural and constructed) and the dynamics that affect processes of interest; (5) provides early warning to enable intervention, if necessary; (6) is subjected to review on a regular basis and adapts to changing circumstances; and (7) archives data in a durable and accessible form. Each of these features is described in Appendix H. The committee used these features as metrics for evaluating the DOE monitoring programs. Because data

[1]Section 3116 of the 2005 National Defense Authorization Act charges the U.S. Nuclear Regulatory Commission, in coordination with the host state, to monitor "disposal actions" to assess compliance with the provisions of that section of the law. There has been some debate (e.g., the U.S. Nuclear Regulatory Commission's public meeting on November 10, 2005) as to what monitoring compliance comprises. Unless otherwise indicated, the term monitoring in this report means monitoring the disposal facility, the environment of the disposal sites, and the surrounding areas, not monitoring waste processing activities or construction of the disposal facilities.

from environmental monitoring play a key role in performance assessments (see Chapter VI), the committee has also evaluated the connection or integration of the long-term monitoring programs to the performance assessment programs.

To evaluate DOE monitoring programs with respect to these features, the committee visited the sites, received input during public meetings, and reviewed documents. Several committee members followed up those information-gathering activities with conference calls with site personnel, which allowed detailed questioning on specific topics. In preparation for these discussions, the committee provided Hanford, Idaho National Laboratory, and Savannah River Site personnel with a list of questions that served as the bases for the discussions. In what follows, the evaluation of DOE's monitoring programs is organized using the features cited above.

Monitoring requirements at sites will shift based on the stage of site development and the evolving state and understanding of the biological, hydrological, and geological features of these complex sites. The committee recognizes that no single monitoring program is suitable for all sites or all stages of site development; rather, monitoring requires careful, site-specific and time-specific planning to ensure that "data of the type and quality needed, and expected for their intended use, are provided, and that decisions involving the design, construction, and operation of environmental technology are supported by appropriate quality-assured engineering standards and practices" (IDQTF, 2005).

DOE'S MONITORING PROGRAMS

Monitoring the Aquifer and Perched Water at the Idaho National Laboratory

DOE has developed long-term monitoring plans under Comprehensive Environmental Response, Compensation, and Liability Act (CERCLA) for waste releases (spills, infiltration, and injection) at the Idaho National Laboratory (INL), including the area in the vicinity of the tank farms (DOE-ID, 2003c, 2004a). This monitoring program for the aquifer and perched water at the Idaho facility, with the associated quality assurance project plan (DOE-ID, 2004b), provides an excellent example of the level of effort and institutional commitment required for site-level monitoring. It illustrates well the linkage that must exist between monitoring and the decision support system for site management. To address the long term, the plan establishes an adaptive process that could extend to 2095. CERCLA monitoring plans emphasize the groundwater pathway; this plan was not integrated with a more broadly based environmental effects element that would include an ecological monitoring plan, although the site has indicated that an ecological monitoring plan may be required when a final cleanup alternative is selected for the tank farm soils. For closure of the tank farms at each of the three sites, it is desirable to have a monitoring plan such as the one in place at the Idaho National Laboratory, but that extends beyond 100 years and explicitly considers the integration with monitoring of other media and the biosphere.

Evaluating Current Monitoring Programs

Goal Oriented

Monitoring at Hanford, the Idaho National Laboratory, and the Savannah River Site is complex, highly variable, and site specific. As the sites' processing and disposal operations have evolved, so has monitoring. Monitoring programs have been developed under CERCLA or RCRA requirements with state and, in some cases, U.S. Environmental Protection Agency (EPA) oversight. Monitoring requirements under these regulations are used widely at sites across the nation and, thus, have been field tested in many different situations. However, it was difficult to evaluate fully what monitoring would be adequate at any of the sites because the interpretation of monitoring program goals varies based on the intricate histories of interactions between each site and its regulators. Nonetheless, the committee's overall impressions are that the sites' monitoring programs are good at fulfilling their current goals, which in most cases are site characterization, operations monitoring, and provision of data to assess compliance with RCRA, CERCLA, and, or state groundwater quality requirements. The programs have been developed recognizing the importance of good data. There is an evolving understanding of the dynamics of contaminant movement on the sites. Investigators continue to be surprised as monitoring finds the unexpected, but these "surprises" are leading to improved understanding of the dynamics and drivers of fluid and contaminant movement (see examples in the section "Provides Early Warning").

As noted at the beginning of this chapter, DOE's plans for post-closure monitoring (including monitoring for compliance with performance objectives) have not, for the most part, been developed at the sites. "DOE Order 435.1 requires that high-level waste facilities "be closed in accordance with an approved closure plan . . . [which] shall include . . . relevant closure controls including a monitoring plan, institutional controls and land use limitations to be maintained in the closure activity" (DOE, 2001a). It is understandable that post-closure monitoring is not DOE's highest priority right now, considering that closure of the tank farms is still decades away. Plans are needed however before closure, because the monitoring systems will have to be built into the closure system as required for low-level waste treatment and storage facilities.[2] DOE Order 435.1 further requires a pre-

[2]DOE's radioactive waste management manual for Order 435.1 states, "Monitoring and/or leak detection capabilities shall be incorporated in the design and engineering of low-level waste treatment and storage facilities to provide rapid identification of failed confinement and/or other abnormal conditions" (DOE, 2001a).

liminary monitoring plan prior to disposal authorization for low-level waste.

Before monitoring systems, or even the monitoring program, can be designed, DOE must know the goals and requirements. The complex regulatory environment for post-closure monitoring at DOE sites results in competing requirements and overlapping jurisdictions.[3] At Hanford, DOE worked with the Environmental Protection Agency and the Washington Department of Ecology to develop a framework that fits together the regulatory goals and strategies for groundwater protection, monitoring, and remediation (DOE-RL, 2004b). When the parties expand this document to address the topics of vadose zone monitoring and long-term stewardship, the pieces will be in place to develop a long-term monitoring plan for all groundwater at Hanford.

The Savannah River Site provided the committee with documents (e.g., WSRC, 2004b) that do not document well the site's strategies and goals. Such goals have to be stated explicitly. Faced with the complex web of regulatory requirements described above, the Savannah River Site has asked the State of South Carolina, the Environmental Protection Agency, and the U.S. Nuclear Regulatory Commission to participate in a meeting to clarify the goals of post-closure monitoring and reach agreement on a single plan for monitoring that will meet the needs of all parties. Although no such meeting had taken place as of January 2006, agreeing on a common set of goals and parameters for the monitoring plans is a sensible first step, in the committee's opinion.

Has an Integrated Vision of Monitoring for the Site

Although there have been efforts to do so, DOE has not yet fully articulated an integrated vision of monitoring for each site. Such an integrated vision would fit together the various monitoring goals for each site and establish an overarching strategy that provides a comprehensive view of the site beyond what is seen as necessary under one or another specific regulatory requirement. Each overall site has many disposal sites, classes of wastes, and contaminated areas that require monitoring, some with different requirements and different people overseeing the monitoring. For example, at Hanford the responsibility for monitoring tank leaks resides with a different contractor than monitoring for contaminants in the saturated zone. One manager is, however, responsible for all monitoring at each site and Hanford has a mechanism for integrating its efforts.

The Hanford groundwater monitoring report for fiscal year 2004 (PNNL, 2005a) shows that Hanford has established a firm foundation and is making good progress in integrating both its various groundwater monitoring efforts and its modeling and monitoring efforts with each other. Hanford is continually updating its monitoring networks, making use of nontraditional well sampling techniques, and using monitoring data to develop a site-wide model of groundwater flow and contaminant movement. Results from the site-wide model provide feedback into the monitoring program. The committee encourages Hanford to continue this progress and to do more to integrate groundwater monitoring with other monitoring activities on the site.

Hanford, the Savannah River Site, and the Idaho National Laboratory each issue annual environmental reports that describe results of monitoring in different media across the overall site (see, e.g., DOE-ID, 2004a; PNNL, 2005b; WSRC, 2005), and these are part of the picture needed for an integrated vision and monitoring plan. The sites described monitoring databases they are developing (the Idaho National Laboratory, at least, has this operational) to make all of the current and historical monitoring data available to scientists and technicians on the sites (see "Archives Information," below). Another piece of an integrated vision is connected to the comprehensive performance assessment for all wastes and contamination on the overall site. Hanford's effort to carry out this assessment is called the System Assessment Capability. Other sites have a composite analysis. The most recent composite analysis for the Savannah River Site, however, does not reflect any recent decisions or proposals for waste disposition and does not cover the overall site, although this concern is important only to the extent that potential impacts from radioactive materials at various locations on the site would overlap in space and time. It is not necessarily a concern in all cases.

Seeks Relevant Information in the Right Places

The sites are currently monitored extensively for both radioactive and nonradioactive constituents. Much of the programs' focus is on groundwater, which is appropriate. The monitoring programs use geostatistical methods (e.g., kriging) to help make decisions about new monitoring well locations and what wells to discontinue using.

Observes the Environment (both Natural and Constructed) and the Dynamics That Affect Processes of Interest

Environmental monitoring at the sites has resulted in widespread site characterization. Each site has operated for periods that have allowed the acquisition of extensive datasets that can serve as baselines to compare against future monitoring data. Characterization data that precede DOE operations (i.e., data obtained during Atomic Energy Commission operations) vary, but in general each site has

[3]The overlapping jurisdictions and regulatory requirements include federal CERCLA requirements for post-closure monitoring, state requirements under RCRA, state and U.S. Nuclear Regulatory Commission (USNRC) needs under Section 3116 (in Idaho and South Carolina) of the 2005 NDAA, and any other requirements agreed to under the federal facility agreements. Also, it should be noted that EPA and state environmental regulatory agencies show great interest in protection of groundwater as a resource (see Appendix C, Table C-3 for a list of requirements DOE considers applicable).

sufficient monitoring to support most site characterization activities. Vadose zone monitoring and investigations of geochemical behavior in the environment may be the exceptions to this observation. Further, each of the sites is supporting long-term ecological monitoring; indeed nationally recognized ecological monitoring programs exist at each site. Knowledge gained mostly through investigations of past releases has yielded a reasonably good understanding of the hydrology at the Savannah River Site, confirmed by following the progress of actual contaminant plumes. Monitoring and analysis of similar releases at Hanford and the Idaho National Laboratory have contributed to the understanding of contaminant transport at those sites, although transport in the subsurface at these sites, particularly the vadose zone, is more complex and not as well understood.

At each of the sites, it appears that there is information adequate to characterize the present environment around the waste tank facilities and to guide the development of the performance assessments. However, as noted in Appendix H, monitoring must not be seen as a static process that is complete and finished at some point. For example, at Hanford, the groundwater flow is changing radically, with indications that the flow may actually reverse direction in some locations because DOE no longer discharges large quantities of contaminated water into the subsurface. To evaluate such events, an adequate long-term monitoring program needs to include a continuing performance assessment to confirm that the site is performing as expected and to point out where additional investigation or explanation is needed.

Questions have been raised about DOE's knowledge of the subsurface flow system beneath Hanford, particularly the locations and extents of clastic dikes and their effect on contaminant migration. It would be difficult to gain detailed knowledge of the subsurface beneath the tanks from the 60 groundwater wells around the tanks and the two slant boreholes that collected high-quality sediment samples and associated contaminants from under the tanks. Site personnel note that the dikes are relatively narrow and, thus, do not have a major effect on contaminant transport, and that there has been no specific case in which the dikes have been cited as the cause of anomalous transport. However, structural discontinuities can have a major impact on both flow direction and sorptive behavior at a local scale, such as beneath an individual tank farm. At the Idaho National Laboratory, the subsurface has proven to be much more complicated than initially thought because of perched aquifers and fracture flow. In recent years, Idaho National Laboratory personnel have created detailed maps of the perched water and interbedded layers to better understand the transport of contaminants from spills at the tank farm.

DOE monitors existing engineered structures (caps and liners for disposal facilities) less consistently than it monitors the environment. At the Savannah River Site, the tank annuli are monitored with resistivity sensors and there are annual visual inspections to the extent that access allows, but fear of damaging piping has prevented the Savannah River Site from monitoring groundwater below the tank farms. Complicating this picture is the fact that the conditions expected after the tanks and disposal sites are closed may be somewhat different, with large capped mounds redirecting precipitation recharge away from areas containing the waste and facilities left on-site. Some monitoring personnel at the site speculate that even now, with the asphalt ground cover currently in place, the top aquifer may have a local depression under the tank farms caused by reduced surface infiltration and continued vertical flow through the aquitard. Current data are insufficient to confirm this idea.

Several noninvasive geophysical options exist for locating buried pipes where exact locations are unknown. In areas where the surface soils are electrically resistive (i.e., low clay content), ground penetrating radar (GPR) provides a cost-effective solution for horizontal location of metal and plastic structures (i.e., pipes) to within a few centimeters and an estimate of vertical location that is typically accurate to within a few tens of centimeters. Ground penetrating radar has been shown to work well at Hanford to depths approaching 10 m (Last and Horton, 2000; Murray, et al., 2005). Ground penetrating radar would also work reasonably well at the Idaho National Laboratory but would likely provide good results to depths of only a few meters at the Savannah River Site because of the partially saturated clay-rich soils. Alternative geophysical methods include magnetic gradiometry surveys and electrical resistivity surveys.

Provides Early Warning to Enable Intervention if Necessary

The committee believes that monitoring within the disposal facilities is the most desirable approach for the early detection of problems, followed by detection in the vadose zone, and finally detection in the nearest aquifer. Hanford has 800 dry wells around its tanks. Detectors in these wells were able to observe some of the leaks from the single-shell tanks. During retrieval operations, the dry wells are checked weekly because retrieval operations are the activities most likely to mobilize waste from the tanks. At other times, the wells are checked at a longer interval. Hanford has tested a capability to detect tank leaks within 24 hours of their occurrence using resistivity sensors (see Sidebar VII-1).

The Savannah River Site has not yet formulated plans for monitoring the covers for its tank farm and Saltstone Vault disposal facilities. Hanford, by contrast, has a prototype barrier system that is outfitted extensively with sensors. DOE indicated that it is premature to decide on the details of the cover system for its tank closures and low-activity waste disposal facility (the Integrated Disposal Facility, or IDF), but Hanford is building lysimeters into multiple layers of the IDF bottom liner. The Idaho National Laboratory has a network of monitoring sites to determine contaminant movement in the vadose zone and the aquifer and is following the

SIDEBAR VII-1
Tank Leak Detection

Early detection of leaks is desirable to stop leaks at the earliest possible time limit the volume of contaminated earth. Geophysical methods have the advantage of integrating the signal from a three-dimensional volume, as opposed to a one-dimensional sample obtainable from individual observation wells.

While several geophysical and geological methods can help in hydrological and geological characterization of possible contaminant flow paths, the electrical resistivity method of geophysical monitoring is most likely to be successful in early detection of tank leaks. In 2002, a 110-day blind test was staged at the Hanford Mock Tank Site (Barnett et al., 2003). The test evaluated electrical resistivity tomography (ERT) with variations using point electrode techniques and the long-electrode technique. These tests resulted in 10 of 12 leaks being detected on the first day.

Another test in the series involved the high-resolution steel casing resistivity technique, which provided leak volume estimates with accuracy within about 30 percent, while detecting 9 of 13 leaks. The other four leaks that were not detected could be associated with identifiable technical problems such as disconnected electrodes or major electrical interference.

Barnett et al. (2003), who compared detection through electrical resistance to other methods, found that these results suggested that electrical methods have one to two orders of magnitude advantage over neutron or gamma logging. They also have the advantage of continuous measurement versus the periodic measurements done with logging methods.

Although these results at Hanford are promising, they were made in an environment with little natural near-surface moisture and relatively low clay content. The geological and hydrological conditions at the Savannah River Site are less favorable for the use of resistivity changes to detect leaks because of higher clay content in the soils and greater interstitial moisture content. With respect to electrical resistivity monitoring, the geological and hydrological conditions at the Idaho National Laboratory are closer to those at Hanford than the Savannah River Site.

Resistivity signatures have temporal variations on seasonal as well as other scales. Consequently, any monitoring using resistivity would have to be done in a way to avoid temporal aliasing (mistaking signals from leaks for the natural temporal fluctuations because they occur with the same frequency or with simple multiples or fractions of the same frequency).

SOURCE: Barnett et al., 2003.

movement of spilled materials as a means of updating knowledge of site dynamics.

Figure VII-1 illustrates the different types of monitoring and where they fit into the physical and regulatory environment. It shows the disposal facility with monitoring ports or access into or near the waste cell that allow monitoring of the engineered facility and early detection of releases, before extensive spread of contamination occurs. The figure illustrates the area for monitoring of releases inside the buffer zone that allows time to correct the problem before it reaches the areas to which the general public would have access and could be impacted. The point of compliance is shown as the line of demarcation for public areas versus the areas inside, in which any person is considered an intruder. The water table is illustrated because it is the entry surface to the saturated zone that is the pathway of most concern for transmission of the waste to a publicly accessible point either by well water or by surface streams to which the groundwater seeps out or gets discharged.

Each site has encountered surprises found by monitoring efforts, such as a discovery of unexpectedly high concentrations of technetium-99, nitrate, and other contaminants at the bottom of the unconfined aquifer downgradient from the T Tank Tarm at Hanford in 2004. Such surprises should be regarded as successes for the monitoring programs, even as they raise concerns about contaminants, including potential undetected releases. However, they also indicate the need for an early warning system to enable decision makers to intervene if appropriate. For example, at the Idaho National Laboratory, site monitoring contributed to a decision to relocate the percolation ponds and to the eventual revision of the site hydrogeological conceptual model (DOE-ID, 2004a). The Savannah River Site provided examples of cases in which contaminants discovered by monitoring led to remediation efforts.

Is Subjected to Review on a Regular Basis and Adapts to Changing Circumstances

The monitoring programs have grown and adapted to changing needs. A certain amount of review is built into the programs as a result of operating under CERCLA or RCRA. CERCLA five-year reviews are required to ensure protectiveness for any remedial action that leaves hazardous sub-

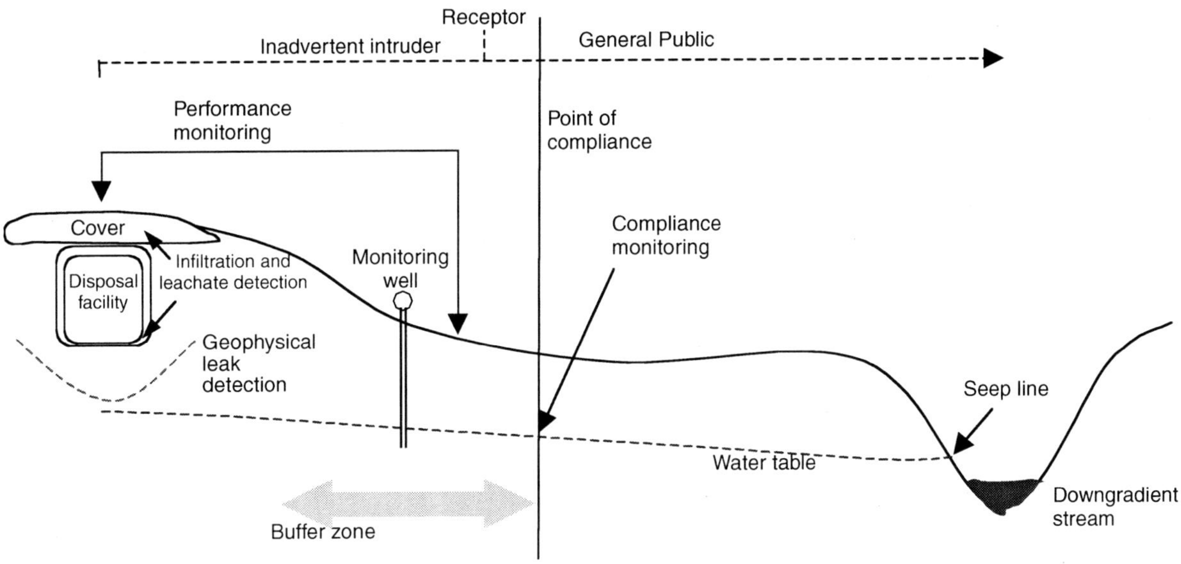

FIGURE VII-1 Different types of monitoring and where they fit into the physical and regulatory environment at a hypothetical site.

stances on a site above levels that allow for unrestricted use by the public.[4] Personnel at each site stated that they conduct regular reviews of site monitoring programs, although the committee did not examine the details of those reviews. As noted in footnote 1 of this chapter, Section 3116 of the 2005 National Defense Authorization Act requires the Nuclear Regulatory Commission, in coordination with the host state, to monitor "disposal actions" to assess compliance with the provisions of that section of the law. The South Carolina Department of Health and Environmental Control requested that DOE provide the state with a long-term monitoring plan for the tank farms at the Savannah River Site. This request prompted DOE to call for the meeting mentioned earlier in the discussion of goals.

Archives Data in a Durable and Accessible Form

Recognizing that DOE might not solve the problem of creating archives that will last and be accessible for centuries, which is a challenge confronting many parts of our society, DOE can at least assemble the information it has so that it is accessible now. The sites have created central databases for accessing monitoring data and have populated the databases with much of the information available, although there is not yet a comprehensive set for any of the sites.[5] The extensive monitoring history provides the sites with an opportunity to create functional archives that provide both data and metadata from each monitoring effort. Unfortunately, none of the sites has a functional, site-level monitoring archive that fully integrates environmental data management. The committee has observed the separation of monitoring procedures and data management by program or project boundaries at each site. Particularly important "data divides" at the sites include divides that exist between process data collection and site characterization, long-term ecological analyses and regulatory compliance monitoring, and a program data focus that fails to integrate media-specific monitoring results (air, surface water, and groundwater).

Except in the case of the Idaho National Laboratory plan for groundwater monitoring described earlier, none of the people to whom the committee talked said that the sites explicitly follow the Uniform Federal Policy for Quality Assurance Project Plans (UFP-QAPP; IDQTF, 2005), which is designed to ensure that a monitoring program is well designed and implemented. It is, of course, possible to have a good monitoring program without explicitly following the policy, but the programs could be improved in terms of interoperability, meeting the needs of the performance assessment for tank wastes, and coordinating not just within the site but at the margins between sites. Perhaps more

[4] A 1999 examination of CERCLA five-year reviews (not specific to DOE sites) found that EPA's backlog of reviews that were past due was increasing (EPA-OIG, 1999). EPA instituted measures to clear the backlog by the end of 2002. A recent Government Accountability Office report found that the five-year interval may be too long for sites that rely on institutional controls as part of their remedies, and the five-year reviews did not consistently review the effectiveness of the institutional controls (GAO, 2005a).

[5] The most complete may be Idaho National Laboratory's Environmental Data Warehouse, a site-wide database for groundwater and perched water data available to site personnel from their desktops.

important is that following the policy may make it more likely that the organizations carrying out monitoring at the sites will coordinate better and continue to do the job well as responsibility is passed from one generation of managers and technicians to the next or from one contractor to the next.

FINDINGS AND RECOMMENDATIONS

Finding VII-1a: The committee's overall impressions are that the sites' monitoring programs are effective at fulfilling their current goals, which in most cases are site characterization and operations monitoring to assess regulatory compliance.

Finding VII-1b: For the most part, the sites have not yet developed plans for post-closure monitoring, including continuing post-closure performance assessment to confirm the performance of on-site waste disposal.

Finding VII-1c: The existing monitoring programs can be improved, and both planning and action are needed to ensure that the programs continue to address the multigenerational challenge posed by long-term monitoring.

Considering the importance of monitoring data in the evaluation activities that support decision making in verifying compliance with performance objectives and in overall protection of human health and the environment, DOE needs to take actions to ensure that the post-closure and long-term monitoring efforts perform well for generations.

Recommendation VII-1: DOE should start planning its post-closure monitoring programs so that provision for monitoring can be built into closure plans and designs. In doing this, each site should implement a process to ensure a high-quality, comprehensive, coordinated, and site-wide monitoring program that meets current and ongoing needs for site evaluation and compliance and is revised as needed to provide long-term monitoring data for continuing evaluation of the site's ability to meet performance objectives. In addition to site-wide quality assurance, this process should also provide guidance for disposal site monitoring and related performance assessments. DOE has guidance presently available in its UFP-QAPP that can assist with this.

VIII

Decision-Making Process

Previous chapters have examined the current state of technology that delimits the options for on-site disposition of tank wastes and analyzed the informational and analytical needs for assessing and monitoring whether a particular disposition plan meets applicable criteria. In this chapter, the committee elucidates the decision-making environment in which the Department of Energy (DOE) operates. Illustrating that environment is important to understanding how the success of the tank remediation program depends on, among other factors, the way DOE approaches its decisions about tank wastes. Specifically, this chapter illustrates the following:

1. There are considerations in high-level waste cleanup and disposition decisions that extend beyond merely meeting the dose limits and other nonquantitative requirements defined in waste determination criteria;
2. Site-specific characteristics translate into different site priorities;
3. DOE operates within an extremely complex and overlapping decision-making structure, which is a source of programmatic risk;[1]
4. DOE's decision-making paradigm has evolved with time; and
5. Given the programmatic risks outlined previously and those identified in this chapter, a more risk-informed, participatory, consistent, transparent, and consultative decision-making process would make DOE's waste management decisions more robust, in the sense that DOE is more likely to succeed in its tank remediation mission.

This chapter is not intended to answer the question, How clean is clean enough? for any particular waste stream, tank, tank farm, facility, or site. Instead, it responds to the statement of task by making recommendations that the committee believes will improve management and decision making so that such questions are answered in a consistent, transparent, and scientifically credible manner that protects public health and the environment now and for generations to come.

The legal authorities that apply to DOE are pivotal in shaping the decision-making framework in which DOE operates. Hanford (not located in a state covered under Section 3116 of the Ronald Reagan National Defense Authorization Act [NDAA] of Fiscal Year 2005) operates in a different management and legal environment than the Savannah River Site and Idaho National Laboratory, which are located in states covered under Section 3116. Nevertheless, the considerations relevant to management and decision making under Section 3116 are fundamentally the same as those that apply to a process for determining an acceptable path forward for Hanford, and DOE Order 435.1 is still in effect at all sites. Whether or not Section 3116 applies, a wide range of site-specific regulatory, economic, institutional, and waste management considerations are part of the decision-making process for waste determinations, as explained in the following section.

MULTIPLE DIMENSIONS OF RISK

The dose limits set out in performance objectives and associated guidance are a fundamental and necessary start-

[1]Programmatic risk is the risk to cost, schedule, and technical performance of a program. It is associated with all uncertainties, including legal uncertainties that can result in delays, cost increases, and failure to reach the established goals. For example, DOE defines the programmatic risk as high for a given project if the technical approach has not been identified for critical or significant portions of the project; key technologies do not exist for critical or significant portions of the project; current investments do not support the resolution of the project's science and technology needs; project end point is not determined or supported by stakeholders and Native American tribal nations; waste/material quantities and characteristics are unknown; process operations are not identified or supported by stakeholders and Native American tribal nations; final disposition location for waste/material has not been identified; activity involves multiple sites; no concurrence has been reached between sites; or a facility does not currently exist and there are no plans for a new facility (DOE, 1998).

ing place for the dialogue about decision making for disposition of tank wastes. Analysis of the site's performance and its ability to meet applicable criteria are part of the legal requirements under Section 3116 of the 2005 NDAA, DOE Order 435.1, and other applicable standards, such as the standards for radionuclides in drinking water that are used as groundwater protection criteria. As explained below, the dose limits in the performance objectives do not fully address the multiple dimensions of risk that characterize tank waste management.

Risks[2] from any disposal facility actually vary over space and time. Examples of different time-space profiles of risks are the following:

- Risks that are very low for a long time and never really increase above low levels (no "amplitude" over time)
- Risks that are very low for a long time, but rise fairly rapidly later in time (albeit to levels still below the performance objectives)
- Risks that are quite elevated near the source (the tank), but decline to acceptable levels at more distant points of compliance

Risks can also differ qualitatively, in terms of their mode of impact on health. For example, some risks are immediate or acute,[3] whereas others are chronic or delayed in their onset (as in the case of carcinogenic compounds). Different compounds also can harm the body or environment in different ways, some affecting reproductive outcomes and others possibly causing harm only to the elderly or already infirm. These qualitative differences in the types of waste in question may also undermine the usefulness of safety established strictly on the basis of performance objectives.

Although each of these hypothetical outcomes might meet the performance objectives, they might engender quite different degrees of concern among the public and decision makers. These kinds of differences in the risk profiles of different waste management choices have important implications for concerns about long-term institutional controls. Some risk profiles may place fewer burdens on assumptions about long-tem institutional controls to ensure that risks are acceptable to the public.

The technical complexities of the disposal sites for tank wastes and their interaction with the geosphere may be included where possible in the performance assessments and in the models they use. Therefore, the performance assessment and sensitivity analyses that are performed using different scenarios can show the impact of various decisions on projected risks. Indeed, DOE's waste determination for the closure of Tanks 19 and 18 (see DOE-SRS, 2005a, Figure 7-14, p. 137) shows how the Savannah River Site used its performance assessment for these two tanks to make tank closure decisions (although the decision tree is somewhat simplified and there is limited supporting information). However, the performance assessment is not intended to provide understanding of the overall nature of the risks that a person at the site would encounter.

For example, each of the three sites considered in this report has additional risks other than those represented by the waste in the tanks (e.g., cribs and previous waste leaks at Hanford). It may not be possible to remediate contamination in soils around the tanks until the tanks are closed. The risks from such preexisting contamination may be substantially greater than any risks that might be created by leaving an incrementally larger (yet still relatively small) amount of waste in the tanks to enable their prompt closure.

Without considering all of these risks and their relationships in time and space in one comprehensive risk-benefit analysis, it is difficult to see how the programmatic goals fit together and what trade-offs should be made in managing risks. Overall, if all potentially related risks for a site are not considered, decisions may be made that do not reduce the risks from the site "system" as a whole to the maximum extent practical. Thus, considering only whether performance objectives are met for a tank closure or a low-activity waste disposal site may not provide a full picture of risks at the site and how they vary in time and space.

Additional risks (e.g., the risk of a leak during waste retrieval, the risk of leaving the tanks emptied for 5 to 10 years before grouting them) and trade-offs among such risks could be identified as part of making a decision for each individual element of the site waste management plan. The ALARA (as low as reasonably achievable) radiation exposure provisions in the performance objectives mentioned in the waste determination criteria require trade-offs among technology limitations; risks to workers, the public, and the environment; and costs. However, the committee has not seen documentation showing detailed risk-benefit analyses of such trade-offs. The waste determinations for Tanks 18 and 19 were considerably better supported than documents the committee reviewed for its interim report, but they still did not contain sufficient information for a complete evaluation of how the trade-offs were made (see discussion in Chapter III on "how clean is clean enough?").

The ALARA analysis required in the performance objectives calls for trade-offs between incremental costs and incremental risk reduction. DOE used such an analysis to decide much tank waste retrieval was enough (see Chapter III). However, the risk-benefit analyses presented in the Idaho National Laboratory and Savannah River Site tank waste determinations were not sufficiently detailed or transparent. The criteria on which trade-off decisions between cost, worker dose, and public dose were made are not

[2]In this chapter, risk is intended in a generic sense as risk to workers, the public, and the environment.

[3]Except in extraordinary circumstances (such as an operational accident), it is not likely that anybody would receive an acute dose of radiation from the tanks.

stated. Finding quantitative cutoff points at which further risk reduction is no longer worth its cost may not be possible in such a multiobjective situation. Similarly, a formal cost-benefit analysis or a comparison of incremental (marginal) risk reduction and cost by reducing both to a single, monetary metric may not be necessary. However, if the options for further action (greater risk reduction) are laid out clearly (i.e., risk-benefit trade-offs are clearly explained), it is possible that a consensus, or at least a majority view, can be developed around a preferred option among a set of diverse stakeholders.

The cost-risk trade-off is the most salient way to show that the performance assessment is only one source of input to a decision. Stakeholders may consider other nonrisk criteria to be important and part of a comprehensive risk-benefit analysis. This analysis would include the distribution of the burden of the risks to different groups, concerns about permanent (or very long term) risks versus temporary or near-term risks, uncertainties affecting the potential risks in time and space, and the potential to detect and mitigate hazards if they emerge. The result of this analysis may alter societal preferences for one management plan over another and impact the likelihood that a plan will achieve its goal. An illustrative and extremely simplified example of an approach for considering tank closure decisions that directly examines trade-offs among these different forms of risk is presented in Chapter X.

SITE-SPECIFIC CONDITIONS CAN LEAD TO DIFFERENT DECISIONS

In a consistent process for decision making, the framework for considering risks and the trade-offs among them is the same for all of the tanks at all of the sites. However, different conditions and priorities may exist at different sites. The considerations influencing the trade-offs among relevant risks may be quite different from tank to tank and from site to site. Thus, it is logical to expect that DOE may make quite different choices about the timing of tank closure and the extent of cleanup to undertake before closure based on the specific details at each site.

In Chapter V, the committee describes the importance of decoupling tank waste retrieval from immediate closure of some tanks to evaluate whether it is desirable and practical to undertake further cleanup before closing a tank (see also Appendix E). Such a trade-off in timing of final closure may be appropriate for some of the more difficult-to-clean tanks at the Savannah River Site whereas tanks at the Idaho National Laboratory may be appropriately closed right away because the amount of remaining radioactive material is small.

An analysis and understanding of the performance objectives is the logical starting place for an analysis of risks posed by tank wastes at the Savannah River Site, Hanford, and the Idaho National Laboratory. To reach decisions about managing tank wastes, DOE should also consider other factors, including the following:

1. The unique risks created by the proposed decision pathways that are not captured in the performance assessment or performance objectives for specific tanks or on-site disposal facilities;
2. Risk considerations at the sites that are altered by cleanup decisions;
3. Changes in, and interactions among, tank residual risks and other waste streams that are associated with the separation of tank wastes into different waste streams;
4. Considerations not captured in risk analyses or performance assessments, such as costs, distributional or equity concerns, and other societal concerns; and
5. The views of the states, Native American tribal nations, and other stakeholders in decision making.

To illustrate more fully how this recommendation impacts decision making, the example in Chapter X shows in an extremely simplified fashion how some of these additional factors can affect the choices about cleanup and waste disposition. The analysis also provides an example of how these additional considerations can be incorporated into the decision-making process in a formal, structured manner that permits consistent approaches to decision making but still allows different outcomes to emerge as a result of site- and tank-specific conditions. Setting out the assumptions and time horizons over which risks can impact health and the environment demonstrates the interplay of factors that complicate risk analysis and shows that performance assessments and performance objectives alone do not capture the full range of possible risks.

PROGRAMMATIC RISK

As shown in previous chapters, the technical challenges that DOE faces in managing tank wastes, tank farm systems, and surrounding environments are difficult, interconnected, and unprecedented. In addition to technical challenges, DOE has faced some legal challenges, due to the litigation concerning provisions for waste determination in DOE Order 435.1. Section 3116 of the 2005 NDAA creates some further complexities in the regulatory environment:

- DOE plans to declare sodium-bearing waste and waste from certain Hanford tanks to be transuranic waste and dispose of them at the Waste Isolation Pilot Plant (WIPP). However, although it is proceeding down this path, DOE does not have agreement from the State of New Mexico that these wastes qualify for disposal at WIPP. The recent regulatory problems that Hanford encountered in its transuranic waste determinations has a real impact on the entire DOE waste management program (see Sidebar VIII-1 and GAO, 2004).
- DOE has wastes for which Section 3116 cannot be used, such as all wastes at Hanford, wastes being sent out of the host state for disposal (e.g., sodium-bearing waste), and

SIDEBAR VIII-1
Recent Government Accountability Office and DOE Inspector General Findings and Recommendations

Recent reports by the Government Accountability Office (GAO, 2005b) and DOE's Office of the Inspector General (DOE-OIG, 2005) point out the legal risks DOE is facing in its accelerated cleanup program. Excerpts from these reports follow.

The types of challenges that could increase cleanup costs at [the Hanford, Idaho National Laboratory, and Savannah River] sites include the following:

- *Delays in disposing of highly radioactive wastes.* In early 2005, DOE reported that a slip in the scheduled opening of DOE's planned repository at Yucca Mountain, Nevada, would delay shipment of waste by at least 2 years—and possibly for as long as 7 years—due to technical and regulatory issues. As a result, sites now storing high-level waste and spent nuclear fuel have been reevaluating their waste disposal plans and associated cost and schedule estimates. The sites potentially affected include Hanford, Idaho National Laboratory, Savannah River, and West Valley. Most sites expect costs to increase as disposal schedules slip. In its fiscal year 2006 budget request, DOE estimated that a five-year delay in opening the Yucca Mountain repository could increase costs by as much as $720 million at its three largest sites. This includes building additional storage buildings and added operating costs.
- *Legal obstacles preventing DOE from implementing aspects of its cleanup approach.* DOE faces challenges to its planned treatment strategy at the Hanford Site that could potentially increase costs. A 2002 lawsuit challenged DOE's plans to separate and determine that a portion of its waste could be treated and disposed of as other than high-level waste, and to DOE's plans to close tanks leaving some radioactive residual in the tanks. In October 2004, a federal appeals court overturned a district court ruling against DOE and held that it was premature to rule on the matter until DOE implemented its strategy. Federal legislation passed in October 2004 provided authority for DOE to carry out its acceleration completion strategy at its Savannah River Site and Idaho National Laboratory. However, the law excluded the Hanford Site. If similar authority is not provided for the Hanford Site, costs at the site could increase significantly—up to $67 billion, according to DOE's estimate. Similarly, uncertainty surrounds Hanford's ability to accept waste from other DOE sites as the result of two ongoing lawsuits: one involving a challenge by the State of Washington to DOE's plan to ship low-level, low-level mixed, and transuranic waste into the state, and one concerning a recent Washington state citizens' initiative that could prohibit Hanford from accepting additional waste until existing waste is cleaned up. Although DOE believes it will ultimately prevail in these lawsuits, some cleanup activities at the other sites may face delays and increased storage costs until the issue is resolved.[a]

The Office of River Protection (ORP) pursued the Transuranic Mixed Tank (TRUM) Waste Project without sufficiently addressing regulatory and permitting issues. [. . .]The Department has not yet completed the regulatory actions required under the *National Environmental Policy Act of 1969* (NEPA) prior to proceeding with the TRUM waste project.[. . .] On December 15, 2003, the Department's ORP approved and issued *Supplement Analysis for Hanford Tank Farm Contact-handled Transuranic Mixed Waste Treatment, Packaging, and Storage (Supplement Analysis)* to the 1996 Environment Impact Statement (EIS). However, the *Supplement Analysis* did not address key issues which the Department's Office of Environment, Safety and Health (EH) considered critical to the public. Specifically, EH noted the analysis did not:

- Clarify the waste classification in light of recent court decisions;
- Address the cost, feasibility, additional waste generation, and timing issues related to reversing the TRUM waste treatment process if the waste is not accepted for disposal at WIPP;
- Consider the environmental impact of reversing the action; and,
- Address potential worker impact for storing the waste above ground.

We recommend that the Principal Deputy Assistant Secretary, Environmental Management, direct the Manager, Office of River Protection to:

1. Mitigate regulatory and permitting risks, including the concerns raised by EH before resuming work on the TRUM tank waste project; and
2. Ensure risk mitigation plans are developed in the future that identify project-specific risks and propose appropriate mitigation strategies before initiating projects and resuming the TRUM waste project.[b]

[a]SOURCE: GAO, 2005b, p. 27.
[b]SOURCE: DOE-OIG, 2005, pp. 1-2.

wastes that are not covered by a state compliance agreement (radioactive wastes being disposed on-site in burial grounds not covered by state compliance agreements). DOE Order 435.1 is presently the only basis for a waste determination for such waste streams, but an attempt to use it may make the suspended litigation "ripe" (see Chapter II) and open the door for resumption of court proceedings, with the potential for further delays.

- The concept of "removal of the highly radioactive radionuclides to the maximum extent practical" does not have a clearly defined meaning and, therefore, is open to interpretation and arguments over interpretation.

• Assessments of whether performance objectives are met can be highly controversial. The choice of point and time of compliance is a decision that has both technical and policy components. This basic fact could render the entire performance assessment outcome subject to controversies, beyond the myriad technical uncertainties and assumptions about human behavior that are inherent in such analyses.

These programmatic uncertainties can stop or delay the tank cleanup program just as surely as technical uncertainties can and with the same result—that highly hazardous wastes remain in aging tanks or in the environmental media for a longer period of time, thus increasing present and future risks and remediation costs.

The committee is not alone in raising these concerns. The Government Accountability Office in a recent report (GAO, 2005b) expressed concern that DOE's failure to clarify the legal and regulatory status of the tank wastes could threaten its ability to accomplish its accelerated cleanup plans. In addition, DOE's Office of the Inspector General recently issued a report (DOE-OIG, 2005) that is critical of the regulatory risk assessment for Hanford's mixed transuranic tank waste program (see Sidebar VIII-1). As with the tank waste program as a whole, the Inspector General found that DOE failed to identify and mitigate "regulatory and permitting risks" prior to initiating waste management plans, despite its awareness of concerns and controversy, including adverse court rulings, regarding these issues. The result was that DOE had to stop work on the project in midstream (as also happened with the Idaho ruling), and resolution of the tank problem is delayed indefinitely.

For waste streams that are to be addressed in the near term, the reasoning that supports waste disposition decisions to be laid out in a manner that is accessible to regulators and stakeholders as part of a transparent, risk-informed, decision-making process. If this reasoning is not laid out properly, there is no way for regulators and stakeholders to evaluate the feasibility of any plans. DOE is taking steps toward this type of process through the waste determinations required by Section 3116 of the 2005 NDAA for the Savannah River Site and Idaho National Laboratory, through the U.S. Nuclear Regulatory Commission consultation process suggested in DOE's Order 435.1 at Hanford, and at the U.S. Environmental Protection Agency's urging in relevant environmental impact statements (EISs; DOE-ID, 2005b, response to comments). The committee urges DOE to continue and expand this practice.

Some of the programmatic risks to which the committee is referring are not easily evaluated, such as congressional appropriation of money. Some of the risks are probably assessable at a technical and engineering level, such as the likelihood that bulk vitrification would work. The regulatory and legal environment under which DOE operates is complex and contested. Therefore, DOE should acknowledge and account for the programmatic risks in its decision making, because there are legal and regulatory challenges that can—as they have in the past—create major barriers to completion of its mission. DOE needs to recognize these programmatic risks that arise from legal and regulatory challenges, just as it needs to recognize the possibility that bulk vitrification may not work at Hanford or that New Mexico may not allow Idaho transuranic waste to be disposed at WIPP. This is simply a matter of good management. As noted earlier, the committee sees an emerging practice of this kind of analysis by DOE in its waste determinations as required by Section 3116 of the 2005 NDAA.

These legal and regulatory difficulties, along with the technical challenges described in previous chapters, increase the programmatic risk that some element of the plan will fail and create problems for other parts of the plan. In the presence of complex technical and regulatory challenges there is the need for a transparent, risk-informed, and participatory decision-making process, as discussed below (NRC, 2005b).

The committee recommends that in its planning, DOE identify sources of programmatic risks as soon as possible so that it can seek ways to mitigate or work around them (see Recommendation VIII-2). For waste streams that do not have to be addressed immediately (e.g., calcine at Idaho, pipes and other ancillary tank systems) or for waste streams that are supposed to be shipped off-site, DOE needs to develop programmatic contingency plans in addition to a disposition pathway.

RISK-INFORMED, PARTICIPATORY, CONSISTENT, AND TRANSPARENT DECISION-MAKING PROCESS

For the past 50 years DOE (or its predecessor agencies) has had the authority and responsibility to manage the high-level waste stored in tanks at the Idaho National Laboratory, the Savannah River Site, and Hanford. During this time, DOE employed available technology and science in addressing treatment, storage, and remediation of waste. As technology and knowledge have evolved and gained in sophistication, the decision making accompanying the use and deployment of nuclear technology and cleanup of wastes has also evolved, becoming more complicated and drawing in parties and partners that were not previously an active part of decision making (see Sidebar VIII-2).

States, local governments, Native American tribal nations and other stakeholders ultimately must live with long-term contamination at sites. Therefore, it is important that these parties also be part of the decision-making process concerning site disposition decisions and the associated choices about long-term stewardship (NRC, 2000c). The views of interested and affected parties can have important effects on the way other contextual factors, such as cost and risk, are treated in site disposition decisions. They may influence site disposition decisions at five levels of generality (NRC, 2000c, pp. 73-74):

SIDEBAR VIII-2
The Evolution of DOE's Decision-Making Paradigm

The Atomic Energy Commission (AEC) was established by the Atomic Energy Act of 1946 to develop nuclear energy for purposes of national defense. In 1954, revisions of the Atomic Energy Act allowed private industry to participate in the development and uses of nuclear technology for peaceful purposes and gave the AEC regulatory powers in the areas of public health and safety and national security as related to nuclear energy.

In response to developing concerns that no single agency should be responsible for promoting and regulating nuclear energy activities, including radioactive waste management and disposal, the Energy Reorganization Act of 1974 replaced the Atomic Energy Commission with the U.S. Nuclear Regulatory Commission and the Energy Research and Development Administration (Public Law 93-438, 88 STAT. 1233). The U.S. Nuclear Regulatory Commission's primary mission is to regulate nuclear reactors, materials, and waste facilities in the commercial sector. The Energy Research and Development Administration was given responsibility for managing nuclear weapon, naval reactor, and energy research and development programs. In 1977 the U.S. Department of Energy was created to provide a comprehensive national energy plan by centralizing the responsibilities of the Energy Research and Development Administration and other energy-related government programs. Part of the mission of the new agency included nuclear weapons research, development, and production. DOE both managed and regulated these facilities, and although an environmental impact statement was required as part of the National Environmental Policy Act (NEPA) process for major federal actions, the public was provided only limited access to information about federal activities at the sites and input into decisions. In 1984, the Legal Environmental Assistance Foundation (LEAF) sued DOE alleging that DOE allowed unpermitted discharges of nonradioactive pollutants from the Y-12 Plant at Oak Ridge National Laboratory in Tennessee (*LEAF* vs. *Hodel*, 1984). In this litigation, the court held that DOE is subject to the provisions of the Resource Conservation and Recovery Act. As a practical matter, states thus gained leverage over some DOE waste management decisions.

In the late 1980s as the Cold War was winding down, DOE began to shift part of its mission from weapons production to environmental remediation of the nuclear weapons complex scattered throughout the United States. In 1989, DOE created the Office of Environmental Restoration and Waste Management, later renamed the Office of Environmental Management, to manage the process of environmental cleanup while protecting the health of workers and the public. The office was also responsible for working with a wide range of stakeholders, including states, Native American tribal nations, other federal agencies, interested and affected members of the public, and public interest groups.

At each of the major sites, DOE, the host state, and the U.S. Environmental Protection Agency signed legally binding Federal Facility Agreements and Consent Orders that describe the roles of each federal and state agency in meeting specific schedules for environmental cleanup (see discussion of policy background in Chapter II).

Native American tribal governments are not parties to these Federal Facility Agreements and Consent Orders and instead turned to the courts as a means to influence DOE decisions that affect tribal peoples, lands, and resources. In addition, public interest groups and other concerned parties have used the courts and the media to try to exert influence over DOE's waste management decisions.

Thus, the decision-making paradigm for DOE and its predecessor agencies gradually changed from operating unilaterally within a self-regulating context of building nuclear weapons and expanding nuclear technologies, to involving multiple federal agencies, states, and other interested and affected parties in the decisions for managing and cleaning up wastes at DOE sites.

1. Stakeholders may help to define risk levels specified in regulations;
2. They may influence priorities about which sites within a facility are addressed first and to what extent (thereby also influencing the management of other waste sites within the facility);
3. They may help specify a desired future state for a site, particularly in terms of its preferred future uses;
4. They may help decide the relative balance of contaminant reduction, contaminant isolation, and stewardship activities to be used in achieving a desired future state for the site; and
5. They may influence choices concerning specific approaches and techniques (e.g., a preference for vitrification over grouting, a desire to have deed restrictions as well as zoning, an objection to the use of on-site incineration).

Given the decision-making environment in which DOE operates now—and in the presence of the programmatic risks mentioned earlier in this chapter—the committee recommends a more risk-informed, consistent, transparent, and participatory process, as recommended in a previous National Research Council (NRC, 2005b) report. This process would make DOE's tank waste management program more robust, in the sense that it is more likely to succeed in its mission, and more transparent, so that regulators, Congress, and the public have a clearer idea of the challenges and choices that DOE faces (see Recommendation VIII-3).

The NRC (2005b) report—the *Risks and Decisions* report—describes the basic elements of a risk-informed approach that is compatible with the needs and legal requirements of this system and is capable of encompassing the nontechnical considerations discussed in Finding VIII-1. The committee authoring that earlier report found that an effective and credible risk-informed decision-making process has the following characteristics:

> It is (1) iterative and participatory; (2) logical; (3) consistent with current scientific knowledge and practice; (4) transparent and traceable; (5) structured with reasonable independence of the decision authority from the petitioner; (6) subjected to thorough, independent peer review; (7) technically credible, with believable results; and (8) framed to address the needs of the decision process.

The *Risk and Decisions* report defines the term "risk-informed approach" as "[a]n approach in which risk is the starting point but still only among several factors in a decision process" (NRC, 2005b, p. 208). More specifically, *Risk and Decisions* points out that:

- Risk assessments (e.g., the scientific evaluation of known or potential hazards) can and should be integrated into risk-informed decision making;
- Because risk-informed decision making cannot be applied in a cookbook fashion, the risk assessment underlying decision making should account for the complexities and uncertainties of the underlying processes being modeled;
- Risk analyses carried out to support exemptions from HLW disposal requirements should be logical, well founded, transparent, and traceable; and
- Risk-informed decision making and the analyses that support it should be structured to inform a specific and well-defined decision based on criteria formulated well in advance of modeling or computation.

A full implementation of a risk-informed approach acknowledges that the *process* of analysis could be more important in achieving transparency, trust, and understanding than an elegant and complicated analysis that is presented as a completed package. Risk-informed decision making thus demands that a suitable process be established and followed. A rote progression through a series of steps would not meet this obligation. Emphasis should be placed on establishing a useful and meaningful process that involves stakeholders from the outset. Using an iterative, staged process brings involved parties along as the complexity and sophistication of the analysis and data increase and is more likely to create confidence in the final disposition decision.

In summary, the critical elements for implementing a risk-informed approach are

- Identify a specific set of options;
- List the information or data needs for deciding among these options;
- Establish a set of criteria for determining the best option (before analysis occurs);
- Carry out rudimentary risk calculations (minimal complexity) to help separate options based on decision criteria;
- Perform risk calculations that take into account alternative views of physical processes, using ranges of parameter values that reflect the state of the science;
- Determine which uncertainties affect the ranking of disposition options, and explain them and their significance for review by experts and stakeholders;
- Make a decision;
- Monitor;
- Compare predicted performance with data collected; and
- Make refinements as necessary through an iterative, staged process.

The *Risk and Decisions* report sets out a six-step process for implementing a risk-informed approach (see Sidebar VIII-3).

In the past there have been concerns about the transparency of DOE's decision-making process. Some of DOE's decisions concerning tank waste management did not have a clear description of risks, alternatives, or rationale for choices. Some decisions had too little supporting information (e.g., choice of glass as a low-level waste form at Hanford), or conversely, while some had an overwhelming amount of supporting information, that was still incomplete in some important areas and so large that time was inadequate to review it in detail (e.g., the first Section 3116 waste determination at the Savannah River Site for salt waste) so stakeholders did not have the opportunity to understand what was being proposed and why. DOE has better chances to reach its programmatic milestones if it adopts a more risk-informed, participatory, consistent, and transparent decision-making process as described in *Risk and Decisions* (NRC, 2005b).

DOE has taken steps to improve its transparency and detail in its most recent waste determinations (Sams, 2004; DOE-SRS, 2005a; DOE-ID, 2005a) and performance objectives demonstration documents (Buice et al., 2005), which describe how DOE reached its decisions and provide supporting data and analyses making it easier for others to understand. The documents supporting Section 3116 waste determinations are a worthy effort and should be pursued for other waste determinations. The states of Idaho and South Carolina also found the Section 3116 process helpful in reviewing DOE's waste determination plans because of the greater transparency in DOE's decision-making process (SCDHEC, 2005a, 2005b; Trever, 2006).

The recent Savannah River Site and Idaho National Laboratory waste determinations, and the Hanford exemption for tank C-106 show the beginning of trade-offs among public risk, worker risk, and cost (e.g., Davis, 1998; Gilbreath, 2005; and Sams, 2004). However, the requests for additional information issued by the U.S. Nuclear Regulatory Commission and state comments show that DOE has further opportunities to improve the waste determination process. For example, in the case of the draft Section 3116 waste determination for Tanks 18 and 19, it was not clear whether waste had been removed to the maximum extent practical from

SIDEBAR VIII-3
The Six Steps of a Risk-Informed, Participatory, Consistent, and Transparent Decision-Making Process

Step 1: Initiate the Process, Laying Out Viable Options and Potential Decisions.
This process initiation phase consists of (1) defining the problem and issues; (2) engaging partners and regulators to discuss and refine issues; (3) defining presumptive and alternative disposition alternatives; (4) defining criteria that are relevant to decision making; (5) developing a process plan by which consultation and analysis will proceed; and (6) seeking review and feedback from stakeholders and partners.

Step 2: Scope Information and Analysis
This step begins the process of estimating the specific risks that will be compared. It includes (1) sketching out the structure of the risk analysis; (2) identifying parameters, datasets, and models required; (3) collecting and reviewing needed information; (4) performing a scoping risk assessment and sensitivity studies to identify critical parameters that require attention; (5) describing data gaps and a data collection plan; (6) conducting review and feedback from experts and stakeholders; and (7) finalizing the work plan and moving forward.

Step 3: Collect Data and Refine Models
This step is a straightforward implementation of the data collection plan developed in Step 2, as refined by the results of the scoping analysis. It consists of the following activities: (1) collecting quality data that describe the waste and the site; (2) describing and collecting data regarding engineering remedies; (3) refining model logic; (4) disclosing new information collected; and (5) submitting the new data and any new risk calculations to external review.

Step 4: Prepare a Refined Risk Assessment
This assessment involves taking the initial risk assessment (see Step 2) and improving it by applying new data, advanced understanding of the underlying processes (i.e., better modeling), or more sophisticated uncertainty analyses. The refined risk assessment is used to inform the disposition decision. Preparing a refined risk assessment involves the following steps: (1) defining the range of uncertainty by, making use of collected data; (2) conducting analysis, including uncertainty analysis, and producing risk estimates; (3) performing a validity check to make sure that the results are reasonable in light of real world experiences; (4) performing a thorough quality assurance-quality control check of model logic and data inputs; (5) summarizing the results of the risk assessment, paying attention to the risk estimates and the uncertainty; (6) obtaining peer review of the model and its results; and (7) releasing the results to the public, in accordance with the agreed upon plan.

Step 5: Conduct Additional Analyses and Data Collection as Needed to Support Decisions
If additional analyses or refinements are needed, they would be done following a plan agreed on with stakeholders. It is likely that such planning will result in analyses and data collection being iterative; additional data collection might be necessary as learning progresses and the need for additional analyses is identified.

Step 6: Finalize the Decision
At this point, DOE has determined that it has sufficient information for decision making and it will use this information to seek final authorization for disposition of the waste stream in question.

Tank 19 (SCDHEC, 2005a). The State of Idaho and the U.S. Nuclear Regulatory Commission raised questions about the residual waste characterization of some of the tanks at Idaho and the state of knowledge of the contamination in the sand pads (Trever, 2006; USNRC, 2006).

Management of Long-Term Risks and Long-Term Stewardship

The committee recognizes that it is not feasible, and does not advocate, removing wastes in tanks "up to the last molecule." This means that in many cases, sufficient amounts of radioactive materials will remain on-site to prevent release of the site for unrestricted use. In the presence of long-term environmental liabilities, a form of "defense in depth" involves establishing institutional controls (also called long-term stewardship). Nevertheless, the committee believes in the importance of (1) not relying on institutional controls exclusively (in lieu of waste removal or other engineered and natural barriers) and (2) assessing the consequences of the failure of long-term institutional management in light of the uncertainties (e.g., present and future behavior of contaminants in the environment, future developments in society and technology, model limitations).

10 CFR Part 61 provides guidance for long-term stewardship of land disposal sites for low-level waste. The U.S. Nuclear Regulatory Commission (USNRC), as provided in 10 CFR 61.59, does not usually allow reliance on active institutional controls for more than 100 years.[4] Therefore, the USNRC requires that DOE demonstrate in its performance assessment that a site meets the performance objectives of 10 CFR 61 and does not depend on active institutional controls beyond 100 years. All three sites rely (at least for portions of the land) on long-term stewardship and institutional controls in perpetuity to help define land use and intruder scenarios.

In addressing its responsibilities to reduce risks and protect the environment over long periods, DOE's long-term stewardship, monitoring, maintenance, and/or solidification or immobilization plans use terms such as "in perpetuity," "1,000 years," and "10,000" years. Long-lived man-made physical structures or institutions are extremely rare. A few obvious examples of long-lived man-made physical structures and a long-lived institution are the Egyptian pyramids (approximately 5,000 years old, although most were violated long ago) and the Roman Catholic Church (approximately 2,000 years old), respectively. Very few structures have lasted intact and can fulfill their initial design purpose at even 1,000 years, such as some of the roads, bridges, aqueducts, and amphitheaters built by the ancient Romans.

In DOE's performance assessments reviewed by the committee, the assumption of no loss of institutional control before 100 years is hardly challenged. Although it is impossible to predict *what* changes will occur centuries from now, it is reasonable to predict *that* changes will occur, by merely considering the significant changes that have occurred in the past 100 years.

The committee acknowledges that DOE is already laying out some of the alternative scenarios in its performance assessments and environmental impact statements (see Chapter VI). An assessment of the consequences of changes in the assumptions (e.g., concerning the effectiveness of institutional controls) used in the performance assessments and an evaluation of the cost, risks, and environmental impact of taking action to mitigate the consequences would be desirable. For instance, if assumptions about land use based on the site being zoned "industrial use, in perpetuity" prove to be incorrect, what would be the consequences in case no action is taken and in case the tanks are re-remediated?

When making tank waste determination decisions, DOE must demonstrate that it has removed the highly radioactive nuclides from the waste to the maximum extent practical. This term refers to technical, safety, and financial considerations. As the committee points out earlier in this chapter, the concept of "maximum extent practical" does not have a clearly defined meaning and, therefore, is open to interpretation and argumentation. DOE takes into account the balance between the costs of worker exposure and the effect on risks of removing additional waste from the tanks or in leaving some waste in the tank as the heel. Financial considerations address the costs of retrieving additional waste from the tanks compared to the corresponding reductions in risks while taking into account additional risks to workers to remove the waste. For example, it is more cost-effective to meet performance objectives by using tank isolation and land-use restrictions than by conducting expensive, more complete contaminant reduction measures. However, a future state that includes stewardship is not the same as a future state reached via more complete contaminant remediation, particularly if the latter would allow unrestricted access. For example, it would be very costly, and perhaps not possible, to remediate the Savannah River Site to allow unrestricted access to the entire site given the existing groundwater contamination. DOE relies heavily on limiting the future use of the Savannah River Site to a "high-security mission" (or at least heavy industrial use) as a rationale for not cleaning up many areas where it intends to maintain active institutional controls for the foreseeable future to levels suitable for unrestricted access.

In addition to considering the costs of using additional retrieval technologies and protecting workers from additional radiation exposure, other cost elements to consider include the costs of monitoring the site as well as the costs of installing and maintaining engineered barriers and sustaining institutional controls. Thus, a decision to end waste removal operations and grout a tank now may appear economically practical, but when long-term stewardship costs are factored in, that may no longer be the case.

A large body of work has been produced on long-term stewardship. The National Research Council published three reports on long-term stewardship relevant to the DOE weapons complex, which includes Hanford, Idaho National Laboratory, and the Savannah River Site (NRC, 2000c, 2003b, 2003c). The main findings and recommendations from the previous NRC committees can be summarized as follows: effective long-term stewardship will likely be difficult to achieve; engineered barriers and institutional controls will eventually fail; great uncertainties remain in assessing the effectiveness of a long-term remediation plan; and no plan developed today is likely to remain protective for the duration of the hazards.

[4]"The period of institutional controls will be determined by the Commission, but institutional controls may not be relied upon for more than 100 years following transfer of control of the disposal site to the owner" (10 CFR 61.59). Note that DOE Order 435.1 has different guidance, "In the intruder assessment, institutional controls should be assumed to be effective in preventing intrusion for at least 100 years following disposal facility closure; longer periods may be assumed with justification (e.g., land-use planning, passive controls)" (DOE, 2001a).

When setting up a long-term stewardship program, DOE should

- Plan for uncertainty;
- Plan for fallibility;
- Undertake scientific, technical, and social research and development; and
- Seek to maximize follow-through using a phased, iterative, adaptive long-term approach, in which monitoring to confirm the performance of the site is a key element.

The National Research Council reports urged DOE to look at the very long term in the stewardship of its environmental legacy. Along with the U.S. Environmental Protection Agency and the U.S. Nuclear Regulatory Commission, other agencies (e.g., Department of Defense, Department of Interior) face similar long-term stewardship challenges and could benefit from a common methodology for dealing with them.

FINDINGS AND RECOMMENDATIONS

Finding VIII-1: Basing tank management and waste disposition decisions only on performance assessment results to demonstrate compliance with performance objectives is inadequate because such assessments do not take into account all of the factors that could be important to decisions such as the evolution of a full-risk profile (risks across the site under different exposure scenarios) over time; compliance with Federal Facilities Agreements; changes in costs and changes in the way people value health and the environment; progress to build confidence in the program; and other site risks. All of these factors become increasingly uncertain as people attempt to anticipate conditions further into the future.

Recommendation VIII-1: Site-specific characteristics such as the type of wastes, tank types, and amounts and location of contamination are important to tank waste management decisions. Differences in these characteristics can lead to different choices among cleanup options at each site and even among tanks at the same site. Despite these differences, DOE's decision-making *process* should be consistent, even if this leads to different outcomes for different tanks. In addition to performance assessment results, the decision-making process should take into account site-specific factors, including the following:

1. The unique risks created by the proposed decision pathways that are not captured in the performance assessment or performance objectives;
2. Risk considerations at the sites that are altered by cleanup decisions;
3. Combined risks associated with tank residues and other waste streams that are associated with the separation of tank wastes into different waste streams;
4. Considerations not captured in risk analyses or performance assessments, such as costs, distributional or equity concerns, and other societal concerns;
5. The views of the states, Native American tribal governments, and other stakeholders in decision making.

Finding VIII-2: DOE operates in an extremely complex and overlapping regulatory and governance structure. Thus, in addition to the technical and health risks that have been detailed to this point in the report, DOE's conceptual approach to tank cleanup poses a number of "programmatic" risks, that is, risks that the cleanup program will not be carried out as planned. Some of these risks are due to purely technical and engineering challenges (e.g., Will bulk vitrification work?), some are budgetary challenges (e.g., Will Congress appropriate sufficient funds to implement the approach?), some are regulatory challenges (e.g., Will New Mexico accept waste from Idaho and Hanford?), and some are legal challenges (e.g., Will DOE be allowed by the courts to rely on DOE Order 435.1 for its waste determinations with respect to locations and materials to which Section 3116 of the 2005 NDAA does not apply?).

Recommendation VIII-2: In its planning, DOE should identify sources of programmatic risks as soon as possible so that it can seek ways to mitigate them. This process will make DOE's planning much more robust, in that success is more likely in its tank cleanup mission; and more transparent, in that regulators, Congress, and the public will have a clearer idea of the challenges and choices that DOE faces.

Finding VIII-3: The National Research Council's *Risk and Decisions* report (NRC, 2005b) describes the basic elements of a risk-informed approach that is compatible with the framework and legal requirements in which DOE operates and is capable of encompassing the nontechnical considerations discussed in Finding VIII-1.

Recommendation VIII-3: DOE should pursue a risk-informed, participatory, consistent, and transparent approach for making decisions under Section 3116 of the Ronald Reagan National Defense Authorization Act of Fiscal Year 2005 and DOE Order 435.1, which governs tank waste decisions at Hanford. DOE is taking steps toward this approach through the waste determinations required by Section 3116 for the Savannah River Site and Idaho National Laboratory. The committee urges DOE to continue and expand this practice.

IX

Focused Research and Development Needs

To reduce long-term risks at the sites and improve the basis for models and assumptions in the performance assessments, the committee recommends that the Department of Energy (DOE) carry out a focused research and development program to support tank closure activities. By "focused research and development activities" the committee means a program that concentrates on improving current technologies or developing technologies that could provide deployable results within 10 years. Although research needs concerning cementitious materials used in tank remediation applications, robotics, and chemical cleaning of tanks (from Chapters III, IV, and V) are discussed in this chapter, long-term research activities have been identified in the monitoring and performance assessment chapters (Chapters VI and VII), these would be carried out in parallel, but they are not the focus of this chapter.

Technologies that are deployable in 10 years could be developed and implemented during the tank remediation program and, in particular, in time to address the most challenging tanks (i.e., those with cooling coils, recalcitrant waste, or leaks), which are the ones most likely to have significant heels.[1] The tank remediation program is a multidecade endeavor and DOE has an opportunity to use this time to its advantage.

The committee believes that there are at least three critical topics warranting focused research and development efforts: (1) in-tank and downstream consequences of existing and advanced chemical cleaning options; (2) technologies to assist in tank waste removal, including robotic devices; and (3) near-term and long-term performance studies on those cementitious materials used to fill tanks and immobilize low-activity waste.

These topics represent the greatest technological challenges (i.e., waste retrieval and tank cleanup) and knowledge gaps (i.e., long-term performance of cementitious material). In addition to the recognized technical challenges in the program, there may be some "unknown unknowns" suggesting additional technological vulnerabilities, that is, areas that warrant additional research and development that cannot be foreseen right now but may become apparent once DOE further progresses in its tank remediation program. DOE should undertake a systematic effort to identify the most important vulnerabilities to reducing programmatic and human health risk as one step to address these "unknown unknowns."

The committee judges that a focused applied research and engineering development program aimed at reducing the amounts of waste left in the tanks or improving waste immobilization could lead to reduced risks on-site. Validating assumptions and improving DOE's knowledge base could increase confidence in its waste management plans or its assumptions about long-term performance of the waste forms disposed of on-site, both of which are desirable outcomes. Moreover, these research and development activities could support the development of contingency approaches to address unanticipated difficulties in baseline processes. Research and development activities to address these topics are discussed below.

IN-TANK AND DOWNSTREAM CONSEQUENCES OF EXISTING AND ADVANCED CHEMICAL CLEANING OPTIONS

The sludge component of tank waste is a sticky, semi-solid material that forms from the agglomeration of oxides and hydroxides of iron, aluminum, and manganese and is a time-dependent consequence of the neutralization of nitric acid processing solutions with sodium hydroxide. Sludge

[1]Previous National Research Council reports contain recommendations on long-term research and development needs for DOE's Environmental Management Science Program and some specifically for high-level waste tanks (NRC, 1996a, 1999a, 1999b, 2000d, 2001a, 2001b, 2001c, 2001d, 2002b, 2003b). The committee recognizes the importance of an ongoing basic and applied research program to support DOE's environmental management mission; however, such a program is not the focus of this chapter.

entrains varying amounts of insoluble actinide and fission products, and its presence therefore contributes to the overall source term of a particular tank. Most of the sludge waste can be mobilized and removed from storage tanks using traditional hydraulic techniques (i.e., mixer and transfer pumps; see Chapter III and Appendix F). Retrieval of residual sludge after the application of hydraulic techniques may require the application of chemical agents, aggressive sluicing methods, or a combination of both. Success with these approaches would further the inventory in, and potential future doses resulting from, individual tanks.

As noted in Chapter III, DOE has demonstrated the efficacy of oxalic acid for the chemical cleaning of waste tanks in certain cases. Two of the three full-scale sludge dissolution trials were performed at the Savannah River Site (West, 1980; Fong, 1985; Adu-Wusu et al., 2003) and one at Hanford (Reddick, 2004). Oxalic acid cleaning was effective in the removal of additional sludge heel in Tank 16 at the Savannah River Site, moderately effective in Tank C-106 at Hanford, and ineffective for treatment or removal of zeolitic materials (used for radiocesium ion exchange) found in Tank 24 at the Savannah River Site. The effectiveness of oxalic acid for removing sludge residues derives from its ability to form stable, soluble oxalate complexes with the iron component of tank sludges. Oxalic acid cleaning was ineffective in the removal of additional sludge heel in Tank 24 at the Savannah River Site because zeolites are primarily aluminum silicates, and these compounds are more stable than aluminum oxalates—they are not "dissolved" by oxalic acid.

Savannah River Site staff believes that oxalic acid has drawbacks associated with criticality safety, downstream processing, and costs (see Chapter III). However, the site recognizes the potential of chemical cleaning for treating sludge residuals in tanks with cooling coils where mechanical technologies for residual waste retrieval may not be effective. Savannah River Site staff has performed some limited research on alternative chemical cleaning agents and approaches to mitigating the potential adverse impact of oxalic acid. In a recent literature survey, Adu-Wusu et al. (2003) compared different chemical cleaning agents.

The committee judges that chemical cleaning is a proven tank cleaning technology that could be effective in tanks with significant obstructions and therefore should be investigated further (see Recommendation IX-1).

Two research and development paths for chemical cleaning can be explored:

1. Cleaning agents other than oxalic acid that would not cause criticality concerns or downstream problems; and
2. Methods to both predict and eliminate criticality concerns and downstream problems if oxalic acid is used as the cleaning agent.

The degree of cleaning is coupled to assessing how much radioactive material is a reasonable amount to remain in the tank. A metric for assessing the need to remove the tank heel is a comparison of radioactivity in the tank to the radioactivity already committed to the site. At the Savannah River Site the total activity in the tanks is 426 MCi (1.58×10^{19} Bq), with a sludge activity estimated to be 203 MCi (7.5×10^{18} Bq; see Table II-1). The radioactivity remaining in the heel is estimated to be 2 percent of the total tank activity (USNRC, 1999),[2] which would imply a heel activity of 8.5 MCi (3.15×10^{17} Bq). The heel isotopic composition will vary with the heel chemical composition. A heel composed of zeolites would be high in cesium-137. If the heel is primarily oxide precipitates, then strontium-90 and actinides would be the main radionuclides.

At the Savannah River Site, around 11 MCi is already committed to the site; 18 KCi (6.70×10^{14} Bq) of transuranic waste and 11 MCi (4.07×10^{17} Bq) of low-level waste. The low-level waste has an isotopic composition that is different than the tank waste and includes a significant contribution from tritium. (The 11 MCi of low level waste is an overestimate since it is not decay corrected.) If the heels contain of 2 percent of the total tank radioactivity, then they will contribute a radioactivity burden to the Savannah River Site comparable to what is currently at the site. This type of assessment can help inform the determination whether additional tank cleaning is needed.

Alternatives to Oxalic Acid

The main problem with oxalic acid is the extremely large quantities that are used to neutralize the residual sludge and dissolve it: 26,000 to 38,000 kg of sodium oxalate per 5,000 gallons of sludge removed. Using a stronger acid, such as nitric acid (HNO_3), to dissolve the sludge would reduce tremendously the amount of oxalic acid needed. Furthermore, nitric acid is an inorganic acid that does not complexate iron compounds (and, thus, does not raise criticality concerns in the waste) and also eliminates downstream problems such as foaming or CO_2 releases that have been seen with oxalic acid. Because less chemical agent is used, the amount of secondary waste generated in the process is also smaller. Consequently, nitric acid would place a smaller burden on compliant tank space at the Savannah River Site, which is in

[2]Another source (DOE, 2002) uses a value 15 times lower but the committee considers the retrieval estimates that underlie that value to be optimistic and unsupported (see Chapter VI). The values used here are estimates based on experience retrieving waste from Tank 16, a tank with cooling coils, before chemical cleaning (USNRC, 1999). DOE estimates that Tanks 18 and 19, which have zeolites but no coils, contain 28,000 Ci (1.0×10^{15} Bq) and 96,000 Ci (3.6×10^{15} Bq), respectively, after waste retrieval (Buice et al., 2005). These are 0.3 percent and 1 percent of the average total radioactivity per tank in tanks that have not undergone waste retrieval at the Savannah River Site, respectively.

scarce supply. Tank corrosion could be made relatively unimportant during tank cleaning because the duration of the chemical cleaning process is brief and the acid would be neutralized by the sludge. Any residual acid could be diluted with a rinse of inhibited water and/or neutralized with additions to the tanks, including the grout.

Savannah River Site personnel indicated that they are in communication with other national laboratories, DOE sites, and Russian experts on tank cleaning technologies, particularly chemical cleaning. Westinghouse Savannah River Company, the main contractor in charge of tank chemistry, is already considering the use of nitric acid for residual sludge removal because of the critical tank space issue (see Appendix E and NRC, 2005a). A team was formed in December 2005 at the Savannah River Site to evaluate the use of nitric acid and its application to sludge removal in the Savannah River Site waste tanks. However, the site recognizes that the current flowsheet does not include nitric acid as a potential tank cleaning agent, so design safety analysis changes would be required if it were used.

Other ligands can be used instead of oxalic acid to dislodge or dissolve the heel and remove the entrained radionuclides. Over the past 30 years, significant advances have been made in metal-specific complexation and ligand design that exploit metal ion speciation in producing selective complexes. A number of these advances are based on biomimetic studies that evaluate specific metal ion-ligand interactions in natural systems (Durbin et al., 1989; Raymond, 1990). Research has been performed on the leaching of actinides and fission elements from synthetic sludges under differing conditions and has been related to metal ion redox and speciation (Nash, 2002; Garnov et al., 2003).

While metal-ligand interactions are understood primarily in the solution phase, ligands produced by bacteria (siderophores) will solubilize iron oxides, will dissolve oxides of uranium and plutonium, and can be exploited in developing metal-specific reactions for solid phases (Brainard et al., 1992). These metal-specific ligand approaches have been used for the selective removal of radionuclides from the human body (see e.g., Gorden et al., 2003). This same selective ligand approach has been applied to the area of radionuclide removal from tank waste (see e.g., Nash et al., 2000). To date, these advances have not been used in waste retrieval operations, but in the future they might offer cost-effective options for removing both solution- and solid-phase radionuclides from the tanks.

Another approach to effect improved removal of entrained radionuclides from residual tank waste is to adjust the chemical oxidation state of the radionuclides. Metal ion solubility in a given aqueous phase varies dramatically with oxidation state. Plutonium is an excellent example with large differences in solubility that are dependent upon the metal ion oxidation state. This property is shared by other actinide elements and is exploited in nuclear fuel treatment for the dissolution of spent fuel and developing methods for tailored separations (Karraker et al., 2001; Thompson et al. 2002).

Additional chemical cleaning agents may also be effective on recalcitrant waste types, such as zeolites at the Savannah River Site (and West Valley) and diatomaceous earths at Hanford. The committee has not identified any specific agents, but it is not clear that this possibility should be dismissed.

Mitigation of Concerns Oxalic Acid

The second possibility for research on chemical cleaning is to use oxalic acid while addressing the nuclear criticality concerns and downstream problems. The committee judges that the probability of a criticality event in a tank is low: It is unlikely that the tank waste processing system would either have a sufficient amount of fissile material in one location or configure it properly to start a chain reaction. Both of these conditions would be needed to achieve criticality. Based on the information provided by Savannah River Site staff on criticality concerns (see Chapter III), the committee was unable to determine whether the criticality concerns with oxalic acid are well founded. Therefore, the committee recommends continuing research on oxalic acid and carrying out a study that shows whether oxalic acid leads to criticality concerns in the tanks or downstream.

Downstream problems could be addressed by destroying oxalic acid and metal oxalates after tank cleaning. The destruction of oxalic acid by oxidation has been investigated and can be used as the basis of further studies. Other oxidative methods have been investigated for the treatment of tank waste, including ozone, chemical oxidation (Patello et al., 1999), and electrochemical oxidation (Nash et al., 2003). While ozone is effective for destruction of oxalic acid in the laboratory, significant quantities would be required for these applications and care would need to be exercised to minimize occupational exposures.

TECHNOLOGIES TO ASSIST IN TANK WASTE REMOVAL, INCLUDING ROBOTIC DEVICES

DOE has been relatively successful in the limited number of waste retrievals undertaken to date. However, such success is tempered by DOE's statement that it initially focused on retrieving waste from less complicated tanks. Future waste retrieval will become more difficult as more complicated situations (e.g., leaking tanks and tanks having substantial internal structures and more recalcitrant solids) are encountered. As a consequence, there is no assurance that previous successes will project into the future (i.e., that current retrieval technologies will be equally successful or even adequate).

In part, DOE is addressing this issue by using a mix of available waste retrieval technologies modified to reflect specific circumstances and experience. As shown in

Chapter III, advanced waste retrieval technologies have been and continue to be developed by DOE or adapted from technology developed by others.

DOE already acknowledges that future efforts to retrieve waste from its tanks will be challenged by the following:

- The environment within the tanks is one of high radiation and harsh chemical conditions.
- Retrieval is complicated by physical obstructions and recalcitrant chemical species.
- The waste characteristics relevant to retrieval are difficult to determine before retrieval because of the heterogeneous and anisotropic nature of the waste so DOE must "learn as it goes" on a tank-by-tank basis.
- The variation in the condition and contents of each tank makes its waste retrieval a unique undertaking.

As shown in Chapter III, DOE faces the need to retrieve waste from many large tanks containing cooling coils and other obstructions, especially at the Savannah River Site. The baseline bulk retrieval approach consists of using water jets from the riser locations to spray material off the internal tank structures onto the bottom of the tank, consolidating it by sluicing, and pumping it from the tank. The potential limitations of bulk retrieval techniques are clear when one notes that the tanks containing a "jungle" of cooling coils are 23 to 26 m (75 to 85 feet) in diameter and up to 9 m (30 feet) in height. The efficacy of bulk retrieval in tanks with coils is uncertain: Good cleaning of tank surfaces appears to have been achieved in zones beneath the risers, but the amount of residual waste remaining in the "dead zones" between risers and in the tank periphery is unknown. Moreover, the limited accessibility of the tanks, the presence of cooling coils, and other obstructions severely limit the size and mobility of retrieval devices. With all of these challenges, bulk retrieval technologies are likely to leave significant amounts of waste in difficult environments that will challenge the capabilities of existing residual waste retrieval technologies.

The committee judges that focused research and development investments in residual waste retrieval technologies suitable for tanks containing cooling coils or other obstructions, recalcitrant waste, or tanks that are leaking appear prudent. In Chapter III and Appendix G, the committee applauds the efforts at Hanford to develop the Mobile Retrieval System, a vacuum retrieval technology complemented by the in-tank vehicle, suitable for leaking tanks, as well as the development of the Salt Mantis, a high-pressure, low-volume water jet to mobilize recalcitrant waste deposits in tanks that have not leaked. Other potentially promising technologies (because they generate little or no secondary waste) include the use of high-pressure steam jets (used at the Savannah River Site to attempt to clean the annulus of Tank 16 and at Hanford in Tank C-106), CO_2, and sodium carbonate pellet blasting (NRC, 2001c).

The concept of an autonomous robot[3] freely roaming inside a tank cleaning various surfaces to certain specifications with minimal human control has an intuitive appeal because of the potential for reduced labor costs and possibly avoiding complications posed by tethers. In general, since the 1970s, advances in robotic technology have been possible thanks to the progress of the microprocessor, although with a lag of some years. The continued increase of processing power, miniaturization, and speed has allowed robotic devices to make major advances in speed, precision, cost-effectiveness, scope of applications, and most critically, reliability. This is especially true in industrial applications that involve repetitive, predictable motions and tasks. More exotic applications such as space, service, military, and security robotics have also seen impressive gains commensurate with the funding levels invested in development (DARPA, 2005; NRC, 1996a).[4]

Because the challenges of DOE tank waste cleanup are unique and the opportunities for deployment have been few due to the pace of the tank waste cleanup program, development and deployment of robotic-like devices for this purpose has been attempted only by a few teams. DOE's previous work and experience on articulated arms (e.g., the Modified Light Duty Utility Arm, in-tank vehicles such as Houdini and ITV at Hanford, and other examples cited in Chapter III and Appendix G) are worthwhile efforts in a necessary, continuing investigation on retrieval technologies. Most of these efforts were done within the Robotics Crosscutting Technology Development Program and Tanks Focus Area, which were both discontinued around 2002.

The environment inside a tank, characterized by the following factors among others, is particularly hostile to untethered semiautonomous robotic technology:

1. A potentially explosive, and of course radioactive, atmosphere;
2. Sludge that impedes the mobility of robots and management of tethers;
3. Physical obstructions such as cooling pipes and debris;
4. Obstructed visibility for necessary vision systems through vapors, sludge, and physical obstructions;

[3]For the scope of this report, a robot is a programmable, multitask manipulator capable of operating within a three-dimensional space and manipulating tools in response to its control programs; the robot may be tethered or untethered. To operate within the tank space the robot would be fitted with on-board sensors whose information may modify its programs. Tele-operated articulated arms or tele-operated mobile vehicles (e.g., tank crawlers) are not robots unless they are programmed to perform their tasks autonomously.

[4]For example, the Defense Advanced Research Projects Agency (DARPA) Grand Challenge is a series of races of autonomous robots through a desert course. Groups compete against each other to create the best autonomous robot that will complete a desert course avoiding all obstacles and following DARPA's preset rules. The grand prize was set at $1 million in 2003 and $2 million in 2004 and 2005 (DARPA, 2005).

5. Contaminants hostile to instruments and sensors;
6. Limited number and size of openings for accessibility (risers);
7. Need for external power source for the robots through tethers or frequent recharging of batteries; and
8. A challenging, unpredictable environment that is contrary to the programming benefits of robots and demanding of unique sensors and intelligence.

The committee judges that the use of autonomous robots (i.e., operating without human control) for waste retrieval operations is not a realistic or practical alternative given the current status of the technology. Any waste retrieval tool for tank cleanup application must have some type of tether for the following reasons:

1. It must be provided with some means to pull it out to safety in case repairs are needed;
2. The tools carried are heavy and generate appreciable forces—hence the need for large actuators and appreciable power make the robot impractically bulky; without a battery, a tether would have to include an electric cable to supply power; and
3, Tooling such as water jets requires a continuous feed.

The management of a tether around pipes and through thick sludge is a challenging task. Moreover, the tank environment is not well defined, especially in tanks with vertical cooling coils, so robotics devices cannot currently be programmed to handle all possible situations that may be encountered.

Given these challenges, developing a deployable and reliable, untethered, semiautonomous robot for waste retrieval purposes may take at least a decade of well-focused research, development, and deployment effort and investment on the order of $10 million per year. A previous National Research Council report (NRC, 2001a) suggests ideas for long-term research needs on untethered, semiautonomous robotic device. Robotic technologies will continue to advance, especially for the control of robots with sensors, and may accomplish in the future what is not possible today.

Nonetheless, robotic technologies may still play a role in enhancing the effectiveness of available retrieval technologies. Existing waste technologies could benefit from "robotization" of some of their functions to simplify their operation, to reduce the level of skill required for their control, and to automate some aspects of the cleanup process. For example, mechanical arms and in-tank vehicles are practical platforms for deploying cleanup tools such as the Salt Mantis, the water mouse, and the wash ball. Robotic enhancements of articulated arms or in-tank vehicles could provide sensory feedback signals (e.g., force, temperature, visual signal) so that the operator can respond accordingly. Some operations, such as "go to point *x,y,z* or advance by 10 mm" could be programmed instead of manually jogging a tool precisely. Having a human in the loop to make operating decisions in response to sensory feedback signals is likely to prove most effective. The trade-off would be in balancing the cost of enhancements against the savings in cleanup time. For tanks with cooling coils, developments are warranted to enable the delivery of cleaning tools, such as high-pressure water jets, to surfaces shadowed by the cooling coils. For example, a directionally compliant tube could be pushed horizontally through the tank; this tube would bend sideways past the cooling coils without overstressing them and still deliver a water jet to the shadowed areas of the tank and pipes.

Such enhancements may improve the rate of retrieval (i.e., reduce cost), but not necessarily increase the amount of waste removed. Other uses of robotics may help reduce exposure of workers to radiation. For example, commercially available industrial robots could be used to insert cleanup tools and retrieve them through tank risers, currently operations that can cause serious worker exposures. A targeted study may find similar uses for industrial robots to reduce worker exposure within the tank farms. Tele-operation with sensory feedback appears to be more amenable than programming to operate in the irregular environment of the tanks.

NEAR-TERM AND LONG-TERM STUDIES REGARDING TANK FILL MATERIALS

For the purposes of performance assessment, certain assumptions must be made about the short-and long-term behavior of cementitious materials used to fill tanks and immobilize low-activity waste. Their behavior with regard to mixing, pumping, and placement can readily be determined in short-term tests such as mockups and can be well informed by the experience of the construction industry. However, the same cannot be said for their long-term grout performance in service. When used for tank stabilization and low-activity waste immobilization for hundreds or thousands of years, cementitious materials are subjected to service conditions well beyond the experience of the construction industry.

The ability to develop meaningful predictions of the behavior of cementitious grout or concrete over the long term requires a good understanding of fundamental mechanisms, such as

- Microstructure formation and degradation mechanisms;
- Pore solution chemistry (including pH and Eh);
- Binding properties (including toxic heavy metals and radionuclides, and also ions such as chlorides and alkalis that participate in deterioration processes);[5]

[5]To some degree, the hydration products of cementitious materials are able to incorporate various foreign materials such as toxic heavy metals, thus rendering them immobile. However, how much and how effectively

- Transport properties (permeability and diffusivity); and
- Mechanical properties.

Much work has already been done on the science of concrete and other cementitious materials to advance our understanding of their microstructure, deterioration mechanisms, transport properties, and mechanical properties as they change over time. However, since most of this work has been done with construction applications in mind, little attention has been paid to their behavior decades after the initial mixing and placement. Also, with the exception of pH, relatively little work has been done with regard to changes in pore solution chemistry and binding properties because they have little relevance to the construction industry. Some work on the use of cementitious materials for the stabilization of toxic heavy metals has been performed and would be relevant to DOE applications. However there remain some gaps in our knowledge that are unlikely to be filled without DOE research and development.

The committee is well aware that DOE is required by law to meet certain milestones in its progress toward closure of the tank farms. However, the committee judges that there is sufficient potential for better decisions about grout applications to warrant a research and development program. The committee believes this can be conducted without adversely affecting the overall schedule for closure of the tank farms. The following section describes the near-term focus of the proposed research and development program and then outlines its features.

Near-Term (5-to-10-Year) Focus of Grout Research and Development

The main focus of the research and development program in the near term would be to provide a sound basis for selecting the best formulations for grout based on the anticipated service environment and performance requirements. In keeping with the committee's recommendation that DOE increase transparency and public involvement in decision making, this program would demonstrate that the grout formulations selected are clearly superior to the alternatives and would provide a quantifiable basis for comparing their performance. The concrete durability research conducted by Atomic Energy of Canada Limited for its near-surface disposal facility indicated that while service life predictions cannot be made with a high level of confidence based on a 5- to 10-year laboratory test program, such a program is helpful in providing a sound basis for comparing the performance of different grout formulations and thus for the selection of candidate grouts (Philipose, 1988; Feldman et al., 1989; Philipose et al., 1990a; 1990b).

Although it is not possible to conduct real-time tests of grout durability for the relevant time frames, deterioration mechanisms can be estimated through accelerated testing. Such tests are performed under conditions that accelerate the degradation of concrete (e.g., elevated temperatures, electrical potentials to accelerate the migration of destructive ions, increased concentrations of destructive chemicals, cycling of temperatures, cycles of wetting and drying). Accelerated durability tests can be problematic however, because conditions imposed to accelerate deterioration may foster different mechanisms than would occur naturally. In addition, durable materials by definition take a long time to deteriorate, which necessitates long-term testing program to obtain results. To avoid these unrealistic test conditions, the proposed research and development program discussed below incorporates only methods that provide a modest degree of acceleration of the deterioration mechanisms, such as mildly elevated temperatures and increased concentrations of the chemicals of interest, similar to the test conditions at Atomic Energy of Canada Limited (Feldman et al. 1991; Philipose et al., 1991, 1992), and the recommended examination methods possess the ability to observe early signs of deterioration.

Research and development on alternative grout placement technologies, such as jet grouting, to improve the degree of mixing of waste with grout could also yield results that are deployable in 5 to 10 years. Research and development work on jet grouting is already in progress at the Los Alamos National Laboratory (AEATES, 2004). Research on grout placement was performed at the three sites of concern in this report; Idaho National Laboratory conducted a mockup test that showing that a five-phase sequential pour would enhance the waste removal process (see Chapter V's section on Tank Grouting at the Idaho National Laboratory).

As mentioned in Chapter V, some short-term testing of grout materials has been conducted at DOE sites, and in particular, at the Savannah River Site. However, the committee has not seen any reports of long-term testing or a comprehensive analysis of basic properties to model long-term behavior.

Experts at the Savannah River Site (Dr. C. Langton and Mr. T. Caldwell) reported that based on their tests, they have concluded there is effectively no mixing of grout with the insoluble tank heel, but the liquid is effectively incorporated into the grout or the dry cementitious materials deposited on top of the grout. Accordingly, DOE's ongoing performance assessment does not take credit for mixing of the grout and heel but, as noted in Chapter VI, assumes that the grout maintains its structural integrity for 500 years and its physico-

they can do so depends on the specific combination of cementitious materials and the pore solution chemistry. For example, slag is particularly effective at binding chloride ions, and in Portland cement the tricalcium aluminate component also binds chloride. However, if the pH of the pore solution is reduced—for example, by carbonation (reaction of the calcium hydroxide with carbon dioxide from the atmosphere)—the capacity of the hydration products to incorporate chlorides is reduced, and some of the bound chlorides will be released. Thus, the ability of the grout to bind foreign materials either incorporated in it at the time of mixing or migrating in from the outside can change over time.

chemical integrity for 10,000 years. The assumption concerning structural integrity is based on an earlier analysis done for the E Area Vaults performance assessment (Martin Marietta Energy Systems et al., 1994). It appears that the values used in the performance assessment for Tanks 18 and 19 for such properties as hydraulic conductivity have been inferred from the literature or from test data on the saltstone test program (see Chapter V). Understanding changes in these properties over time (decades, centuries, and millennia) is important to model waste releases to the environment in the performance assessment. Therefore, the committee cannot assess the assumptions of physical and chemical durability of the cementitious material used to fill the tanks and to immobilize low-activity waste. However, the committee believes that fundamental understanding of parameters that affect grout stability is not adequate to support the expected 500- to 10,000-year performance period.

Outline of Proposed Grout Research and Development Program

The U.S. Nuclear Regulatory Commission (USNRC) is cosponsoring research at the National Institute of Standards and Technology (NIST) on degradation mechanisms, mixing formulations, durability, and modeling of cementitious materials (Garboczi et al., 2005).

DOE could provide the necessary input parameters for NIST's microstructure-based model, such as

- The various environmental conditions present and anticipated at the three sites (e.g., soil chemistry, conditions of wetting and drying);
- The chemicals within the tanks or the low-activity waste;
- The ingredients for formulating grout available at each site (e.g., cementitious materials such as cement—ordinary Portland, sulfate resisting, and others), fly ashes, slag, silica fume; admixtures such as sodium thiosulfate (reducing agent); and
- A range of water-cementitious materials ratios.

The DOE laboratories have the most intimate knowledge of the particular environment and performance requirements of the grouts and can provide essential insights into the best surrogates to simulate the chemical environment for safe testing. Universities bring a rigorous approach to research and often possess unique or specialized laboratory equipment. Industry has practical knowledge that can keep the program tied to the real world, as well as the capability to conduct a large-scale test program within the limits of budget and schedule while maintaining a high level of quality.

A series of grout samples of different formulations could then be subjected to a succession of simulated environmental conditions over the duration of the program. Periodically, specimens can be taken out of the solution baths and "sacrificed" to locate the reaction front to determine the depth of ionic ingress into the specimens. Variation of depth of ingress with time for the various formulations is an indication of their transport properties and hence an indication of durability. Examples of examination methods would include the following:

- *Petrographic examination.* Petrography uses mainly optical microscopy, supplemented by scanning electron microscopy and sometimes chemical analyses or other methods. It is concerned primarily with determining spatial relationships among the hydration products of the cementitious materials and can identify early signs of changes in microstructure, the products of degradation reactions, and the mechanisms of degradation. The NIST model is based on microstructure; thus, the information provided by petrography is essential both to establish a starting point and to validate and refine the model's predictions. An additional advantage of petrography is that the most informative specimens it employs are thin sections, which are slices of concrete approximately 20μm thick mounted on microscope slides. These specimens are extremely stable and remain usable for 100 years or more. Thus they could be archived to provide opportunities for examination and comparison many decades in the future. Such basic information would be an asset to both model validation and the site's long-term monitoring efforts discussed in Chapter VII.
- *Bulk chemical analysis.* Where the mechanism of interest is the ingress of some specie, a series of thin slices of grout can be analyzed (e.g., with an electron microprobe) for that chemical to determine concentration gradients, from which the transport properties of the chemical can be determined.

A study of the changes in the characteristics of leachate over time would inform the performance assessment. In this example, monitoring would continue beyond 10 years to allow comparisons with and updating of the performance assessment to increase confidence that the grout is working as designed.[6]

More basic research activities that could be performed in the same 5- to 10-year time frame include identifying and evaluating oxidation pathways and kinetic mechanisms for grout degradation. Some of these activities could also be conducted in parallel with saltstone to compare retention capabilities for the mobile radionuclides such as technetium-99. Many of these research and development needs were

[6]The tanks or vaults and cementitious materials within them would have to degrade before water could get in, leach contaminants, and leak out of them. The time at which this might occur is uncertain, and there may not be any water leaching out of these tanks for decades. Hence, this monitoring activity would have to be coordinated with the long-term monitoring activity (see Chapter VII).

identified by Westinghouse Savannah River Company at the committee's meetings.

USE OF TEST BEDS FOR THE STUDY OF RETRIEVAL TECHNIQUES

As noted in Chapter III, waste retrieval (bulk or residual) has not yet occurred in most tanks at DOE sites. Many of the tanks contain internal features such as cooling coils or other debris that promise to impede waste retrieval. In these situations, new technology or adaptations of existing technology may be desired or required. Adapting an existing retrieval technology or deploying a new retrieval technology in a radioactive environment can cost millions of dollars, and failures can cost even more. Thus, it is technically prudent and cost-efficient to test retrieval technologies in nonradioactive test facilities (test beds) before attempting to deploy them. The usual approach to ensuring that a radioactive waste process will work consists of two steps: tests with the actual material (hot tests) at the laboratory scale to ensure that the process fundamentals are understood (e.g., sludge dissolution, pumpability), and tests with nonradioactive simulants (cold tests) at large or full scale to prove the design and the equipment.[7]

The purpose of a test bed is to:

- Test all equipment prior to field deployment,
- Provide hands-on experience with retrieval systems under simulated conditions,
- Troubleshoot retrieval system design and operations,
- Train equipment operators;
- Support development of operating procedures and maintenance protocols, and
- Test off-normal (outside of planned operation or behavior) and recovery activities.

In 2002, the Hanford Site built a test bed called the Cold Test Facility (CTF) at a cost of $2.87 million (Dodd, 2005). The CTF consists of a 75-foot diameter open-top tank to simulate a single-shell tank or a double-shell tank. The CTF can hold up to 600,000 gallons of simulated waste. A steel bridge or primary superstructure spans the open tank to accommodate full-scale mixer pumps and transfer pump system mockups along with waste retrieval equipment. Although the structure is open, it is capable of simulating customized constraints such as hanging interference mockups and risers pits. Tests are sometimes conducted at night to simulate the lack of visibility inside the tank. Retrieval technologies demonstrated at the CTF include the in-tank vehicle crawler, an off-riser sampler, and the hydrolaser water lance (Salt Mantis), and technologies used at other sites (see Table F-2 and Appendix G).

The Savannah River Site has operated a test bed at the TNX facility, but it is not clear that this test bed will be available in the future. The Pump Test Tank is a partial Type IV tank mockup at the mostly decommissioned TNX facility, used for testing and equipment before deployment.

The Idaho National Laboratory does not have a test bed on its facility, but it conducted two major mockup activities: (1) to test grout placement techniques (see Chapter V); and (2) to test waste retrieval equipment (see Chapter III). The first mockup facility was constructed in 1999 at an Idaho Falls industrial facility. It consisted of a full-scale horizontal slice of the bottom of a tank only a few feet in height, but 50 feet in diameter, with cooling coil structures and simulated residual solids. Grout mixtures were tested to validate assumptions of flowability in both the tank and the vault areas and the ability of the grout to move the in-tank solids toward the steam jet (INEEL, 1999).

The second mockup facility was constructed and used during 2000-2001 at another Idaho Falls industrial facility. This mockup was used to test the ability of the spray wash system to clean solids from the tanks. A full-height tank mockup was constructed, with only half the circumference. Simulated solids were placed in the tank, and the spray system components were operated to validate assumptions of their cleaning ability. In addition, assumptions regarding the ability of the steam jets were tested to ensure that heavy solids loadings could be removed from the tanks without plugging the steam jets (INEEL, 2001).

The benefits of test beds have long been recognized by DOE staff and contractors. Often cited among waste retrieval lessons learned is the importance of a cold test facility to test all equipment before deployment (Caldwell, 2005a; Dodd, 2005; Lockie, 2005a). The following is a quote from a Pacific Northwest National Laboratory review of lessons learned in tank waste remediation at Oak Ridge National Laboratory (Bamberger et al., 2001; p. 7.1):

> First and foremost, cold testing is extremely beneficial prior to any first-time field deployment. Not only does this initial testing in a clean environment allow any significant design flaws to be identified and reworked before contamination controls become a significant issue, but it also provides valuable training for the operators and craft personnel by providing them with an opportunity to become familiar with the equipment from the inside out. In addition, integrated cold testing allows development of procedures that reflect how operations are actually conducted and allows multiple operators to receive training under low pressure conditions rather than "on the front lines." Finally, cold testing can provide important opportunities to demonstrate readiness as part of a phased readiness review process.

[7]Laboratory-scale tests and tests with simulants did not reveal the difficulties that emerged when DOE used the in-tank precipitation process to remove cesium from waste in Tank 48. That process, however, is rather different in nature than the type of waste removal and tank cleanup technologies that the committee describes here.

The committee recommends that one or more radioactive test beds for retrieval technologies that can be adapted to simulate a variety of tank situations (e.g., recalcitrant heels, cooling coils, debris) be maintained and made available for other DOE sites.

LEVEL OF RESEARCH AND DEVELOPMENT INVESTMENT

Savannah River Site staff is collaborating with DOE's Office of Cleanup Technologies to develop technologies for the following:

- Chemical cleaning methods for heel removal
- The impact of oxalic and other acids on carbon steel tanks during heel removal
- Selective dissolution of targeted radionuclides during heel removal
- Mechanical methods for heel removal that minimize water usage
- Removal of fast-settling solids (e.g., zeolites) and hardened sludge
- Efficient removal of salt heels
- Criticality control during heel removal
- Characterizing and mapping residual material after heel removal
- Cleaning methods for tank annular space
- Methods for verifying waste removal from tank annular space

Another example of a technology development initiative is the tank cleaning technology exchange meeting that is scheduled for the first quarter of 2006. The purpose of this meeting is to share technology development and deployment efforts among DOE sites and vendors to identify equipment and systems applicable to tank closure that have been successful within DOE and in commercial applications. New technologies that do not require as much high-level waste tank space are also being pursued (Ross, 2006b).

The committee recommends a centralized, focused, 10-year bench-scale research and development program with a budget amounting to a minimum of $10 million per year and, more desirably, $50 million per year to support DOE's tank cleanup program. This is based on experience with the Environmental Management Science Program (EMSP; prior to its transfer to DOE-Office of Science), the Office of Science and Technology Tanks Focus Area (TFA), and Robotics Crosscutting Technology Development Program, and similar programs such as, Defense Advanced Research Program Agency (DARPA) programs. The magnitude of these investments is justified as follows. The initial EMSP budget was approximately $50 million per year at the beginning in 1996 and decreased to about $30 million in 2003 when the program was moved to the DOE-Office of Science. The average size of EMSP grants is about $500,000 for three years.[8] In 2000 there were 306 research projects, of which 76 were categorized as high-level waste problem areas (i.e., relevant to the tanks), which represented at yearly budget of $38 million. The Tanks Focus Area's yearly budgets[9] before it was disbanded were $28.5 million (FY 1997), $30.1 million (FY 1998), $29.1 million (FY 1999), $47.9 million (FY 2000), $41.5 million (FY 2001), and $43.125 million (FY 2002).

In addition to bench-scale research effort, new technologies may require studies using real waste (hot tests) and pilot-scale studies using test beds (e.g., tank mockups equipped with coils and surrogate heels) to test their effectiveness before full-scale deployment. Hot and pilot-scale studies will require additional funds, realistically $50 million per year and possibly more.

A budget of $50 million per year would correspond to approximately 3 percent of the FY 2007 DOE-Environmental Management (EM) budget request for tank waste storage, retrieval, treatment, immobilization, and disposal at the three sites (roughly $1.6 billion) (OMB, 2006).[10] A higher research and development investment as a percentage of the program's budget would bring the program more in line with research and development investment in industry and other federal agencies.[11] A $50 million per year budget for 10 years ($500 million total) also represents one percent or less of DOE's estimated cost for the tank cleanup

---

[8]Projects and budgets are available on the EMSP web site at: *http://emsp.em.doe.gov*.

[9]These numbers do not include the management costs to support the Tanks Focus Area activities.

[10]The President's 2007 proposed budget requests $104.5 million for radioactive liquid tank waste stabilization and disposition at Idaho National Laboratory, $571 million for tank waste stabilization and disposition at the Savannah River Site, and $964 million for Hanford's Office of River Protection (which is responsible for the storage, retrieval, treatment, immobilization, and disposal of tank waste). The proposed budget for research on high-level waste treatment and storage is zero dollars (termination of all research), and the proposed budget for technology development in DOE-Environmental Management (for which tank wastes are listed as the top priority) is approximately $21 million (OMB, 2006).

[11]A 2001 report by the National Research Council found that research and development efforts in other federal agencies are 9 percent for EPA, 15 percent for DOD, and 40 percent total for DOE, but only 4 percent for Environmental Management (NRC, 2001d). Although the research and development budgets in private industry are not directly comparable to research and development in the federal government, the research and development intensity (research and development funding as a percentage of net sales) "is highest [12-13.7 percent] in knowledge-intensive industries such as software and pharmaceuticals, whereas research intensity is lowest [0.66-0.73 percent] for such mature industries as petroleum and construction" (NRC, 2001d). The report goes on to conclude that DOE's environmental quality mission is fairly knowledge intensive. It is beyond the scope of this present study to update the statistics from six years ago, but these numbers do give a sense of the scale of research and development investment that other organizations deem appropriate. DOE's tank waste mission is arguably not as knowledge intensive as the software industry, but the tank waste program needs research and development more to fulfill its mission than do the mature industries cited.

program ($50 to $75 billion[12]). The research and development program would fund applied research and engineering development projects in the high-priority areas identified in this chapter and those identified through an assessment of the engineering vulnerabilities in DOE's tank remediation program (i.e., cases in which performance poorer than planned causes significant problems for the program). Projects should be selected using a competitive process similar to that used for the EMSP and would be in addition to or in place of any directed (noncompetitive) awards. The tank remediation program at the sites should be responsible for supporting field testing and deployment of technologies resulting from this program. However, program management should be centralized to ensure prioritization and coordination of the projects.

The Tanks Focus Area showed the effectiveness of centralizing research and development activities at DOE tank sites. Despite variations in site and tank characteristics, many technological issues were common among two or more sites. The research and development needs outlined in this chapter and elsewhere in the report are common to all three sites so it seems more cost-effective to develop common solutions for the sites, where appropriate. According to the last Tanks Focus Area report, the application of technical solutions from the program accounted for more than $250 million in cost savings (or avoidance) with a projected life-cycle savings of more than $5 billion (DOE-TFA, 2001).

The Tanks Focus Area was also an effective platform to centralize and coordinate results from other components of DOE's Office of Science and Technology, such as the Environmental Management Science Program; DOE's Environmental Management Accelerated Site Technology Deployment Program; crosscutting programs (e.g., efficient separations and processing, robotics, characterization monitoring, sensor technology); and DOE-Environmental Management's cooperation with industrial partners, universities, and national laboratories. Between the program's inception in 1996 and its disbandment in 2002, the Tanks Focus Area studied more than 200 technology applications that resulted in approximately 160 deployments, including the wash ball used to clean the interior of the tanks at the Idaho National Laboratory, the caustic-side solvent extraction proposed for the Salt Waste Processing Facility at the Savannah River Site, and the Light-Duty Utility Arm and the Houdini in-tank vehicle used for waste retrieval at the Idaho National Laboratory. Numerous other technologies were demonstrated at DOE sites and showed promise for implementation, including saltcake dissolution, residual waste mapping, and sampling technologies (DOE-TFA, 2001).

[12]In 2003, DOE estimated that by implementing its plan for accelerated cleanup, the department could reduce the projected $105 billion cost and 70-year time frame for cleanup of tank wastes at the Savannah River Site, Hanford, and Idaho National Laboratory to $76 billion and 35 to 50 years. The General Accounting Office (now the Government Accountability Office) found opportunities for further cost and schedule savings and errors in DOE's estimates on the order of billions of dollars (GAO, 2003). Yet even with these corrections and allowing for uncertainties, the tank waste cleanup program is expected to cost at least $70 billion over several decades.

FINDINGS AND RECOMMENDATIONS

Finding IX-1: Based on experience with the Environmental Management Science Program (prior to its transfer to DOE-Office of Science), the Office of Science and Technology, Tanks Focus Area, Robotics Crosscutting Technology Development Program, and similar programs such as the Defense Advanced Research Program Agency programs, a centralized, focused, 10-year bench-scale research and development program with a budget amounting to a minimum of $10 million per year, and more desirably, $50 million per year seems reasonable, if pilot-scale studies are included.

Recommendation IX-1: DOE should fund applied research and engineering development projects in the high-priority areas identified in this chapter and those identified by the vulnerability analysis. Projects should be selected using a competitive process similar to that used for the DOE Environmental Management Science Program. These projects and their funding would be in addition to or in place of directed (noncompetitive) awards, if any. The tank remediation program at the sites should be responsible for supporting field testing and deployment of technologies resulting from this program. However, the program management should be centralized to ensure prioritization and coordination of the projects.

Finding IX-2: Oxalic acid has proven to be an effective chemical cleaning technology in Tank 16 at the Savannah River Site and, to a certain extent, in Tank C-106 at Hanford. The committee believes that oxalic acid can be a helpful cleaning tool in tanks with significant obstructions. There are signs of interest within DOE or its contractors at the Savannah River Site, but currently there is no active research and development on chemical cleaning.

Recommendation IX-2: DOE should fund research and development partnerships among universities, national laboratories, and industry focused on options for chemical cleaning of tanks to find alternative cleaning agents or to mitigate the criticality and downstream processing problems that Savannah River Site staff pointed out to the committee.

Finding IX-3: Untethered semiautonomous robotic devices are not likely to add value to the retrieval process in tanks, given the current state of the technology. Proven technologies, such as an articulated arm with a water jet and remotely controlled vehicles with pusher blades, provide for retrieval tool deployment and visual feedback to human operators and represent the limit of useful retrieval technology at this time. Robotic enhancements to proven human-controlled techniques are likely to yield more effective performance than untethered, semiautonomous robotic devices.

Recommendation IX-3: During tank cleanup and closure operations, DOE should continue the investigation and development of more effective technologies for waste retrieval from tanks with emphasis on tanks with obstructions and recalcitrant waste. In particular, DOE should continue to use, adapt, and improve the most effective suite of available waste retrieval technologies on a tank-by-tank basis, including consideration of technologies at other DOE sites, in industry, and internationally. Additionally, if retrieval beyond the capabilities of existing physical methods is required, DOE should

- Reevaluate the use of oxalic acid to determine whether criticality and downstream problems are a real concern;
- If they are a real concern, DOE should investigate methods to mitigate these shortcomings or develop alternative acceptable chemical cleaning approaches; and
- If chemical cleaning proves to be inapplicable to some types of recalcitrant deposits (such as zeolites), DOE should develop alternative mechanical methods, such as delivery tools and techniques for waste mobilization that enhance waste accessibility to water jets.

Finding IX-4: In the near term, decisions about the formulation of grouts for tank fill and immobilizing low-activity waste are being made on the basis of experience in very different applications and in some cases also on data from short-term tests on saltstone. The committee has not seen any reports of long-term testing or of more fundamental research directed at the unique aspects of DOE applications, particularly the binding capacity of grouts and the changes in various properties over the extended times contemplated by the DOE.

Recommendation IX-4: DOE should initiate a focused research and development program over a 5- to 10-year period, and longer where necessary, to improve the fundamental understanding of the long-term performance of cementitious material and to tailor different formulations of grout to different tanks or groups of tanks, and different low-activity waste compositions. The program should involve collaboration among government laboratories, universities, and industry.

Finding IX-5: In addition to the research and development areas identified in Findings 1-3, future research and development needs may become apparent as DOE progresses in its tank cleanup program.

Recommendation IX-5: DOE should support an independent assessment of its tank remediation program for the purpose of comprehensively identifying and prioritizing any additional technical vulnerabilities in the program as a basis for funding additional research and development. The vulnerability assessment should be independent of, but rely heavily on, information obtained from the tank remediation programs at the sites.

Finding IX-6: The benefits of test beds have long been recognized by DOE staff and contractors at sites where waste tanks need remediation. Often cited among waste retrieval lessons learned is the importance of a cold test facility to test all equipment before deployment.

Recommendation IX-6: DOE should maintain one or more radioactive test beds for retrieval technologies that can be adapted to simulate a variety of tank situations (e.g., recalcitrant heels, cooling coils, debris) and make them available for use by other DOE sites.

X

Illustrative Example of the Recommended Decision-Making Process

The technical complexities of the tanks and their interactions with the geosphere are included where possible in the Department of Energy's (DOE's) performance assessments and in the models they use. Therefore, the performance assessment and sensitivity analyses that are performed using different scenarios can indicate the impact of various decisions on projected risks. Indeed, DOE's draft Section 3116 waste determinations for the closure of Tanks 18 and 19 (see DOE-SRS, 2005a, Figure 7-14, p. 137) and for the Idaho National Laboratory tank farms show how these two sites used performance assessments to make tank closure decisions. However, as discussed in Chapter VIII, the trade-offs between risks and benefits were not always transparent.

EXAMPLE OF HOW TO STRUCTURE DECISIONS ABOUT WHETHER A TANK IS SUFFICIENTLY CLEAN TO GROUT AND CLOSE

The following is an example of a risk-informed, transparent decision-making framework for tank closure. This example illustrates how a single, consistent approach to thinking about both waste removal to the maximum extent practical and tank closure can lead to very different choices from site to site and even from tank to tank within a single site. It also shows that the choice of the point and time of compliance impacts the decision to close a tank or to carry out additional waste removal.

However, the following are important caveats about this example. The example uses grossly simplified (indeed hypothetical) assumptions about waste retrieval methods, scenarios, and groundwater flow patterns to focus on an illustration of the decision-making framework and to avoid distracting the reader with the assumptions used or the numbers obtained.

Also, the example does not account for the increase in risks to workers associated with decisions to clean tanks further. The trade-off between increasing health risks to workers living today in favor of decreasing these risks to residents and intruders living 10,000 years from now is a major consideration in tank closure decision making that could not be dealt with in this example.

The example is intentionally simple so that readers can gain an appreciation of the value of a consistently structured approach to thinking about tank closure decisions and gain better insight into the merits of some of the committee's recommendations regarding tank closure decisions. Application to a real tank would require far more technical detail and consideration of more decision-relevant factors, but it would not have to differ in any fundamental way from the framework presented in this chapter.

The Illustrative Tank and Its Hypothetical Risk Profile

Risks associated with future release of radionuclides in the tank heels vary over time and space, which is called the "risk profile" to distinguish it from the narrower concept of risks at a single "point of compliance," as required for establishing whether performance objectives have been met.[1] Figure X-1 summarizes the illustrative example's risk profile for grouting and closing a tank that already has some soil contamination around it, which will start to be cleaned up as soon as the tank has been closed. To economize on data presented in what is intended as a simple illustrative example, the committee summarizes this profile by showing how the dose level would change over time at three specific locations around the tank (identified in Figure X-1 as shaded circles) since any plume of radionuclides released from the tank after closure would move through the surrounding vadose zone

[1]The term "risk" is used here to describe the doses at a particular point in time and space, under the standard assumptions of exposure of individuals at that location. This is not the usual concept of risk, which would account for the probability of such exposure occurring and also for the health effects of such exposure. In this example, the committee also assumes for simplicity of exposition that the performance objective is that the risk may not exceed 1.0 at the point of compliance.

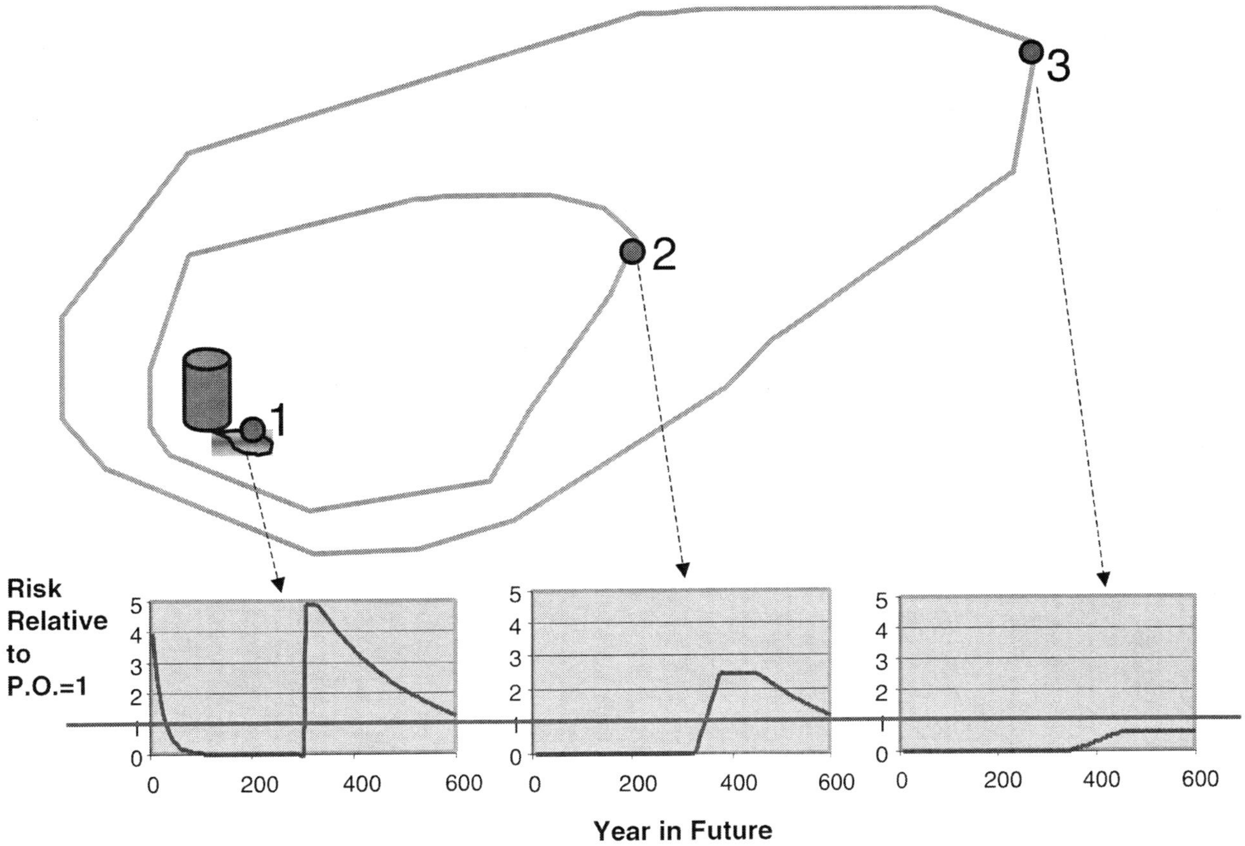

FIGURE X-1 Schematic of a tank site and locations where risk profile is summarized.
NOTE: P.O. = performance objective.

and groundwater. The three graphs at the bottom show the assumed projected risk levels at each of the three locations. (The gray concentric lines encircling the tank represent loci of approximately equal risk levels as estimated at each of the three illustrative locations.)

Location 3 is intended in this example to be the designated "point of compliance." The associated figure for Location 3 shows that radionuclides released from the closed tank would not reach that location until almost 400 years from now. The simplified example assumes that no other plume of contamination will reach the point of compliance before then. At its maximum (about 500 years from now), the risk from that contamination would not exceed 1.0, consistent with this example's starting assumption, which is that if the tank in question could be closed with its residual heels as-is, it would meet the legally required performance objectives. Risks are not below the level of the performance objectives at all locations, however. Closer to the tank, the risks are larger and occur earlier in time. Location 1 reflects the temporal pattern of risks in the immediate vicinity of the tank, and Location 2 is a point about half-way between Location 1 and the point of compliance, Location 3.

The hypothetical and grossly simplified spatial-temporal risk pattern in Figure X-1 should not be used to draw conclusions about the risks at the three receptor locations at any specific site. The example is hinged on a simplified situation where transport through the vadose zone flow is assumed to be approximately vertical and flow in the saturated zone is assumed to be approximately horizontal. Although these assumptions are not inconsistent with groundwater flow at the Savannah River Site, they are not representative of the conditions at Idaho National Laboratory or the Hanford Site, where horizontal transport in the vadose zone can mean that the highest dose from exposure to groundwater does not occur at the point closest to the tanks.[2]

Figure X-1 illustrates the committee's statement that an entire profile of risks is relevant to decisions and not just the legally required meeting of performance objectives at a point of compliance. It also illustrates the committee's statements

[2]At the Idaho National Laboratory, the point of calculation was moved farther away from the facility boundary because estimated radionuclide concentrations are higher several hundred meters away from this boundary.

about the significance of the policy decision regarding what location will serve as the point of compliance in estimating whether a closure plan can meet performance objectives. If Location 2 were designated the point of compliance, the tank could not legally be closed under Section 3116 without further removal of residual radionuclides. The committee is not suggesting where the point of compliance should be placed because this is a policy decision made by the state and DOE. However, the committee believes that the full spatial and temporal variations of risks are not explicitly indicated in the performance assessment for tank closure.

Figure X-1 also shows some hypothetical risks associated with the present soil contamination near this tank. It is reflected as high current risks at Location 1. If this soil contamination is remediated once the tank is closed, it never appears as far from the tank as Locations 2 or 3. Although this source of contamination is not related to the tank heels, it is not necessarily included in a performance assessment for tank closure. However, decisions that might affect the timing of tank closure might be affected by the interaction of this timing with the ability to rapidly reduce near and present contamination.[3] For this reason, existing soil contamination in the vicinity of the tanks is included in this decision example.

Alternatives to Immediate Tank Closure with As-Is Residual Heel Quantities

In this example, once the bulk of the waste has been removed and the heel reduced to the limit of a given technology, the options are either to grout and close the tank or to proceed with further heel removal with alternative technologies. Two types of alternative technologies are chosen in this example based on considerations in the waste retrieval chapter (Chapter III):

1. Further waste removal using an established technology, such as chemical cleaning with oxalic acid; or
2. Further waste removal using advanced technologies that are still in the development stages but might become available within the next 10 years and have the ability to substantially reduce the heel in the tank before it is actually closed (see Chapter IX for examples).

Both options involve trade-offs. Additional cost is one consideration, but this example relies solely on the offsetting risks that each alternative may create while offering a potential to further reduce the risk from heels that remain in a tank. The assumption is that the tank is still connected to the rest of the tank farm so additional removal does not involve reconnecting it or removing abandoned equipment from inside a tank that was prepared for final closure.

The example arbitrarily introduces one risk from using oxalic acid:[4]

> If the tank to be washed has a crack that is not an active leak site because it is plugged by the solid sludge matrix, the plug might be cleared during the washing process and some of the mobilized radionuclides may be released through the crack before the wash liquids can be pumped out of the tank.

The example also takes into account two risks of waiting for an emerging new waste removal technology:[5]

1. The new technology may never work as expected, making the wait fruitless.
2. The sludges in the ungrouted tank may become mobilized in the intervening years. The sludges are in a solid form at present and therefore are unlikely to "leak" from the tank under current conditions. However, if water (e.g., from rain or flood) were to enter the tank despite current precautions to prevent its ingress, mingle with the residuals, mobilize some of the radionuclides, and then exit again before evaporating (e.g., through a crack in the tank), some of the residuals could enter the environment before they were stabilized. Another risk scenario might be a catastrophic tank failure, due to subsidence or possibly an earthquake.

Therefore, once the bulk of the waste has been removed and the heel reduced to the limit of a given technology, DOE faces three basic choices:

1. Grout the tank immediately, with the residual heels at their current levels.
2. Delay closure for about two years to perform an oxalic acid wash to reduce the heels to lower levels.
3. Delay closure for about 10 years in the hopes that a specific emerging cleanup technology will allow heels to be reduced to lower levels during that interval.

Each of these three options produces a different profile of "risks" over space and time around the tank site. Figure X-2 shows a decision tree that summarizes the trade-offs in this decision. To keep the example simple, the tree includes only the risk of sludge leaking from the tank during the wash process for oxalic acid washing. Under Option A, there is a

[3]For example, staff at Idaho National Laboratory indicated that they feel urgency to close their emptied high-level waste tanks because they cannot start to clean up very high dose rate soil contamination from earlier transfer pipe leaks until the tanks are closed.

[4]Other potential or perceived drawbacks of using oxalic acid, such as the criticality concerns and downstream problems described in Chapter III, are not considered in this example.

[5]Once a specific technology has been identified as the one worth waiting for, a more specific set of risks can be added to the list that follows and included in the risk analysis that supports this decision.

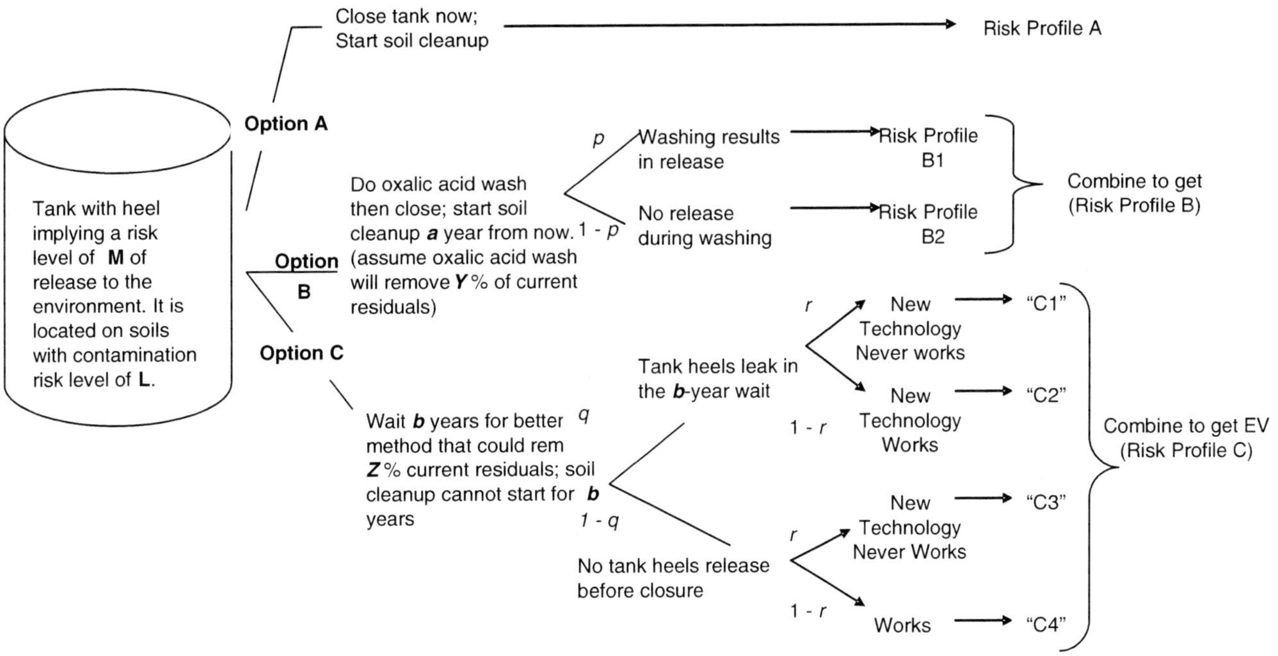

FIGURE X-2 Illustrative tank decision tree structure. Decisions are made by comparing risk profiles EV(A), EV(B), EV(C), which combine risks now in soils (near-term) with risks possible if tanks leak (long term).
NOTE: EV = expected value.

particular risk profile that is consistent with the results of the performance assessment that finds the performance objectives will be met at the point of compliance at a time when exposure would be at a maximum. This was illustrated in Figure X-1 for a particular tank.

Under Option B, the risk profile (including risk at the point of compliance) will be changed in one of two ways. If there is no leakage as a result of the acid washing, the main effect will be that risks at each point in time and space will be reduced from those under Option A by the amount by which the washing step reduces the overall radionuclide content that will be grouted in the tank. With some probability p, however, there would be a release from the tank at the time of the washing, and this will increase potential exposures at each location much earlier than in the case of immediate grouting. These two risk profiles can be combined probabilistically to obtain a single expected risk profile for Option B by weighting each profile by its probability and then adding them together.

Option C in this illustrative analysis can be characterized by four different profiles. The four outcomes reflect the combinations of the tank leaking or not leaking before it can be grouted, and whether the technology becomes available or not. In the event that the tank does not leak before grouting at year 10, but the technology never becomes available during that time, the risk profile is identical to that of Option A. However, it now has a probability less than one, equal to $(1 - q) \times r$. This profile is then probabilistically combined with the three other profiles to obtain a single expected risk profile for Option C. To the extent that there is a high risk of leakage in the 10 years of waiting, the expected risk profile for Option C will have higher risks than Option A in the early decades. Additionally, the greater the probability that the technology will never become available to reduce the quantity of heels in the tank, the less is the expected benefit from Option C in terms of reducing risks relative to those estimated for Option A in later years when risk would be at its maximum at the point of compliance.

For a strictly risk-based comparison, these three options create some trade-offs between higher risks in the near term, and potentially lower risks in the long term, which is the time that is the focus of the performance objectives. These trade-offs can be assessed only by considering the full temporal and spatial profile of risks, because any comparison based merely on the long-term risk at the point of compliance would always suggest that an alternative to Option A would be "better" solely on a risk basis. Of course a more complete risk-benefit analysis would consider cost, worker exposures, and other important factors for decision making.

There is another risk-based consideration that adds further richness to the comparison of the alternatives. At many of the tank sites, there is soil contamination already present around the tank. In some cases, the existing contamination implies very high potential doses now, if any individuals were to be exposed to it. Cleanup of these soils cannot proceed safely until the tanks in the vicinity have been stabi-

TABLE X-1 Assumptions for Options B and C in Figure X-2

Variable in Figure X-2	Full Interpretation of Variable	Numerical Assumption for Example
Variables Related to Oxalic Acid Wash (Option B)		
P	Probability that the oxalic acid wash will cause some release of mobilized radionuclides due to existing crack in tank underneath sludge cake	.01
Y	Percentage of residual sludge now in tank that will be removed as a result of the oxalic acid washing	75%
A	Delay before tank will be closed due to oxalic acid wash	2 years
Variables Related to Waiting for Emerging Technology (Option C)		
Q	Probability that radionuclides will be released from ungrouted and unstabilized tank during 10-year wait for new technology	.05
R	Probability that the emerging technology in question fails to become a working method of cleaning within the 10-year wait	.50
Z	Percentage of residual sludge now in tank that will be removed if the new technology does become workable	90%
B	Delay before tank will be closed due to waiting for the new technology	10 years

lized and closed. Thus, any delay in tank closure to limit projected risks in the long term can cause delays in the reduction of large and known risks in the present. This trade-off has also been incorporated into the illustrative example that follows.

Illustration of Impact of Alternative Tank Options on Risk Profiles

To illustrate how the decision tree in Figure X-2 can actually be used to provide a risk-based evaluation of the three alternatives, specific numerical assumptions have to be introduced in addition to those that generated the initial risk profile shown in Figure X-1. Table X-1 summarizes the key assumptions needed to assess the risk profiles for Options B and C, given a starting point that is the risk profile under Option A (i.e., close the tank with no further reduction of the residual radionuclides presently in it). The numerical assumptions shown in Table X-1 are intended for a single specific tank, with its specific forms of residuals and specific new technology needs. As will be shown later, the numerical assumptions will vary from tank to tank, as will the Option A initial risk profile. These tank-specific factors affect the relative desirability of each of the alternative options.

Table X-1 lists pessimistic assumptions about the efficacy of the oxalic acid wash, namely that the process would remove only about 75 percent of the remaining radioactive material. The probability that washing would create a leak is assumed to be fairly small, .01 (1 percent).[6] Table X-1 also gives a fairly pessimistic assumption about the likelihood that the new technology will ever function, with only a 50-50 chance that it will become a viable option. It would, however, allow 90 percent heel removal.

Figure X-3 shows the expected risk profiles for Options A, B, and C when using the tank example in Figure X-1 and the assumptions in Table X-1 about the alternative options. On the basis of a pure risk-risk comparison, it would appear that the oxalic acid wash (Option B) would provide the best overall outcome, even after considering that this would delay cleanup of the substantial existing soil contamination. Since the performance objectives are met at Location 3 (the designated point of compliance) even under Option A, a comparison of risks based solely on the ability to meet performance objectives does not demonstrate a convincing case to stakeholders for the extra residual waste reduction with oxalic acid washing unless the risk-benefits trade-offs required by ALARA (as low as reasonably achievable) are laid out explicitly. However, Figure X-3 shows how much overall benefit to the risk profile may be obtained from Option B. The performance objectives could be met at a wider range of locations and even are almost met right near the tank itself (i.e., at Location 1). This would not be true under the expected outcomes for Options A or C for this tank situation.

Additionally, the two-year delay in reducing the current contamination affecting Location 1 has relatively little impact on the overall timing and level of risks. The near-term risks at Location 1 would have to be deemed a very significant hazard for actual individual exposures—and thus given very high weights relative to the longer-term, more hypothetical risks for the delay in cleaning up present site contamination—to affect the trade-offs in favor of waiting for an oxalic acid wash to occur.[7]

[6]This assumes that the quantity of residuals and their constituent elements are such that there is no concern with criticality in this tank; otherwise the oxalic acid wash would not be an option for the tank.

[7]Alternatively, one would have to believe that the two-year delay in initiating remediation of the existing soil contamination would create risks that contamination would never be cleaned up at all and would cause a plume of risk to appear at Locations 2 and/or 3 in the next several decades. Given that Options B and C imply relatively brief delays in cleanup relative to the life span of these DOE site cleanup programs, the committee considers this too unlikely a concern to be incorporated into the risk profiles of locations distant from the tank.

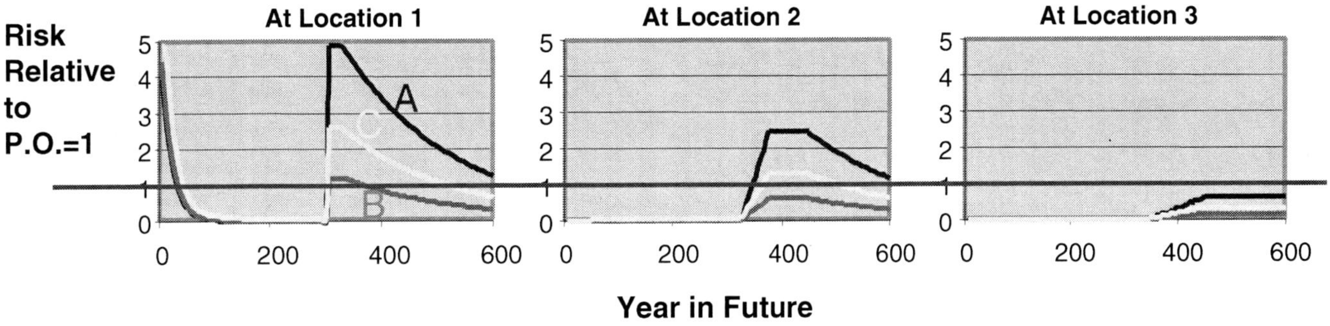

FIGURE X-3 Expected risk profiles for illustrative tank under Options A, B, and C. NOTE: P.O. = performance objective.

This illustrative example of a decision has been based strictly on risk considerations, although the concept of risk has been extended beyond the single point-in-space-and-time analysis that is currently the modus operandi for tank waste determinations. It has also been extended to incorporate risks from contamination other than that associated solely with future releases from the heels left in the tank. Just these extensions have added insights to the case for considering alternatives that might further reduce the heels. It is important to remind ourselves that there are other important considerations in choosing among the options, including costs of alternative options, worker risks, and so forth. These would also be important to incorporate into the decision, and doing so could lead to a different choice than has been suggested might make sense on a purely risk-based comparison. However, it is the committee's view that this comparison of options would be greatly enhanced for decision makers and the public if it starts from a rigorous and structured decision tree such as has been highlighted here.

Different Decisions Will Be Warranted for Different Tanks

The example in the preceding section has shown how a structured analysis of risks in time and space can be used to assess alternatives to the immediate closure of a tank that may contain some amount of residual heels. This was a single hypothetical tank situation, and there is nothing in the way of a conclusion from it that can be applied to all tanks generally. Site-specific and tank-specific conditions can substantially alter the comparison among options, even within the risk-only framework illustrated thus far. This section, demonstrates how tank-specific conditions can alter the appearance of the best option by extending the initial example to a wider range of tank situations.

Table X-2 shows a set of assumptions for three different tanks. All are assumed to have the same general geological conditions (e.g., all exist in the same tank farm), so we do not alter the assumptions for these tanks are not altered regarding the timing and spatial patterns of risk once radionuclides are released from each tank. However, in the real world, these conditions also will vary, further adding to the ways in which decisions may differ from tank to tank.

Tank X in Table X-2 is the same as the illustrative tank already used in the preceding section. Tank Y differs from Tank X in two ways. First, Tank Y has far less contamination of the surrounding soils, although it has a comparable degree of risk with the heel remaining in the tank. Second, the materials in the Tank Y heel are not expected to be as easily removed via oxalic acid washing, because some of those residuals are in zeolite form. Tank Z has very little residual but is in a location that has intense existing surrounding soil contamination. Also, Tank Z's small remaining heel will not likely benefit from oxalic acid wash (because, for example, the materials may have never been in sludge form). Additionally, it seems unlikely any new mechanical or chemical cleaning method will be able to further reduce the small amounts that remain in the tank, simply because so little physical material remains to be removed. Of the three illustrative tank examples, Tank Z might be viewed as relatively more like the situation with Idaho National Laboratory's liquid high-level waste or sodium-bearing waste tanks.

Figure X-4 shows the relative risk profiles for all three options at each of the three hypothetical tanks. It can be seen that the best choice, based solely on comparisons of risk profiles, is likely to vary. Whereas oxalic acid wash (Option B) appears to be a preferred choice for the original example Tank X, waiting for the new technology becomes preferred for Tank Y, and immediate closure seems to be a reasonable choice for Tank Z. In the case of Tank Z, the long-term risks are very low to start with and hardly vary for Options B or C. At the same time, there is substantial near-term risk at Location 1, and the delay in reducing this near-term risk under Options B and C presents the only visible way in which risks differ from option to option.[8]

[8]There are real differences in the risk profiles for Options A, B, and C at Tank Z, but they require that a smaller scale be used on the graph. However, all of the graphs in Figure X-4 purposely were drawn with the same vertical scales, to allow the differences among the three tanks to be more apparent.

TABLE X-2 Assumptions for Options B and C in Three Different Tanks

Variable in Figure X-4	Full Interpretation of Variable	Tank X	Tank Y	Tank Z
Conditions of Tank Heels and Surrounding Soil Contamination				
L	Risk levels at Location 1 due to existing soil contamination (1.0 = performance objective risk level)	5	0.5	5
M	Maximum risk levels that will occur at Location 1 if existing heels are released to environment (1.0 = performance objective risk level)	5	5	1
Variables Related to Oxalic Acid Wash (Option B)				
P	Probability that the oxalic acid wash will cause some release of mobilized radionuclides due to existing crack in tank underneath sludge cake	.01	.01	.01
Y	Percentage of residual sludge now in tank that will be removed as a result of the oxalic acid washing	75%	30%	10%
a	Delay before tank will be closed due to oxalic acid wash	2 years	2 years	2 years
Variables Related to Waiting for Emerging Technology (Option C)				
q	Probability that radionuclides will be released from ungrouted and unstabilized tank during 10-year wait for new technology	.05	.05	.05
r	Probability that the emerging technology in question fails to become a working method of cleaning within the 10-year wait	.50	.50	.90
Z	Percentage of residual sludge now in tank that will be removed if the new technology does become workable	90%	90%	10%
b	Delay before tank will be closed due to waiting for the new technology	10 years	10 years	10 years

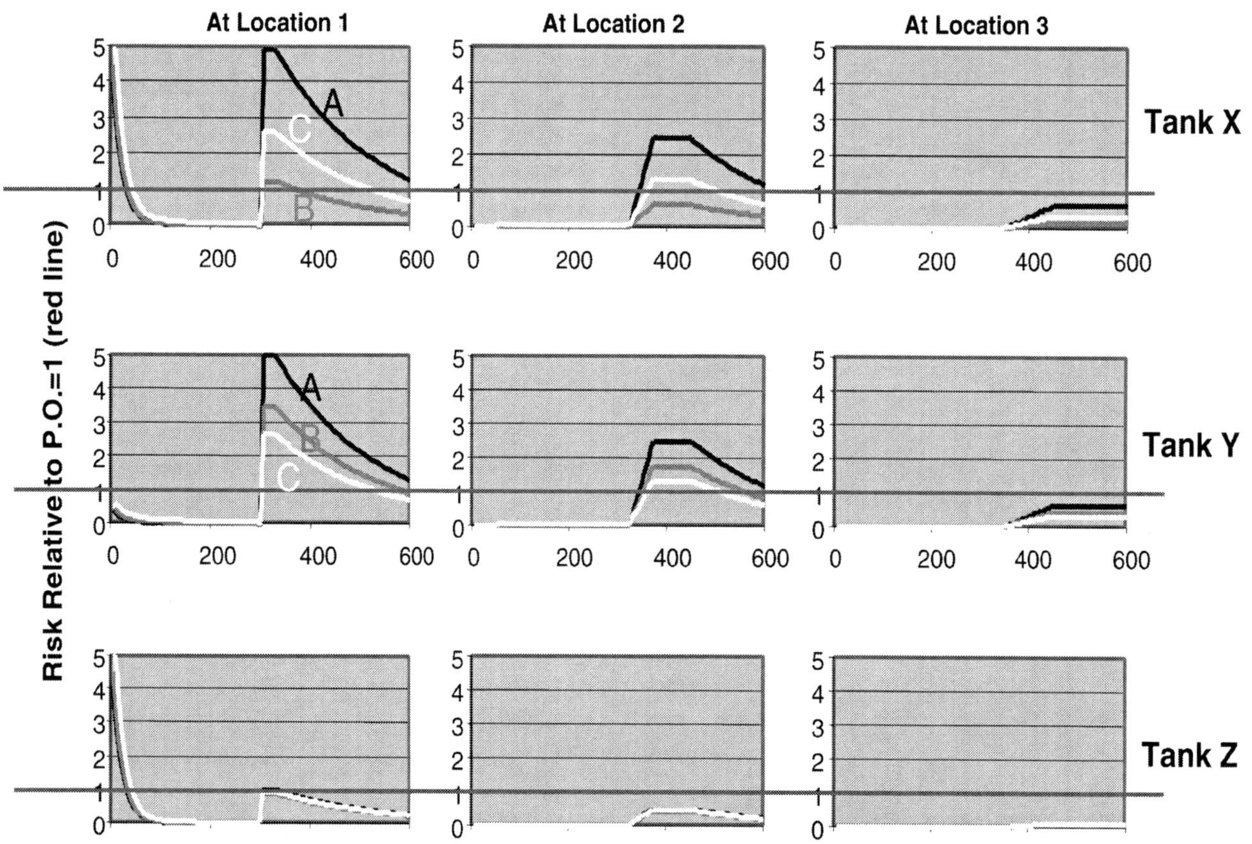

FIGURE X-4 Risk profiles for Options A, B, and C for three hypothetical tanks facing different situations. NOTE: P.O. = performance objective.

These comparative risk figures also bring to mind some issues related to the need for reliance on long-term institutional controls and the ways in which tank management decisions can lessen that concern. The situation for Tank Z is one in which risks at many locations on the site from the long-term potential release of radioactivity in tank heels are small. Long-term institutional controls are therefore of lesser concern than strong action in the near term to reduce present risks. Tanks X and Y, on the other hand, impose substantial risks throughout the site over the very long term, even though performance objectives are "met." Without additional tank cleanout, there may be unacceptable levels of risk at some locations unless institutional controls are assumed to be effective for hundreds of years. In the case of Tank Y, near-term risks are small relative to the long-term risk from the tank heels, so it may make sense to delay tank closure long enough to obtain a reasonably large reduction in the area that would have long-term contamination above levels deemed acceptable at assumed points of compliance. Tank X faces a more complex trade-off between near-term needs for remediation and long-term risks that could be quite high in the absence of institutional controls. In this example, however, the slight delay to perform an oxalic wash (combined with the fact that the wash could be very effective for this particular tank) may be worth the substantial reduction in need for long-term institutional control to ensure that future exposures do not exceed limits.

Again, the additional decision-relevant considerations of cost, worker exposure, et cetera, might further modify these choices and should not be ignored in a full analysis for actual tank decisions. Additionally, decisions to weight near-term or long-term risks more heavily, or to weight the risks at one location more heavily than another, could affect the choices. Nevertheless, the point remains that different choices may be reasonable for different tanks and that the decision to close a tank immediately may not be the best option even when a tank is projected to be able to meet its performance objectives. (Note that all three of these hypothetical tanks would meet the performance objectives with Option A—immediate closure—if the point of compliance is Location 3.) Table X-3 provides an example of an approach that DOE could adopt to decide when to stop waste retrieval in a specific tank. The table requires that DOE quantify the various inputs to the risk-benefit analysis described in the table, such as public risk, worker risk, costs, compliance with Federal Facility Agreements, effect on nuclear power receptivity, and how these risks vary with time (e.g., in 30, 300, and 1,000 years). The table takes into account whether additional time has been spent to achieve additional waste reduction (second column). The third column entries would be the assumed rate of monetary inflation, and the fourth column represents the projected change in level of public trust in DOE. The fifth column lists times at which the inputs are measured. Each parameter is attributed a weighing factor according to DOE's

TABLE X-3 Determining How Much Waste to Leave in a Tank

ATTRIBUTE	START DELAY, YEARS		INFLATION RATES	PUBLIC PERCEPTION CHANGES YEARS,		TIME, YEARS			WEIGHTING				TOTAL	OTHER SITE RISKS
	5	10		30	300	30	300	1,000	PUBLIC RISK	WORKER RISK	COSTS	FFA COMPLIANCE		
PUBLIC RISK														
WORKER RISK														
COSTS														
FFA COMPLIANCE														
EFFECT ON NUCLEAR POWER RECEPTIVITY														

NOTE: FFA = Federal Facility Agreement.

priorities. In the next-to-last column there is a number that captures all of these risks and benefits for each amount of waste proposed to be left in the tanks. The table also takes into account the potential presence of other site risks with radioactivity already committed to the ground. In this scenario, if the residual left in the tanks is one or more orders of magnitude smaller than the radioactivity already committed to the site, pursuing additional waste retrieval may not be a good use of resources and workers.

Summary of the Illustrative Example

In summary, the radioactive materials that pose the risks in tank wastes have widely varying levels of activity, half-lives, and physical properties. This situation results in risks that are extremely long term and highly variable over both space and time. Because risks can change over space and time, a risk-informed approach would help in considering the long-term consequences and risks of tank waste management and disposition decisions. Each of the three sites should undertake a separate examination of how the tank risks vary over space and time (see Chapter VIII, Recommendation VIII-1).

An example discussed in this chapter shows how

- A structured approach to decision making about tank closure helps clarify the trade-offs, making it a more transparent process;
- Each decision has implications for an entire risk profile and adds to the information provided by the performance assessment;
- The interaction between tank closure decisions and the need for long-term institutional controls can be better informed by a structured assessment of risk profiles; and
- "Maximum extent practical" could be communicated more transparently to regulators and stakeholders in the context of changes in risk profiles than in the context of the limit of a waste retrieval technology.

As explained in Chapter VIII, the structured assessment of risk profiles is still only a part of the information needed for risk-informed decision making. Costs and worker risks are not found in the illustrative example but a real-world analysis would include these considerations as well. The structuring activity helps identify the additional concerns that are relevant for a full risk-informed approach and can promote the kind of discussion about choices that is a hallmark of a participatory decision process. Different end points for waste removal and final tank closure can emerge for different types of tank, sites, and waste types.

XI

Conclusions

As discussed in the previous chapters, there is uncertainty in some of the major phases of the Department of Energy's (DOE's) tank waste remediation program. DOE has only recently made public its justification for tank waste disposition decisions and many of its quantitative examinations of questions relevant to the committee's charge. Even in these documents, however, there are no clear, definitive answers to some of the questions that Congress posed to the committee in its request for this study (see discussion of congressional charge to the committee, below). In what follows, the committee summarizes the messages it means to convey by its general and site-specific findings and recommendations (detailed findings and recommendations can be found in Chapters III through IX).[1] The committee also provides abbreviated answers to the questions that Congress asked.

FINDINGS AND RECOMMENDATIONS

Summarized below are the committee's general and site-specific findings and recommendations (see Chapters III through IX for details).

1. DOE's overall approach is workable but there are technical and programmatic challenges in reaching the goals of the tank remediation program.

DOE's overall approach for managing its tank wastes and the framework in which this must be done is workable: to the maximum extent practical, retrieve the waste from the tanks and separate the recovered waste into high- and low-activity fractions; and dispose of both waste remaining in tanks and recovered low-activity waste on-site in a manner that protects human health and the environment. Nonetheless, DOE faces technical and programmatic challenges in implementing this approach. Examples of technical challenges include retrieving waste from tanks with significant obstructions at the Savannah River Site and from tanks with leaks at the Hanford Site and assessing the uncertainties in the performance of planned waste processing approaches, such as the deliquification, dissolution, and adjustment (DDA) process at the Savannah River Site and the bulk vitrification process at Hanford.[2] Programmatic challenges are those affecting the success of the tank cleanup program, such as budgetary challenges and regulatory challenges (see Chapter VIII).

2. Decisions about planned disposal activities require multiple inputs and should not be dictated solely by schedule conformance.

Basing tank management and waste disposition decisions only on performance assessments to demonstrate compliance with performance objectives is inadequate because such assessments do not take into account all of the various factors that could be important to decisions such as the evolution of the full risk profile (risks across the site under different exposure scenarios) over time; compliance with federal facilities agreements; changes in costs and changes in how people value health and the environment; progress to build confidence in the program; and other site risks, among others. All of these factors are increasingly uncertain the further into the future that people attempt to anticipate conditions.

DOE uses conformance to a schedule (i.e., meeting milestones) as one of three criteria in determining what constitutes removal of waste to the maximum extent practical in the Section 3116 draft waste determination for Tanks 19 and 18 at the Savannah River Site. While meeting agreed-upon schedules and milestones is important and in many cases legally enforceable, a schedule-driven approach could lead

[1]This chapter is essentially the same as the précis of the report's findings and recommendations found in the summary.

[2]Other technical challenges concerning waste retrieval, processing, immobilization, and monitoring are described in Chapters III, IV, V, and VII.

to retrieval and closure actions that may later be judged insufficient.

In addition to DOE, several other parties play an important role in the ultimate success of DOE's program: Congress, the Environmental Protection Agency, the Nuclear Regulatory Commission, host states, American Indian nations, local governments, and other stakeholders. A risk-informed, transparent, participatory, and consistent decision-making process facilitates the involvement of these parties and enhances the effectiveness of the process.

3. Tank cleanup is a multidecade project allowing opportunities to improve the efforts to retrieve waste from the tanks

The milestones to close all tanks are at least a decade away (ranging from 2016 to 2032).[3] Of all 246 high-level waste tanks, only 2 have been closed (i.e., grouted) so far, and about 14 are being prepared for closure.[4] However, the cleanup of each tank will provide DOE with experiences that can be used to improve cleanup of the others. Following closure, modification of stabilized tank waste that is left on-site will be difficult, meaning that the form, concentration, and mass of tank waste left on-site will basically be fixed. Retrieval of waste from each tank will be somewhat different because each tank's specific combination of waste type, tank design and construction, and operation history is unique. Complete closure of all tanks will involve many one-of-a-kind and first-of-a-kind endeavors that would be carried out more effectively by building on experience with tank waste remediation. DOE is learning from its experience to date, but there still are substantial opportunities to continue to improve its program with respect to waste retrieval, processing, immobilization, and disposal, monitoring compliance and performance assessment, decision making, and research and development. Each of these opportunities is discussed in greater detail below.

Waste Retrieval *(see Chapter III)*

Depending on the particular requirements of individual tanks, DOE should use the most effective sequence and combination of waste retrieval tools to ensure that waste is removed to the maximum extent practical. When the limit of a given technology is reached, DOE should utilize, when necessary, other waste retrieval tools (already available or to be developed) so that the maximum extent practical is not contingent solely upon what technology has already been deployed or the proposed tank cleanup schedule. Reaching the limit of a given technology does not in itself demonstrate that all practical efforts for retrieval have been made.

The committee continues to believe that DOE should decouple the schedule for tank waste removal from the schedule for tank closure on a case-by-case basis, particularly in the case of tanks with significant heels (radioactive material remaining after planned retrieval operations are complete), as is likely in tanks with obstructions and/or with recalcitrant waste. In these tanks, more time may be needed to implement additional waste retrieval methods. DOE should not make decisions based solely on schedule conformance. Decoupling will enhance future opportunities to remove additional radioactive material from these tanks as retrieval technologies are improved. If implemented properly, decoupling for individual tanks need not delay the final closure of the tank farms. There is little technical advantage in the accelerated closure of the tanks.

Waste Processing *(see Chapter IV)*

When selecting a waste processing technology, DOE should take into account the impacts, flexibility, and robustness of processing facilities and waste forms. The Savannah River Site uses the DDA process to free up tank space, but the committee has concerns about the amount of radioactive material that the DDA process would allow to be disposed as low-activity waste on-site. The committee also has concerns about the bulk vitrification option for Hanford's supplemental low-activity waste treatment (see site-specific concerns, below). The cost and risks to workers, members of the public, and the environment if the processes should fail to perform acceptably, along with schedule uncertainties, need to be taken into account in making decisions among alternatives.

Waste Immobilization *(see Chapters V, VI, and IX)*

The committee agrees with DOE's selection of cementitious material (grout) as the most appropriate material for tank closure and does not foresee the development of better alternatives. However, the committee has concerns about DOE's understanding of the very-long-term performance of the grout used to inhibit water flow and immobilize waste in closed tanks. As a result, the committee recommends further short- and long-term research and development on the performance of cementitious materials. These efforts should be tailored to the formulations of grout planned for use in tank closures and waste immobilization and to the demands DOE places on their long-term performance in its performance assessments.

[3]Currently, the closure schedule for Hanford is 2024 for single-shell tanks and 2032 for double-shell tanks; at the Savannah River Site, the closure milestones are 2022 for Type I, II, and IV and 2028 for Type III tanks; at Idaho, the six-phase tank closure process began in 2005 and will reach completion in 2016. No milestone has been selected for closing the Idaho bins.

[4]DOE has officially submitted a waste determination to close the following tanks: Tanks 18 and 19 at the Savannah River Site, Tanks WM-180 through 186, and tanks WM-103 through WM-106 at the Idaho site (see Table F-1 in Appendix F). A separate state and USNRC review required under the Hanford Federal Facility Agreement is underway for Hanford's Tank C-106.

Performance Assessments and Monitoring Compliance (see Chapters VI and VII)

The committee views monitoring programs and performance assessments as iterative, interrelated, evolutionary activities that require updating as new information becomes available and as changes occur at the site. The sites have not yet, for the most part, developed plans for post-closure monitoring so the committee is not able to comment on them. DOE Order 435.1 requires that plans for closure of high-level waste facilities include a monitoring plan and further requires that iterations of performance assessments for low-level waste disposal facilities continue through facility closure and beyond, as needed (DOE, 2001a). It is understandable that post-closure monitoring is not DOE's highest priority right now, given that closure of the tank farms is still decades away. Plans are needed, however, before closure because some of the components of monitoring systems should be built into the closure system. DOE should begin to build provision for monitoring into its tank closures and disposal facilities and develop plans for a post-closure monitoring program, ensuring that post-closure monitoring and the updating of performance assessments are given appropriate attention as the site progresses toward closure and beyond.

External and independent peer review of DOE's draft waste determinations and performance assessments introduced by Section 3116 of the NDAA has led to demonstrable improvement in DOE's analyses (such as incorporating sensitivity studies) and the technical documents that are being prepared. This in turn has sharpened the understanding of DOE's rationale, assumptions, analysis, and conclusions. DOE should continue to seek transparent, independent peer review of critical data and analyses used to support decisions about tank waste retrieval, processing, and disposal even if review is not required under the NDAA.

Decision Making (see Chapter VIII)

Determining how clean is clean enough for tank waste retrieval, separation, and disposal is a decision in which DOE must consider a range of technical and nontechnical factors. The question does not have a unique, numerical solution. In such decisions, DOE should take into account, in addition to the performance assessments results for specific locations at specific times, how the risks from the materials left on-site vary over space and time; technical capabilities for waste retrieval and radionuclide separation from the removed wastes; cost, both in terms of dollars spent and worker doses incurred per increment of risk reduction achieved; and the potential risks from other wastes to be left on-site. Given the technical and programmatic challenges in DOE's waste management environment, one way to improve decision making is to adopt a more risk-informed, participatory, transparent, and consistent decision-making process. Such a process, as recommended in a previous National Research Council report (NRC, 2005b), would give regulators, Congress, the public, and especially DOE a clearer idea of the challenges and choices that DOE faces. It also will make DOE's planning more robust, in the sense that it is more likely to succeed in its mission. DOE has taken steps to improve its transparency in its most recent draft waste determinations and performance objectives demonstration documents, which describe how DOE reached its decisions and provide supporting data and analyses for understanding the rationale for its decisions (Sams, 2004; Buice et al., 2005; DOE-ID, 2005a; DOE-SRS, 2005a; Rosenberger et al., 2005). The committee commends this improvement and encourages DOE to continue to increase transparency, accessibility, participation, and peer review in all aspects of its tank waste management program.

Research and Development (see Chapter IX)

As DOE is in the initial stages of retrieval and closure, and as the committee continues to see delays in key pieces of the tank program (e.g., Salt Waste Processing Facility at the Savannah River Site and Waste Treatment Plant at Hanford; see below), it is increasingly clear that there is more time for implementing a research and development program that could improve waste retrieval, tank stabilization, and low-activity waste immobilization. DOE should initiate a targeted, aggressive, collaborative research and development program focused on (1) options for chemical cleaning of tanks; (2) emerging technologies to assist in tank waste removal, including robotic enhancements to current waste retrieval technologies; and (3) near- and long-term performance and monitoring of tank fill materials as they interact with the environment. Based on experience with the Environmental Management Science Program,[5] Tanks Focus Area,[6] and similar programs, a 10-year program on the order of $50 million per year would seem appropriate to generate the technological know-how needed for continuous improvement of tank waste management.

Site-Specific Findings and Recommendations

Savannah River Site

Compliant Tank Volume for Processing Needs

Tank wastes at the Savannah River Site occur in three different physical forms: a salt solution, a water-soluble

[5]DOE's Environmental Management Science Program (EMSP) was created by the 104th Congress to stimulate basic research and technology development for environmental cleanup of the nation's nuclear weapons complex (NRC, 1997a).

[6]The Department of Energy's Tanks Focus Area (TFA) was funded by DOE's Office of Science and Technology to provide technical assistance with issues related to tank wastes at the Savannah River Site, Hanford Site, Idaho National Laboratory, Fernald Site, and West Valley Demonstration Project.

saltcake, and an insoluble sludge. All phases contain radioactive materials. For the last 10 years, the site's Defense Waste Processing Facility has immobilized sludge in glass and poured the glass into steel canisters, which are stored pending shipment off-site for disposal in a high-level radioactive waste repository. DOE has stated that it needs open volume in compliant waste tanks[7] for secondary wastes from sludge treatment to ensure that sludge removal from noncompliant tanks continues apace and the Defense Waste Processing Facility continues to operate at full capacity. DOE ultimately plans to use the Salt Waste Processing Facility to remove radionuclides from most of the salt solution and saltcake phases of the tank waste, which occupy most of the volume in compliant waste tanks. To obtain open tank volume for waste inputs before the Salt Waste Processing Facility is operational and for efficient operation of that facility, DOE plans to use two interim processes: DDA and a separate, low-throughput chemical processing unit. Because of the recently announced 26-month delay in startup of the Salt Waste Processing Facility, DOE has been forced to reexamine its alternatives in obtaining that open tank volume.

The committee reemphasizes its concern[8] that too much waste will be processed through the DDA if it is a standalone process. There are two principal reasons for this concern. First, as described in DOE's plans as of 2005 (when the committee's interim report was prepared), DDA would send large amounts of radioactive material to the Saltstone Vaults, orders of magnitude more than was originally envisioned. Second, because of the 26-month delay in operation of the Salt Waste Processing Facility, it is possible that additional radioactive material could be disposed in saltstone if DDA has to operate longer than previously expected. In its salt waste determination (DOE, 2006), DOE said it intends to put no more radioactive material in the saltstone than described in the draft salt waste determination (DOE-SRS, 2005b); but DOE has not yet proposed any solutions to the tank space problem arising from the 26-month delay. Thus, the committee could not evaluate the problem further. The committee reiterates, however, that DOE should seek alternatives to the DDA process, either by slowing waste inputs (slowing operations or gaining efficiencies) or by finding storage alternatives for the least hazardous of the tank wastes to free up storage space.

Point of Compliance

DOE's point of compliance (the location where compliance with performance objectives is determined) for its F Tank Farm is 1.6 kilometers (1 mile) away from the facility boundary, rather than the standard 100 meters (109 yards) away. When DOE uses a nonstandard point of compliance, it should state clearly the potential exposures closer to the disposal facility in case assumptions about human behavior and institutions do not turn out to be true. The selection of the point of compliance has both policy and technical dimensions. The committee believes that those technical dimensions should be stated clearly and prominently so that the policy decision is well informed.

Estimated Doses from the Predicted Waste Residuals in the F Tank Farm

In estimating the residual tank inventories for its performance assessment calculations for the F tank farm, DOE assumes that future efforts to clean out tanks will be much more effective than they were for most of the tanks that have already been cleaned out. The committee views this assumption as both optimistic and unsupported. Without a technical basis for the inventory estimates, the committee does not have confidence in the results of the performance assessment for the F Tank Farm.[9]

Hanford

The challenges DOE faces at Hanford are significant and varied, but given that the revised performance assessment for the single-shell tank farms at Hanford had not been issued by the end of 2005, the committee was unable to evaluate DOE's plans with respect to several elements of the charge from Congress. However, the committee has concerns about the technical performance and safety features of the bulk vitrification option for supplemental low-activity waste treatment. Each site is pursuing different technologies for immobilizing its non-high-level tank waste—grout at the Savannah River Site, steam reforming at the Idaho site, and vitrification in the Waste Treatment Plant and bulk vitrification[10] at Hanford. The Hanford bulk vitrification process is currently less well developed technically than either the Savannah River Site saltstone or the Idaho National Laboratory steam reforming. Before selecting an immobilization technology at Hanford, DOE should sponsor a detailed, transparent, independent, technical review of bulk vitrification versus other options, focusing on process risks and uncertainties.

Idaho National Laboratory

Idaho faces smaller, simpler challenges than either the Savannah River or Hanford Sites in cleaning out the tanks. Less spent fuel was reprocessed at the site—all of it with the

[7]Compliant waste tanks meet the Resource Conservation and Recovery Act requirements for storage of hazardous waste. Noncompliant tanks do not meet those requirements (e.g., full secondary containment).

[8]See the committee's interim report (NRC, 2005a). The summary of that report can be found in Appendix E.

[9]No tank closures have been proposed yet for the H Tank Farm.

[10]Bulk vitrification is the lead candidate for the Hanford supplemental low activity waste process; however, the preferred technology has not yet been officially selected.

same process; most of the tank waste was calcined and resides as granular solids in bins; and what liquid waste was stored in the tanks did not separate substantially into sludge and salt because the waste was left in its acidic state. The Idaho National Laboratory is making good progress in dealing with its liquid wastes. However, it remains to be seen whether the solidified waste (the calcine) stored in bins will be as easy to remove as projected by the site.

CONGRESSIONAL CHARGE TO THE COMMITTEE

The committee's charge from Congress contains six specific topics. Each topic is presented below and is followed by the committee's response.

Topic 1: The "Department's understanding of the physical, chemical, and radiological characteristics of the waste referred to above, including an assessment of data uncertainties."

The committee believes that DOE has reached a point in its analysis of the physical, chemical, and radiological characteristics of the waste in the tanks where further understanding would not change its overall approach substantially. DOE's knowledge of the waste in the tanks is sufficient for waste retrieval. DOE needs to know the waste composition in greater detail for processing purposes and to confirm compliance with performance objectives, but this must be done after waste retrieval when mixing makes representative sampling of the retrieved waste possible and when samples of the tank heels can be taken. Even then, the waste composition need only be known sufficiently for reliable and efficient processing to take place. Some processing methodologies, such as steam reforming and grouting, do not require detailed feed characterization, whereas others, such as vitrification, may require greater knowledge and control of the waste characteristics. When these requirements are very stringent, it may be necessary to look for processing and immobilization technologies that accommodate a wider range of feed characteristics. The costs and the risks to workers, members of the public, and the environment if the processes should fail to perform acceptably have to be taken into account for each processing option. For different processes and different locations, the knowledge required may be different.

Topic 2: "Any actions additional to those contained in current plan that [DOE] should consider to ensure that the plan will comply with the performance objectives of Part 61 of Title 10, Code of Federal Regulations".

10 CFR 61.41 states: "Reasonable effort should be made to maintain releases of radioactivity in effluents to the general environment as low as is reasonably achievable." After DOE shows that its plans meet the dose limits, DOE should further demonstrate how its plans for waste retrieval and immobilization meet ALARA requirements to protect workers, the public, and the environment now and in the future. In Section 3116(a)(2) of the NDAA, a criterion for on-site disposal is that the waste "has had highly radioactive radionuclides removed to the maximum extent practical." The risks posed depend on the assumed location and time at which the performance criteria must be met. DOE has issued only a few documents detailing how it determined what amounts of material left in a tank would be acceptable. There is not sufficient information available to evaluate whether all of the components of importance to such decisions have been taken into account. However, it would be advantageous to have a common process that illustrates how risks under each option are likely to change over space and time, which would be useful in determining what wastes can be disposed on-site. An illustrative and extremely simplified example of such a process is given in Chapter X.

Topic 3: "The adequacy of the Department's plans for monitoring disposal sites and the surrounding environment to verify compliance with those performance objectives."

Monitoring is important in performance assessment model validation exercises and for early detection of failures. It also allows remedial actions to be taken at the earliest possible time, thereby minimizing human and environmental impact and cost. The committee judges that monitoring within the disposal facility is the most desirable location for the early detection of problems, followed by detection in the vadose zone, and finally by detection in the nearest aquifer. The committee's overall impressions are that the sites' monitoring programs are satisfactory at fulfilling their current goals, which in most cases are site characterization and operations monitoring to assess regulatory compliance. However, the committee believes that the DOE plans for monitoring should go beyond the requirement of verifying compliance with performance objectives. DOE should start planning its post-closure monitoring programs so that provision for monitoring can be built into closure plans and designs.

Topics 4 and 5: "Existing technology alternatives to the current management plan for the waste streams mentioned above and, for each such alternative, an assessment of the cost, consequences for worker safety, and long-term consequences for environmental and human health;" and any "technology gaps that exist to effect improved efficiency in removal and treatment of waste from the tanks at the Hanford, Savannah River, and Idaho sites."

The committee had to operate within schedule constraints, which limited the extent to which it could evaluate technology alternatives. Such evaluation was particularly difficult because DOE was able to provide only limited information on cost, worker safety, and long-term human and environmental health consequences of technology alternatives for tank waste management. However, it is apparent that DOE should continue to adapt existing and develop new technologies for effective waste retrieval, with emphasis on tanks with obstructions and recalcitrant waste. Additionally, the committee recommends a targeted aggressive, collaborative research and development program on chemical cleaning of the tanks, mechanical waste retrieval, and tank filling materials for tank stabilization. The committee recommends support at approximately $50 million per year to focus on technologies that could become available in the near term (within 10 years) in time to be implemented during the tank cleanup program.

Topic 6: "Any other matters that the committee considers appropriate and directly related to the subject matter of the study."

Following are issues that the committee believes are important, but either DOE's plans are not detailed enough at this time to make specific recommendations or the issues are independent of tank management plans:

- Remediation of pipelines, leaking underground pipes and interwall spaces in double-walled tanks, and other auxiliary equipment in the tank farms could be challenging, particularly at Hanford where there are about 100 plugged pipelines (see Chapter III).
- Although the Idaho National Laboratory should focus on tank wastes first, some consideration should be given to the calcine waste and bins and their disposition (see Chapter III).
- DOE needs regulatory approvals for the off-site disposal of some Hanford tank waste and Idaho sodium-bearing tank waste.
- The philosophy and methodology for post-closure monitoring needs to be developed and articulated.
- Attention should be paid to long-term stewardship, post-closure monitoring, and the meaning of "in perpetuity," and rigorous (within the limits of long-term prediction) planning for these activities should commence. A focus on these issues is important and would incorporate, but extend beyond, DOE's tank waste management plans (see Chapters VII and VIII).

FINAL WORDS

DOE is progressing slowly, sometimes through no fault of its own, to remediate its high-level waste tanks at Hanford, Savannah River, and Idaho. Because this is largely a first-of-a-kind activity, some problems with chosen remedies should be expected. This should be seen as part of the learning curve for the process. However, the process can be improved by a more focused, experimental (adaptive) approach and by following a risk-informed, consistent, and transparent methodology, such as the one outlined in a previous National Research Council report *Risk and Decisions About Disposition of Transuranic and High-Level Radioactive Waste* (NRC, 2005b). This methodology involves analyses of risks, benefits, costs, and alternatives in a transparent environment that would allow meaningful stakeholder involvement, adaptation to surprises that will be encountered along the way, and continuous learning both from experience and from an ongoing focused R&D program.

References

Adu-Wusu, K., M.J. Barnes, N.E. Bibler, J.R. Cantrell, F.F. Fondeur, B.A. Hamm, C.C. Herman, D.T. Hobbs, E.T. Ketusky, M. Singleton, M.E. Stallings, W.E. Stevens, and B.J. Wireman. 2003. Waste Tank Heel Chemical Cleaning Summary, WSRC-TR-2003-00401, Rev. 0. September 9. Aiken, S.C.: Westinghouse Savannah River Company.

AEATES (AEA Technology Engineering Services, Inc.). 2004. Demonstration of a System for In-Situ Tank Waste Heel Immobilization at LANL. Final Report. ESI Document No: 2173-6-003 Rev 0. January. Mooresville, N.C.: AEA Technology Engineering Services, Inc.

AEATES. 2005a. Consolidated Tank Closure Demonstration. September 7. Mooresville, N.C.: AEA Technology Engineering Services, Inc.

AEATES. 2005b. Calcine Bins Retrieval and Transfer System: Test Report. ESI Document No.: 2156-6-004. January. Mooresville, N.C.: AEA Technology Engineering Services, Inc.

Allison, J.F. 2005. Letter from J.F. Allison, Manager, U.S. Department of Energy Savannah River Operations Office to The Honorable A.J. Eggenberger, Chairman, Defense Nuclear Facilities Safety Board, re: Design Approach for Providing Performance Category 3 (PC-3) Confinement for the Salt Waste Processing Facility (SWPF). November 23.

Badheka, N.P., and N.S. Elraheb. 2004. Annulus Cleaning Technical Plan. CBU-LTS-2004-00109, Rev. 0. September 22. Aiken, S.C.: Westinghouse Savannah River Company.

Bamberger, J.A. and G.F. Boris. 1999. Oak Ridge National Laboratory Old Hydrofracture Facility Waste Remediation Using the Borehole-Miner Extendible-Nozzle Sluicer. June. PNNL-12225/UC-721. Richland, Wash.: Pacific Northwest National Laboratory.

Bamberger, J.A., B.L. Burks, K.D. Quigley, S.W. Butterworth, and D.D. Falter. 2001. Technical Review of Retrieval and Closure Plans for the INEEL INTEC Tank Farm Facility. PNNL-13651 UC-721. September. Richland, Wash.: Pacific Northwest National Laboratory. Available at *http://www.pnl.gov/main/publications/external/ technical_reports/ pnnl-13651.pdf.*

Barnes, C.M., C.B. Millet, and V.J. Johnson. 2004. Feed Composition for the Sodium-Bearing Waste Treatment Process. INEEL/EXT-2000-01378, Rev. 4. June. Idaho Falls, Idaho: Idaho National Engineering and Environmental Laboratory.

Barnett D.B., G.W. Gee, M.D. Sweeney, M.D. Johnson, V.F. Medina, D.P. Mendoza, B.G. Fritz, F. Khan, W.D. Daily, J.B. Fink, A. Ramirez, M.T. Levitt, A. Binley, and C.A. Baldyga. 2003. Results of Performance Evaluation Testing of Electrical Leak-Detection Methods at the Hanford Mock Tank Site—FY 2002-2003. February. PNNL-14192. Richland, Wash.: Pacific Northwest National Laboratory.

Barton, B. 2005a. S-112 Retrieval Progress. Savannah River/Hanford Technical Exchange Meeting, Savannah River Site, Aiken, S.C. March 22-24.

Barton, B. 2005b. C-106 Waste Retrieval. Savannah River/Hanford Technical Exchange Meeting. WSRC-MM-2005-00044, Savannah River Site, Aiken, S.C., March 22-24.

Barton, B. 2005c. High-Resolution Resistivity Leak Detection. WSRC-MM-2005-00044. Savannah River/Hanford Technical Exchange Meeting, Savannah River Site, Aiken, S.C. March 22-24,

Bechtold, D.B., G.A. Cooke, D.L. Herting, J.C. Person, R.S. Viswanath, and R.W. Warrant. 2003. Laboratory Testing of Oxalic Acid Dissolution of Tank 241-C-106 Sludge, RPP-17158, Rev. 0, July 9. Richland, Wash.: Fluor Hanford.

Beckman, S. 2006. E-mail communication from S. Beckman (Fernald) to B. Pastina, Re: Calcine at Fernald. February 16.

Bernero, R.M. 1993. Letter from R.M. Bernero, Director, Office of Nuclear Materials Safety and Safeguards, U.S. Nuclear Regulatory Commission, to J. Lytle, Office of Environmental Restoration and Waste Management, U.S. Department of Energy. March 2. Available at *http:// www.nrc.gov/reading-rm.html.*

Bogen, G.J., G.C. DeWeese, T.H. May, and W.W. Pickett. 1999. Alternatives Generation and Analysis C-104 Single-Shell Tanks Waste Feed Delivery. TWR-4454, Revision 1. Richland, Wash.: Numatec Hanford Company.

Boomer, K.D., S.K. Baker, A. Boldt, J.D. Galbraith, J.S. Garfield, C.E. Golberg, B.A. Higley, L.J. Johnson, M.J. Kupfer, R.M. Marusich, R.J. Parazin, A.N. Praga, G.W. Reddick, J.A. Reddick, E.J. Slaathaug, L.M. Swanson, T.L. Waldon, and C.E. Worcester. 1993. Tank Waste Technical Options Report. WHC-EP-0616, Revision 0. March. Richland, Wash.: Westinghouse Hanford Company.

Bradley, R.F., and A.J. Hill, Jr. 1977. Chemical Dissolving of Sludge from a High Level Waste Tank at the Savannah River Plant. DP-1471. November. Aiken, S.C.: E.I. du Pont de Nemours and Company, Savannah River Laboratory.

Brainard, J.R., B.A. Strietelmeier, P.H. Smith, P.J. Langston-Unkefer, M.E. Barr, and R.R. Ryan. 1992. Actinide binding and solubilization by microbial siderophores. Radiochimica Acta 58-59:357.

Bredehoeft, J.D. 2003. From models to performance assessment: The conceptualization problem. Ground Water 41:571-577.

Buice, J.M., R.K. Cauthen, R.R. Haddock, B.A. Martin, J.A. McNeil, J.L. Newman, and K.H. Rosenberger. 2005. Tank 19 and Tank 18 Closure Performance Objective Demonstration Document (PODD). CBU-PIT-2005-00106, Revision 0. September. Aiken, S.C.: Westinghouse Savannah River Company.

Burks, B. 2005. Tank Waste Retrieval Using Remotely Operated Systems. Presentation during the 162nd meeting of the Advisory Committee on Nuclear Waste, August 2-3, Rockville, Md.

Cahn, L. 2005. Tank Farm Contaminated Soils and Groundwater Cleanup. Presentation to the Committee on the Management of Certain Radioactive Waste Streams Stored in Tanks at Three Department of Energy Sites, July 25, Idaho Falls, Idaho.

Caldwell, T. 2005a. Waste Removal and Treatment Technology. Presentation to the Committee on the Management of Certain Radioactive Waste Streams Stored in Tanks at Three Department of Energy Sites, April 14, Augusta, Ga.

Caldwell, T. 2005b. Tank 4 Interior Photo Study—1954 Pre-Startup. CBU-PIT-2005-00111. May 5. Aiken, S.C.: Westinghouse Savannah River Company.

Camper, L. 2005. NRC's Role in Waste Determinations. Presentation to the Committee on the Management of Certain Radioactive Waste Streams Stored in Tanks at Three Department of Energy Sites, March 7, Washington, D.C.

Cantrell, J.R. 2004. Heel Removal and Annulus Cleaning Technology Development Suspension Plan. CBU-FTF-2004-0094. Aiken, S.C.: Westinghouse Savannah River Company.

CH2M Hill Hanford Group, Inc. 2003. Tank Closure Activities at U.S. Department of Energy Sites. RPP-11762, Revision 0. June. Richland, Wash.: CH2M Hill Hanford Group, Inc.

CH2M Hill Hanford Group, Inc. 2005. Single-Shell Tank Performance Assessment. Predecisional Partial Draft. DOE/ORP-2005-01, Revision Draft. Richland, Wash.: CH2M Hill Hanford Group, Inc.

Choho, A.F., and K.A. Gasper. 2002. Evaluation of Low-Activity Waste Feed Supplemental Treatment Options by the C3T Mission Acceleration Initiative Team for the Office of River Protection. RPP-11306, Revision 0. Richland, Wash.: CH2M Hill Hanford Group, Inc.

Clark, B. 2005a. E-mail communication from Bill Clark (DOE-SRS) to Micah Lowenthal, Re: Fact check on rad waste quantities. November 18.

Clark, B. 2005b. E-mail communication from B. Clark (DOE-SRS) to M. Lowenthal, Re: Follow-up questions for SRS, attachment, CBU-PIT-2005-00092 abridged.

Cole, M.G. 1992. Underground Storage Tanks Installation and Management. Chelsea, Mich.: Lewis Publishers, Inc.

Cook, J.R. 2005. Estimated All Pathways and Inadvertent Intruder Doses from Saltstone Disposal Using Updated Salt Waste Compositions. Revision 0. February 24. WSRC-RP-2005-01402. Aiken, S.C.: Westinghouse Savannah River Company.

Cook, J.R., E.L. Wilhite, and L.B. Collard. 1999. Addendum to the Composite Analysis for the E-Area Vaults and Saltstone Disposal Facilities. September 23. WSRC-RP-99-00844. Aiken, S.C.: Westinghouse Savannah River Company.

CRESP (Consortium for Risk Evaluation with Stakeholder Participation). 2005. Preliminary Risk Evaluation of Calcined High-Level Waste Disposition at the Idaho Site. June. Switzer, C., K.G. Brown, D.S. Kosson, J.G. Clarke, F.L. Parker, C.W. Powers, H.J. Mayter, and M. Greenberg. Piscataway, N.J.: CRESP.

Daily, D.E. 2005a. Tank 5 Sludge Volume Estimation After the First Phase of Bulk Sludge Removal. M-ESR-F-00107. Rev. 0. November 3. Aiken, S.C.

Daily, D.E. 2005b. Tank 5 Sludge Volume Estimation After the Second Phase of Bulk Sludge Removal (U). M-ESR-F-00109. Rev. 0. December 12. Aiken, S.C.

DARPA (Defense Advanced Research Projects Agency). 2005. Grand Challenge. Arlington, Va. Available at *http://www.darpa.mil/grandchallenge/*.

Davis, N. R. 1998. Tank 19 Heel Removal Systems Engineering Evaluation. Waste Removal Programs. PIT-MISC-0040. Revision 0. Aiken, S.C.: Westinghouse Savannah River Company.

DNFSB (Defense Nuclear Facilities Safety Board). 1993. DNFSB Recommendation 93-5 to the Secretary of Energy pursuant to 42 U.S.C. § 2286a(5) Atomic Energy Act of 1954, as amended. Letter from J.T. Conway, Chairman, Defense Nuclear Facilities Safety Board to U.S. Department of Energy Secretary, Hazel O'Leary. Available at *http://www.deprep.org/ archive/rec/93-5.asp*.

DNFSB. 2005. Letter from A.J. Eggenberger, Chairman, Defense Nuclear Facilities Safety Board to U.S. Department of Energy Assistant Secretary for Environmental Management forwarding August 9 Staff Issue Report, Re: Demonstration Bulk Vitrification Project, Hanford Tank Farms. September 7. Available at http://www.*dnfsb.gov/pub_docs/hanford/cor_2005_hd.html*.

Dodd, R.A. 2005. Waste Retrieval Technologies. Presentation to the Committee on the Management of Certain Radioactive Waste Streams Stored in Tanks at Three Department of Energy Sites, July 22, Richland, Wash.

DOE (U.S. Department of Energy). 1998. Accelerating Cleanup Paths to Closure. DOE/EM-0362. June. Washington, D.C.: U.S. Department of Energy Office of Environmental Management. Available at *http://web.em.doe.gov/closure/final/onlinenp.html*.

DOE. 1999. Radioactive Waste Management. DOE Order 435.1 Approved July 9. Washington, D.C.: DOE.

DOE. 2001a. Radioactive Waste Management Manual. DOE M 435.1-1. Approved June 19. Washington, D.C.: DOE.

DOE. 2001b. Summary Data on the Radioactive Waste, Spent Nuclear Fuel, and Contaminated Media Managed by the U.S. Department of Energy. April. Washington, D.C.: Office of Environmental Management, U.S. Department of Energy.

DOE. 2001c. Radioactive Waste Management. DOE Order 435.1 Change 1. August 28. Washington, D.C.: DOE.

DOE. 2002. Final Environmental Impact Statement for a Geologic Repository for the Disposal of Spent Nuclear Fuel and High-Level Radioactive Waste at Yucca Mountain, Nye County, Nevada. Volume II, Appendixes A through O. DOE/EIS-0250. Office of Civilian Radioactive Waste Management Available at *http://www.ocrwm.doe.gov/documents/feis_2/vol_2/apndx_a/appndx_a.htm*.

DOE. 2005. Summary Report of Performance Monitoring Workshop Cosponsored by MSE Technology Applications, Inc. and Office of Cleanup Technologies, Office of Environmental Cleanup and Acceleration, Office of Environmental Management, U.S. Department of Energy, June 21-23, Butte, Mont.

DOE. 2006. Basis for Section 3116 Determination for Salt Waste Disposal at the Savannah River Site. DOE-WD-2005-001. Washington, D.C.: DOE.

DOE-EM (Office of Environmental Management). 2001a. DOE/OR/01-1953&D2. Removal Action Report for the Bethel Valley Main Plant Inactive Liquid Low-Level Radioactive Waste Tanks U.S. Department of Energy Office of Environmental Management.

DOE-EM. 2001b. DOE/OR/01-1953&D2/A2. Addendum to the Removal Action Report for the Bethel Valley Main Plant Inactive Liquid Low-Level Radioactive Waste Tanks. U.S. Department of Energy Office of Environmental Management, Washington, D.C.

DOE-EM. 2005. Notice of preferred sodium bearing waste technology. Federal Register 70(148):44598-44600. Washington, D.C. Available at *http://www.eh.doe.gov/NEPA/notices/44598.pdf*.

DOE-ID (Idaho Operations Office). 2002 Idaho High-Level Waste & Facilities Disposition Final Environmental Impact Statement. DOE/EIS-0287. September. Idaho Falls, Idaho: DOE.

DOE-ID. 2003a. INTEC Injection Well: Summary of Historical Information and Groundwater Quality Trends. EDF-3943. December 18. Idaho Falls, Idaho: DOE.

DOE-ID. 2003b. Performance Assessment for the Tank Farm Facility at the Idaho National Engineering and Environmental Laboratory. DOE/ID-10966, Revision 1. April. Idaho Falls, Idaho: U.S. Department of Energy Idaho Operations Office.

DOE-ID. 2003c. Long-Term Monitoring Plan for Operable Unit 3-13, Group 5, Snake River Plain Aquifer. DOE/ID-10783 Revision 2. March 2003. Idaho Falls, Idaho: DOE.

DOE-ID. 2004a. Long-Term Monitoring Plan for Operable Unit 3-13, Group 4 Perched Water. DOE/ID-10746. February. Idaho Falls, Idaho: DOE.

DOE-ID. 2004b. Quality Assurance Project Plan for Waste Area Groups 1, 2, 3, 4, 5, 6, 7, 10, and Inactive Sites. DOE/ID-10587, Revision 8. March. Idaho Falls, Idaho: DOE.

DOE-ID. 2004c. Risk-Based End State Vision for the Idaho National Engineering and Environmental Laboratory Site (Draft). January 2004. DOE/ID-1110 Revision D. Idaho Falls, Idaho: DOE.

DOE-ID. 2005a. Draft Section 3116 Determination Idaho Nuclear Technology and Engineering Center Tank Farm Facility. DOE/NE-ID-11226. Revision 0. September 7. Idaho Falls, Idaho: DOE.

DOE-ID. 2005b. Record of Decision for the Idaho High-Level Waste and Facilities Disposition Final Environmental Impact Statement. DOE/EIS-0287. December. Washington, D.C.: DOE.

DOE-OFO (Ohio Field Office). 2006. Fernald Closure Project. Silo 3 Project. Available at *http://www.fernald.gov/Cleanup/Silo3.htm*.

DOE-OIG (Office of Inspector General). 2005. Management Controls over the Hanford Site Transuranic Mixed Tank Waste. OAS-O-06-011. November. Washington, D.C.: DOE.

DOE-ORNL (Oak Ridge National Laboratory). 2005. FY 2005 Cleanup Progress. Annual Report to the Oak Ridge Community. DOE/OR-2210. Oak Ridge, Tenn.: ORNL. Available at *http://www.bechteljacobs.com/pdf/CleanProg2005.pdf*.

DOE-RL (Richland Operations Office). 2002. Waste Encapsulation and Storage Facility (WESF) Fact Sheet. REG-0275. Richland, Wash.: DOE.

DOE-RL. 2003. Waste Encapsulation and Storage Facility. Richland Operations Office Fact Sheet. RL-D02-013 DOE. September. Richland, Wash.: DOE.

DOE-RL. 2004a. Final Hanford Site Solid Waste Program Environmental Impact Statement. DOE/EIS-0286F. January. Richland, Wash.: DOE. Available at *http://www.hanford.gov/doe/eis/eis-0286D2/2003_FINAL.cfm*.

DOE-RL 2004b. Hanford Site Groundwater Strategy: Protection, Monitoring, and Remediation. DOE/RL-2002-59. Richland, Wash: DOE.

DOE-RL. 2005. Conference call about waste retrieval among members of the Committee on the Management of Certain Radioactive Waste Streams Stored in Tanks at Three Department of Energy Sites with Hanford Site staff. October 24.

DOE-SRS (Savannah River Site). 1997a. Industrial Waste Water Closure Module for the High-Level Waste Tank 17 System. Revision 2. PIT-MISC-0004. August 26. Construction Permit Number: 17,424-IW. Aiken, S.C.: Department of Energy Savannah River Operations Office.

DOE-SRS. 1997b. Industrial Wastewater Closure Module for the High-Level Waste Tank 20 System. January 8. Construction Permit Number: 17,424-IW. Aiken, S.C: Department of Energy Savannah River Operations Office.

DOE-SRS. 1999. Savannah River Site Residual Waste Characterization System. February 23. Aiken, S.C.: DOE.

DOE-SRS. 2002. Savannah River Site High-Level Waste Tank Closure, Final Environmental Impact Statement. DOE/EIS-0303. May. Aiken, S.C.: DOE. Available at *http://www.eh.doe.gov/nepa/eis/eis0303/eis0303.html*.

DOE-SRS (Savannah River Site). 2005a. Draft Section 3116 Determination for Closure of Tank 19 and 18 at the Savannah River Site. DOE-WD-2005-002. Revision 0. September. Aiken, S.C.: DOE.

DOE-SRS. 2005b. Draft Section 3116 Determination. Salt Waste Disposal Savannah River Site. DOE-WD-2005-001. February 28. Aiken, S.C.: DOE.

DOE-SRS. 2005c. F/H Area High Level Waste Tank Status Report for CY2004. CBU-ENG-2005-00005. Aiken, S.C.: Westinghouse Savannah River Company.

DOE-SRS. 2005d. Conference call about waste retrieval among members of the Committee on the Management of Certain Radioactive Waste Streams Stored in Tanks at Three Department of Energy Sites with Savannah River Site staff. October 21.

DOE-SRS. 2005e. Conference call on oxalic acid among committee members Sattelberger, Levenson, and Czerwinski and Savannah River Site staff. October 19.

DOE-SRS. 2005f. Savannah River Site End State Vision. July 26. Aiken, S.C.: DOE. Available at *http://sro.srs.gov/pubact1.htm*.

DOE-SRS. 2005g. Response to information gaps identified in tank wastes planned for on-site disposal at three Department of Energy sites: The Savannah River Site—Interim Report. CBU-PIT-2005-00248. Revision 0. October. Aiken, S.C.: Westinghouse Savannah River Company.

DOE-SRS. 2006. Basis for the Section 3116 Determination for Salt Waste Disposal at the Savannah River Site. DOE-WD-2005-001. January. Available at *http://apps.em.doe.gov/swd/ salt/DOE-WD-2005-001.pdf*.

DOE-TFA (Tanks Focus Area). 2000a. Heel Retrieval Technology Guide. Tanks Focus Area. Available at *http://www.tanks.org/ttgdoc/heel/heel_retrieval_technology_guide. htm*.

DOE-TFA. 2000b. TFA Multiyear Program Plan FY01-FY05. PNNL-13339. September. Richland, Wash.: Pacific Northwest National Laboratory.

DOE-TFA. 2001. Tanks Focus Area FY01 Annual Report. Delivering Technical Solutions for Radioactive Tank Waste Remediation. November 15. Richland, Wash.: DOE.

DOE-WVDP (West Valley Demonstration Project). 2006. High-Level Waste Tank Closure. U.S. Department of Energy Ohio Office West Valley Demonstration Project. Available at *http://www.wv.doe.gov/High-Level%20Waste%20Tank%20Closure.htm*.

Durbin, P.W., D.L. White, N. Jeung, F.L. Weitl, L.C. Uhlir, E.S. Jones, S. Bruenger, and K.N Raymond. 1989. Chelation of 238 PU(IV) in vivo by 3,4,3-LICAM(C): Effects of Ligand Methylation and pH. Health Physics 56:839.

Elmore, M.R. and C. Henderson. 2001a. Post Retrieval and Pre-Closure High-Level Radioactive Liquid Waste Tank Lay-Up Strategies—Part II Implementation Plan. PNNL-13900. May 15. Richland, Wash.: Pacific Northwest National Laboratory.

Elmore M.R. and C. Henderson. 2001b. Functions and requirements for West Valley Demonstration Project Tank Lay-Up. PNNL-13898. March 15. Richland, Wash.: Pacific Northwest National Laboratory.

Elmore M.R., and C. Henderson. 2002a. Review of Tank Lay-Up Status at U.S. Department of Energy Radioactive Waste Tank Sites. PNNL-13908. March 5. Richland, Wash.: Pacific Northwest National Laboratory.

Elmore M.R., and C. Henderson. 2002b. Summary of High Level Waste Tank Lay-Up Activities Supporting the Tanks Focus Area, Fiscal Years 2001-2002. PNNL-13901. May. Richland, Wash.: Pacific Northwest National Laboratory.

Emison, J.A., B.B. Spencer, and B. E. Lewis. 2002. Gunite and Associated Tanks Waste Conditioning System: Description and Operational Summary. ORNL/TM-2001/149. Oak Ridge, Tenn.: Oak Ridge National Laboratory.

EPA-OIG (U.S. Environmental Protection Agency, Office of Inspector General). 1999. Superfund: Backlog of Five-year Review Reports Increased Nearly Threefold. 1999P-218. September 30. Washington, D.C.: U.S. Environmental Protection Agency.

Feldman, R.F., J.J. Beaudoin, and K.E. Philipose. 1989. Durability of concrete for a waste repository—Measurement of ionic ingress. Presented at the Fall Meeting of the Material Research Society, Boston, Massachusetts, November 27.

Feldman, R.F., J.J Beaudoin, and K.E. Philipose. 1991. Effect of cement blends on chloride and sulphate ion diffusion in concrete. Il. Cemento 88 (Jan.-March): 3-18.

Flanders, S. 2005. NRC's Previous WIR Reviews. Presentation to the Committee on the Management of Certain Radioactive Waste Streams Stored in Tanks at Three Department of Energy Sites, March 7, Washington, D.C.

Florida International University. 2001. Case Studies of Pipeline Plugging and Unplugging Experiences at Russian Nuclear Defense Material Production Sites. February. Available at *http://www.tanks.org/ttgdoc/ARES-Russian-Pipe-Plugging.pdf*.

Fluor Fernald. 2003. Access and Retrieval Strategy for the Silo 3 Project. 40430-PL-0002, Revision 1. September. Cincinnati, Ohio: Fluor Fernald Inc.

Fogle, R.F. 2002. Tank Riser Suspension System Conceptual Design. WSRC-TR-2002-00362, Revision 0. September 15. Aiken, S.C.: Westinghouse Savannah River.

Fong, M.C.H. 1985. Oxalic Acid Cleaning of Tank 24H. DSPT-85-782. September. Aiken, S.C.: E.I. du Pont de Nemours and Company, Savannah River Laboratory.

Fowler, J.R. 1981. Radiochemical Analysis of Samples from Tank 16 Cleanout. DPST-81-441.May 18. Aiken, S.C.: E.I. du Pont de Nemours and Company.

GAO (General Accounting Office). 1997a. Department of Energy: Management and Oversight of Cleanup Activities at Fernald. GAO/RCED-97-63. March. Washington, D.C. GAO. Available at *http://www.gao.gov/archive/1997/rc97063.pdf.*

GAO. 1997b. Nuclear Waste: Department of Energy's Project to Clean up Pit 9 at Idaho Falls Is Experiencing Problems. GAO/RCED-97-180. July. Washington, D.C.: GAO. Available at *http://www.gao.gov/archive/1997/rc97180.pdf.*

GAO. 1999. Nuclear Waste: Process to Remove Radioactive Waste from Savannah River Tanks Fails to Work. GAO/RCED-99-69. April. Washington, D.C.: GAO. Available at *http://www.gao.gov/archive/1999/rc99069.pdf.*

GAO. 2002. Waste Cleanup. Status and Implications of DOE's Compliance Agreement. GAO-02-567. May. Washington, D.C.: GAO.

GAO. 2003. Nuclear Waste: Challenges to Achieving Potential Savings in DOE's High-Level Waste Cleanup Program. GAO-03-593. June. Washington, D.C.: GAO. Available at *http://www.gao.gov/new.items/d03593.pdf.*

GAO. 2004. Nuclear Waste: Absence of Key Management Reforms on Hanford's Cleanup Project Adds to Challenges of Achieving Cost and Schedule Goals. GAO-04-611. June. Washington, D.C.: GAO. Available at *http://www.gao.gov/new.items/ d04611.pdf.*

GAO (Government Accountability Office). 2005a. Hazardous Waste Sites: Improved Effectiveness of Controls at Sites Could Better Protect the Public. January. GAO-05-163. Washington, D.C.: GAO.

GAO. 2005b. Better Performance Reporting Needed to Assess DOE's Ability to Achieve the Goals of the Accelerated Cleanup Program. July. GAO-05-764. Washington, D.C.: GAO.

Garboczi, E.J. 2005. Status of Technology for Predicting the Durability of Cementitious Materials, August 2-4, 162nd ACNW meeting, Rockville, Md.

Garboczi, E.J., J.W. Bullard, and D.P. Bentz. 2004. Virtual testing of cement and concrete—USA 2004. Concrete International 26(12):33-37.

Garboczi, E.J., D.P. Bentz, K.A. Snyder, N.S. Martys, P.E. Stutzman, C.F. Ferraris, and J.W. Bullard. 2005. An electronic monograph: Modeling and measuring the structure and properties of cement-based materials. June 13. Gaithersburg, Md.: National Institute of Standards and Technology.

Garnov, A.Y., L. Rao, K.L. Nash, A.H. 2003. Bond: Leaching of U(VI), Am(III), and Sr(II) from simulated tank waste sludges. Separation Science and Technology 38(2):359-374.

Gasper, K.A. 2003. Hanford Tank Farm Overview. Savannah River/Hanford Salt Retrieval Technical Exchange meeting, April 22-23, Savannah River Site, Aiken, S.C.

Gasper, K.A. 2005. Status of Hanford Activities. Savannah River/Hanford Technical Exchange, March 22-24, Savannah River Site, Aiken, S.C.

Gephart, R.E. 2003. Hanford: A Conversation about Nuclear Waste Cleanup. Columbus, Ohio: Battelle Press.

Gilbreath, K.D. 2005. Risk Benefit Evaluation of Residual Heel Removal in Tanks 19 and 18. CBU-PIT-2005-00169. September. Aiken, S.C.: Westinghouse Savannah River Company.

Glass S.W., and F.C. Klahn. 2001. ReTRIEVR, A Segmented, Long-Reach Robot for Tank or Silo Waste Retrieval. Presented at the ANS 9th International Robotics and Remote System Conference, March 4, Seattle, Wash, American Nuclear Society Paper 073. Available at *http://www.tanks.org/ ttgdoc/ f073.pdf.*

Glassell, L., B.L. Burks, and W.H. Glover. 2001. System Review of the Modified Light Duty Utility Arm After the Completion of the Nuclear Waste Removal from Seven Underground Storage Tanks at Oak Ridge National Laboratory. Providence Group, Knoxville, Tennessee, Proceedings of the American Nuclear Society Ninth International Topical Meeting on Robotics and Remote Systems, March 4-8, Seattle, Wash. Available at *http://www.tanks.org/ttgdoc/f046.pdf.*

Gorden, A.E.V., J. Xu, K.N. Raymond, and P. Durbin. 2003. Rational design of sequestering agents for plutonium and other actinides. Chemical Reviews 103(11):4207-4282.

Hamel, W.F., and F.W. Damerow. 2001. Completing HLW Vitrification at the WVDP; The Approach to Final Retrieval, Flushing and Characterization. Waste Management '01 Conference, February 25-March 1, Tucson, Ariz.

Hamel, W.F., C. L. McMahon, and D.C. Meese. 2000. Waste Removal from the West Valley Demonstration Project High-Level Radioactive Waste Storage Tanks. Waste Management '00 Conference, February 27-March 2, Tucson, Ariz. Available at *http://www.tanks.org/ttgdoc/heel/documents/ tucon2000paper.pdf.*

Hanford FFACO (Federal Facility Agreement and Consent Order). 2003. Hanford Federal Facility Agreement and Consent Order by Washington State Department of Ecology, United States Environmental Protection Agency, United States Department of Energy as amended through September 1, 2003. Available at *http://www.hanford.gov/tpa/ agreement-6/tpadoc.pdf.*

Harbour J.R., 2005. Summary of grout development and testing for single-shell tank closure at Hanford. WSRC-TR-2005-00195. Rev. 0. April 28. Aiken, S.C: Savannah River National Laboratory.

Harbour, J.R., T.B. Edwards, T.H. Lorier, C.A. Langton, R.C. Moore, J.L. Krumhansl, K.C. Holt, F. Salas, and C.A. Sanchez. 2004. Stabilizing Grout Compatibility Study. WSRC-TR-2004-00021, Rev. 0. Aiken, S.C.: Westinghouse Savannah River Company.

Hartman, M.J., L.F. Morasch, and W.D. Webber. 2005. Hanford Site Groundwater Monitoring for Fiscal Year 2004. PNNL-15070. March. Richland, Wash.: Pacific Northwest National Laboratory. Available at *http://hanford-site.pnl.gov/envreport/ 2004/15070toc.htm.*

Hatchell, B., T. Alberts, B. Barakov, and B. Lewis. 2001. Russian Pulsating Mixer Pump Deployment in the Gunite and Associate Tanks at ORNL. 2001 ANS Paper 1. Available at *http://www.tanks.org/ttgdoc/f001.pdf.*

Hewitt, B. 2005a. E-mail communication from B. Hewitt (DOE-ORP) to F. Parker, Re: Hanford responses to NAS questions. December 7.

Hewitt, B. 2005b. E-mail communication from B. Hewitt (DOE-ORP) to F. Parker, Re: Hanford Information. December 14.

Hewitt, B. and T. Sams. 2005. Tank Waste Residuals. Presentation to the Committee on the Management of Certain Radioactive Waste Streams Stored in Tanks at Three Department of Energy Sites, July 22, Richland, Wash.

Hill, A.J., Jr. 1978. Chemical Dissolving of Tank 16H Sludge, DPST-78-208. January 9.

Hill, P.J. 2005. Summary of the call on oxalic acid between Savannah River Site staff and selected committee members. October 19.

Hill, P.J. 2006. E-mail from P.J. Hill (Westinghouse Savannah River Company) to M. Lowenthal, Re: Inventory Uncertainty. March 21.

Hintze, D. 2005. Tank Waste Removal Processes. Presentation to the Committee on the Management of Certain Radioactive Waste Streams Stored in Tanks at Three Department of Energy Sites, April 14, Augusta, Ga.

Hobbs, D.T. 2003. Impacts of Sodium Oxalate on High-Level Waste Processing at the Savannah River Site. WSRC-TR-2003-00120, Revision 0. Aiken, S.C.: Westinghouse Savannah River Company.

Hoitink, D.J., K.W. Burk, J.V. Ramsdell, Jr., and W.J. Shaw. 2005. Hanford Site Climatological Summary 2004 with Historical Data. PNNL-15160. May 2005. Richland, Wash.: Pacific Northwest National Laboratory.

Honeyman. 2005. Understanding of Hanford Tank Wastes. Presentation to the Committee on the Management of Certain Radioactive Waste Streams Stored in Tanks at Three Department of Energy Sites, July 22, Richland, Wash.

IAEA (International Atomic Energy Agency). 1999. Near Surface Disposal of Radioactive Waste. Safety Requirements. Safety Standards Series. WS-R-1. Vienna: International Atomic Energy Agency.

IAEA. 2000. Predisposal Management of Radioactive Waste, Including Decommissioning. Safety Requirements. Safety Standards Series. WS-R-2. Vienna: International Atomic Energy Agency.

IAEA. 2004. Long Term Behaviour of Low and Intermediate Level Waste Packages Under Repository Conditions. TECDOC Series No.1397. Vienna: International Atomic Energy Agency.

Idaho FFACO. 1991. Idaho National Engineering Laboratory (INEL) Federal Facility Agreement and Consent Order, December 9, 1991. The United States Environmental Protection Agency, Region 10, State of Idaho, Department of Health and Welfare, and the United States Department of Energy. Available at *http://web.em.doe.gov/2001_Agreements/Idaho/INEEL_FFA_CO_12-9-91.pdf.*

Idaho SACO (Idaho Settlement Agreement and Consent Order). 1995. Settlement Agreement and Consent Order, October 1985. State of Idaho, United States Navy, the United States Department of Energy reached agreement (most often called the Settlement Agreement. Available at *http://www.deq.state.id.us/inl_oversight/contamination/settlement_agreement_entire.cfm.*

IDQTF (Intergovernmental Data Quality Task Force). 2005. Uniform Federal Policy for Quality Assurance Project Plans Evaluating, Assessing, and Documenting Environmental Data Collection and Use Programs. Part 1: UFP-QAPP Manual. Final Version 1. March. EPA-505-B-04-900A; DoD: DTIC ADA 427785. Washington D.C.: U.S. Environmental Protection Agency and U.S. Departments of Defense and Energy. Available at *http://www.epa.gov/fedfac/pdf/ufp_qapp_v1_0305.pdf.*

INEEL (Idaho National Engineering and Environmental Laboratory). 1999. Idaho Nuclear Technology and Engineering Center Tank Farm Facility (TFF) Closure TFF WM-182 Grout Mock-Up October. INEEL/EXT-99-01067. Idaho Falls, Idaho: Idaho National Engineering and Environmental Laboratory, Bechtel BWXT Idaho, LLC.

INEEL. 2001. *INTEC* Tank Farm Facility Closure Mockup Test Report Project File No. 01572. INEEL/EXT-2001-00299. Idaho Falls, Idaho: Idaho National Engineering and Environmental Laboratory, Bechtel BWXT Idaho, LLC.

ITSR (Innovative Technology Summary Report). 1999a. Pipe Explorer ™ Surveying System. Innovative Technology Summary Report. Industry Programs Crosscut and Deactivation and Decommissioning Focus Area. DOE/EM-0440. Available at *http://apps.em.doe.gov/OST/pubs/itsrs/itsr74.pdf.*

ITSR. 1999b. Pipe Crawler® Internal Piping Characterization System. Innovative Technology Summary Report. Deactivation and Decommissioning Focus Area. DOE-EM-0355. OST reference 1810. May. Available at *http://apps.em.doe.gov/OST/ pubs/itsrs/itsr1810.pdf.*

ITSR. 1999c. Pipe Inspection Using the Pipe Crawler. Innovative Technology Summary Report. May 1. Deactivation and Decommissioning Focus Area. DOE/EM-0425. May. Available at *http://apps.em.doe.gov/OST/pubs/itsrs/itsr1810-may99.pdf.*

ITSR. 1999d. Pipe Inspection Using the BTX-II. Deactivation and Decommissioning Focus Area. OST reference 1811. May. Available at *http://apps.em.doe.gov/ OST/pubs/ itsrs/itsr1811.pdf.*

ITSR. 1999e. Pipe Cutting and Isolation System. Tanks Focus Area. DOE/EM-0448. OST Reference 2093. August. Available at *http://apps.em.doe.gov/OST/pubs/itsrs/itsr 2093. pdf.*

ITSR. 1999f. AEA Fluidic Pulse Jet Mixer. Tanks Focus Area. DOE/EM-0447. OST Reference 1511. August. Available at *http://apps.em.doe.gov/ost/pubs/itsrs/itsr1511. pdf.*

ITSR. 2001. Heavy Waste Retrieval System. Innovative Technology Summary Report. OST/TMS-ID 2194. Robotics Crosscutting Program and Tanks Focus Area. DOE-EM-0595. July. Available at *http://apps.em.doe.gov/OST/pubs/itsrs/itsr2194.pdf.*

Karraker, D. G. 1998. Solubility of Uranium, Plutonium, Iron and Manganese in Weak Nitric Acid and Oxalic Acid Solutions: A Literature Survey, WSRC-RP-98-00091. March 5. Aiken, S.C.: Westinghouse Savannah River Company.

Karraker, D.G., T.S. Rudisill, and M.C. Thompson. 2001. Studies of the Effect of Acetohydroxamic Acid on Distribution of Plutonium and Neptunium by 30 Vol Percent Tributyl Phosphate. WSRC-TR-2001-00509. November 28. Aiken, S.C.: Westinghouse Savannah River Company.

Kosson, D., H. Grogan, K. Higley, R. Maddalena, and C. Whipple. 2004. Merit Panel Review of the C-Tank Farm Closure Performance Assessment. Final Report. April 20. Richland, Wash.: CH2M-Hill Hanford Group, Inc. Available at *http://www.hanford.gov/orp/?page=61&parent=128.*

Lambert, S.L. 2005. Residual Waste Inventories in the Plugged and Abandoned Pipelines at the Hanford Site. RPP-25113. September 30. Richland, Wash.: DOE.

Langton, C. 2005. Tank Closure Grouts. Presentation to the Committee on the Management of Certain Radioactive Waste Streams Stored in Tanks at Three Department of Energy Sites, April 14, Augusta, Ga.

Langton C., R. Spence, and J. Barton. 2001. State of the Art Report on High-Level Waste Tank Closure (U). July 31. WSRC-TR-2001-00359, Revision 0. Aiken, S.C.: Westinghouse Savannah River Company.

Langton, C.A., J.R. Harbour, D.H. Miller, W.L. Mhyre, C.N. Shramek, J. Carroll, E. Johnson, and A.L. Furgeson. 2003. Grout Placement and Property Evaluation for Closing Hanford High-Level Waste Tanks–Scale-Up Testing. December 15. WSRC-TR-2003-00556, Rev. 0. Aiken, S.C.: Westinghouse Savannah River Company.

Last, G.V., and D.G. Horton. 2000. Review of Geophysical Characterization Methods Used at the Hanford Site. PNNL-13149. March. Richland, Wash.: Pacific Northwest National Laboratory.

Latchum, J.W., A.L. Biladeau, E.S. Brown, D.D. Deming, B.L. Schmalz, F.M. Warzel, and L.J. Weber. 1962. Report of the Investigating Committee CPP Waste Tank WM-187 Leakage of March 17, 1962. LA-53-62A. April 17. Phillips Petroleum Company. Atomic Energy Division. Available at *http://www.id.doe.gov/InsideNEID/3116_Links. htm.*

LEAF v. Hodel (*Legal Environmental Assistance Foundation* v. *Hodel*). 1984. 586 F. Supp. 1163 (ED Tenn. 1984). April.

Lewis, B., P. Gibbons, and K. Manke. 2002a. Annex US(1). Waste Mobilization and Removal from Oak Ridge National Laboratory Gunite and Associated Tanks. P02-114917. September 11. Available at *http://www.tanks.org/ttgdoc/GAAT-IAEA-us1. doc.*

Lewis, B., P. Gibbons, and D. Short. 2002b. Fluidic Pulse Jet Mixer at the Bethel Valley Evaporator Service Tanks. P02-114813. August 29. Oak Ridge, Tenn.: Oak Ridge National Laboratory. Available at *http://www.tanks.org/BrowseDocuments.asp? SearchText=.*

Lockie, K. 2005a. Overview of Reprocessing, Waste Generation, Tank Farm, and Calcine Storage. Presentation to the Committee on the Management of Certain Radioactive Waste Streams Stored in Tanks at Three Department of Energy Sites, July 25, Idaho Falls, Idaho.

Lockie, K. 2005b. E-mail communication from K. Lockie (DOE-ID) to B. Pastina, Re: I need a picture of the high-pressure water jet. October 25.

Lockie, K. 2006. E-mail communication from K. Lockie, (DOE-ID) to D. Thompson, Re: INL map that is better than this one for NAS Certain Tank Wastes committee report, with attachment, groundwater flow regional.jpg. March 27.

Lockie, K., K. Quigley, and N. Stanisich. 2005. Plans for Cleaning and Closure of Idaho Tank Farm. Presentation to the Committee on the Management of Certain Radioactive Waste Streams Stored in Tanks at Three Department of Energy Sites, July 26, Idaho Falls, Idaho.

Lokken, R.O. P.F.C. Martin, and, J.W. Shade. 1992. Durability of Double-Shell Tank Waste Grouts. December. PNL-7835. Richland, Wash.: Pacific Northwest National Laboratory.

Lukens, W., J. Bucher, D.K. Shuh, and E.N. Edelstein. 2005. Evolution of technetium speciation in reducing grout. Environmental Science & Technology 39:8064-8070.

Mahoney, M. 2005 High Level Waste System Analysis. Presentation to the Committee on the Management of Certain Radioactive Waste Streams Stored in Tanks at Three Department of Energy Sites, May 5, Augusta, Ga.

Mahoney, M.J., and P.D. d'Entremont. 2004. Interim Salt Processing Strategy Planning Baseline. CBU-PED-2004-00027, Revision 0. August 27. Aiken, S.C.: Westinghouse Savannah River Company.

Makhijani, A. 2004. The Savannah River at Grievous Risk: Analysis of the Proposal to Allow the Department of Energy to Leave a Significant Portion of Its High-Level Radioactive Waste at the Savannah River Site in the Savannah River Watershed. May 17. Takoma Park, Md.: Institute for Energy and Environmental Research. Available at *http://www.ieer.org/reports/srs/hlwanalysis.html.*

Makhijani A., and M. Boyd. 2004. Nuclear Dumps by the Riverside: Threats to the Savannah River from Radioactive Contamination at the Savannah River Site. March 11. Takoma Park, Md.: Institute for Energy and Environmental Research. Available at *http://www.ieer.org/reports/srs/fullrpt.pdf.*

Makhijani, A., R. Alvarez., and B. Blackwelder. 1986. Deadly Crop in the Tank Farm: An Assessment of the Management of High-Level Radioactive Wastes in the Savannah River Plant Tank Farm, Based on Official Documents, July. Takoma Park, Md.: Institute for Energy and Environmental Research. Available *at http://www.ieer.org/reports/index.html.*

Mann, F.M. 2003. Annual Summary of the Immobilized Low-Activity Waste Performance Assessment for 2003, Incorporating the Integrated Disposal Facility Concept. DOE/ORP-2000-19, Revision 3. Richland, Wash.: CH2M Hill Hanford Group, Inc. Available at *http://www.hanford.gov/docs/gpp/fieldwork/ilaw/ORP-2000-19_r3.pdf.*

Mann, F.M. 2005. Immobilized Low-Activity Waste (ILAW) Properties and Disposal in Integration Disposal Facility. Presentation to the Committee on the Management of Certain Radioactive Tank Waste Streams Stored in Tanks at Three Department of Energy Sites, July 22, Richland, Wash.

Mann, F.M., and M. Connelly. 2003. Preliminary Performance Assessment for Waste Area Management Area C at the Hanford Site, Washington. DOE/ORP-2003-11, Rev. 0. Richland, Wash.: CH2M Hill Hanford Group, Inc. Available at *http://www.hanford.gov/orp/?page=61&parent=128.*

Mann, F.M., C.T. Kincaid, and W.J. McMahon. 1999. Computer Code Selection Criteria for Flow and Transport Code(s) To Be Used in Vadose Zone Calculations for Environmental Analyses in the Hanford Site's Central Plateau. HNF-5294, Rev. 0. Richland, Wash. Available at *http://www.hanford.gov/docs/gpp/library/gwvz/ CodeSelect.pdf.*

Mann, F.M., R.J. Puigh II, S.H. Finfrock, E.J. Feeman, Jr., R. Khalelle, D.H. Bacon, M.P. Bergeron, B.P. McGrail, and S.K. Wurstner. 2001. Hanford Immobilized Low-Activity Waste Performance Assessment: 2001 Version. DOE/ORP-2000-24, Rev. 0. Richland, Wash.: U.S. Department of Energy.

Martin Marietta Energy Systems, EG&G Idaho, Inc., and Westinghouse Savannah River Company. 1994. Radiological Performance Assessment for the E-Area Vaults Disposal Facility (U). WSRC-RP-94-218. April. Aiken, S.C.: Westinghouse Savannah River Company.

Murray, C., G. Last, and M. Truex. 2005. Review of geophysical techniques to define the spatial distribution of subsurface properties or contaminants, Letter report prepared for Fluor Hanford Inc. Richland, Wash.: Pacific Northwest National Laboratory.

Murray, P. 2005. Status of Fluidic Technology for Retrieving Waste from Tanks. Presentation during the 162nd meeting of the Advisory Committee on Nuclear Waste, August 2-3, Rockville, Md.

Nash, K.L., R.E. Barrans, R. Chiarizia, M.L. Dietz, M.P. Jensen, P.G. Rickert, B.A. Moyer, P.V. Bonnesen, J.C. Bryan, and R.A. Sachleben. 2000. Fundamental investigations of separations science for radioactive materials. Solvent Extraction and Ion Exchange 18(4):605-631.

Nash, K.L., A.V. Gelis, M.P. Jensen, A.H. Bond, J.C. Sullivan, L. Rao, and A. Garnov. 2002. Actinides in alkaline media: Dissolution, mineral associations, and speciation in Hanford waste tank sludge simulants. Journal of Nuclear Science and Technology 512.

Nash, K.L. M. Hancock, I. Laszak, and M. Borkowski. 2003. Oxidative leaching of plutonium from simulated Hanford tank-waste sludges. Abstracts of Papers, 225th ACS National Meeting, March 23-27, New Orleans, La., American Chemical Society.

NCRP (National Council on Radiation Protection and Measurements). 2002. Risk-Based Classification of Radioactive and Hazardous Chemical Wastes. NCRP Report No. 139. December 31. Bethesda, Md.: NCRP.

Neuman, S.P., and P.J. Wierenga. 2003. A Comprehensive Strategy of Hydrogeologic Modeling and Uncertainty Analysis for Nuclear Facilities and Sites. NUREG/CR-6805. Washington, D.C.: U.S. Nuclear Regulatory Commission.

NRC (National Research Council). 1994. Alternatives to Ground Water Cleanup. Washington, D.C.: National Academy Press.

NRC. 1995. Interim Report of the Committee on Research and Peer Review in EPA. Washington, D.C.: National Academy Press.

NRC 1996a. Environmental Management Technology-Development Program at the Department of Energy: 1995 Review. Washington, D.C.: National Academy Press. Available at: *http://books.nap.edu/catalog/5172.html.*

NRC. 1996b. Nuclear Wastes: Technologies for Separations and Transmutation. Washington, D.C.: National Academy Press.

NRC. 1997a. Building an Effective Environmental Management Science Program: Final Assessment. Washington, D.C.: National Academy Press.

NRC. 1997b. Peer Review in the Department of Energy-Office of Science and Technology: Interim Report. Washington, D.C.: National Academy Press.

NRC. 1998. Peer Review in Environmental Technology Development Programs. Washington, D.C.: National Academy Press.

NRC. 1999a. An End State Methodology for Identifying Technology Needs for Environmental Management, with an Example from the Hanford Site Tanks. Washington, D.C.: National Academy Press. Available at *http://books.nap.edu/ catalog/6366.html.*

NRC. 1999b. Alternative High-Level Waste Treatments at the Idaho National Engineering and Environmental Laboratory. Washington, D.C.: National Academy Press.

NRC. 1999c. The State of Development of Waste Forms for Mixed Wastes. Washington, D.C.: National Academy Press.

NRC. 2000a. Alternatives for High-Level Waste Salt Processing at the Savannah River Site. Washington, D.C: National Academy Press.

NRC. 2000b. Strengthening Science at the U.S. Environmental Protection Agency: Research-Management and Peer-Review Practices. Washington, D.C.: National Academy Press.

NRC. 2000c. Long-Term Institutional Management of U.S. Energy Department of Energy Legacy Waste Sites. Washington, D.C.: National Academy Press.

NRC, 2000d. Seeing into the Earth: Noninvasive Characterization of the Shallow Subsurface for Environmental and Engineering Applications Washington, D.C.: National Academy Press.

NRC. 2000e. Research Needs in Subsurface Science. Washington, D.C.: National Academy Press. Available at: *http://books.nap.edu/catalog/9793.html.*

NRC. 2001a. Research Needs for High-Level Waste Stored in Tanks and Bins at U.S. Department of Energy Sites: Environmental Management Science Program. Washington, D.C.: National Academy Press. Available at *http://books.nap.edu/ catalog/10191.html.*

NRC. 2001b. Science and Technology for Environmental Cleanup at Hanford. Washington, D.C.: National Academy Press. Available at *http://books.nap.edu/catalog/10220.html.*

NRC. 2001c. Research Opportunities for Deactivating and Decommissioning Department of Energy Facilities. Washington, D.C.: National Academy Press. Available at *http://books.nap.edu/catalog/10184.html.*

NRC. 2001d. A Strategic Vision for Department of Energy Environmental Quality Research and Development. Washington, D.C.: National Academy Press. Available at *http://www.nap.edu/catalog/10207.html.*

NRC. 2001e. Conceptual Models of Flow and Transport in the Fractured Vadose Zone. Washington, D.C.: National Academy Press. Available at *http://fermat.nap.edu/catalog/10102.html.*

NRC. 2002a. Review Procedures for Water Resources Project Planning. Washington, D.C.: National Academy Press.

NRC. 2002b. Research Opportunities for Managing the Department of Energy's Transuranic and Mixed Wastes. Washington, D.C.: National Academy Press. Available at *http://books.nap.edu/catalog/10513.html.*

NRC. 2003a. Improving the Scientific Basis for Managing DOE's Excess Nuclear Materials and Spent Nuclear Fuel Washington, D.C.: The National Academies Press. Available at *http://www.nap.edu/catalog/10684.html.*

NRC. 2003b. Long-Term Stewardship of DOE Legacy Waste Sites: A Status Report. Washington, D.C.: The National Academies Press.

NRC. 2003c. Environmental Cleanup at Navy Facilities: Adaptive Site Management. Washington, D.C.: The National Academies Press.

NRC. 2005a. Tank Wastes Planned for On-site Disposal at Three Department of Energy Sites. The Savannah River Site. Interim Report. Washington, D.C.: The National Academies Press.

NRC. 2005b. Risk and Decisions About Disposition of Transuranic and High-Level Radioactive Waste. Washington, D.C.: The National Academies Press.

NRC. 2005c. Improving the Characterization and Treatment of Radioactive Wastes for the Department of Energy's Accelerated Site Cleanup Program. Washington, D.C: The National Academies Press.

NRDC (Natural Resources Defense Council). 2003. DOE Attempting Legislative End-Run Around Court Ruling on Nuke Waste. July 17. Available at *http://www.nrdc.org/BushRecord/nuclear.asp.*

NRDC, 2004a. Senate Passes Graham Amendment to Reclassify Radioactive Waste to Avoid Cleaning It Up. NRDC Press Archives. Statement of Karen Wayland. June 3. Available at *http://www.nrdc.org/media/pressreleases/ 040603.asp.*

NRDC. 2004b. Backroom Deal: Congress OKs Abandoning Highly Radioactive Waste in South Carolina and Idaho, Threatening Drinking Water Supplies. NRDC press archives. Statement of Geoffrey Fettus. October 8. Available at *http://www.nrdc.org/media/pressreleases/041007a.asp.*

NRDC. 2005. Energy Department Cannot Leave Highly Radioactive Waste in Tanks Without EPA or NRC Approval, National Academy of Sciences Panel Says. NRDC Calls for Congress to Ensure Oversight of DOE to Protect Public Health and the Environment. March 1. Available at *http://www.nrdc.org/media/pressreleases/050301.asp.*

NRDC, Snake River Alliance, and the Confederated Tribes and Bands of the Yakama Nation v. Abraham, Secretary, U.S. Department of Energy. 2003. Memorandum Decision. Civ. No. 01-043-S-BLW. U.S. District Court for the District of Idaho. 271 F. Supp. 2d 1260. July 2.

NRDC, Snake River Alliance, Confederated Tribes and Bands of the Yakama Nation, and Shoshone Bannock Tribes v. Abraham, Secretary, U.S. Department of Energy. Opinion. U.S. Court of Appeals for the Ninth Circuit. 388 F.3d 701. November 5.

O'Brien, R.F., P.G. Heasler, and L. Rowell. 2001. West Valley Tank 8D-1 and 8D-2 Inventory Estimation Methodology. PNNL-13585. July. Richland, Wash.: Pacific Northwest National Laboratory.

Olson, A. 2005. Origin and Characteristics of Idaho Tank Waste. Presentation to the Committee on the Management of Certain Radioactive Waste Streams Stored in Tanks at Three Department of Energy Sites, July 26, Idaho Falls, Idaho.

OMB (Office of Management and Budget). 2006. Budget of the United States Government, Fiscal Year 2007. Available at *http://www.whitehouse.gov/omb/budget/fy2007/.*

Pacquet, E.A., and G.A. Leshikar. 2001. Evaluation of Flygt™ Propeller Mixers for Double-Shell Tank High-Level Waste Auxiliary Solids. RPP-6421, Rev 0. June 1. Richland, Wash.: CH2M Hill Hanford Group, Inc. Available at *http://www.tanks.org/Browse Documents.asp?SearchText=.*

Paperiello, C.J. 1997. Classification of Hanford Low-Activity Tank Waste Fraction. Letter from, Director, Office of Nuclear Material Safety and Safeguards, U.S. Nuclear Regulatory Commission, to J. Kinzer, Assistant Manager, Office of Tank Waste Remediation System, Richland Operations Office, U.S. Department of Energy. Washington, D.C. June 9. Available at *http://www.nrc.gov/reading-rm/doc-collections/commission/secys/1997/secy1997-083/1997-083scy.html#ATTACHMENT%202.*

Patello G.K., R.L. Russell, and L. Renee. 1999. Oxalate destruction in a simulated high-level tank residual waste using nitric acid. Ceramic Transactions 93:179.

Patterson, M. 2005. Calcine Characteristics and Plans for Calcine Retrieval and Disposal. Presentation to the Committee on the Management of Certain Radioactive Waste Streams Stored in Tanks at Three Department of Energy Sites, July 26, Idaho Falls, Idaho.

PCCRARM (Presidential-Congressional Commission on Risk Assessment and Risk Management). 1997a. Framework for Risk Assessment: Final Report, Volume 1. Available at *http://www.riskworld.com/Nreports/1996/risk_rpt/Rr6me001.htm.*

PCCRARM. 1997b. Risk Assessment and Risk Management in Regulatory Decision-Making: Final Report, Volume 2. Available at *http://www.riskworld.com/Nreports/ 1996/risk_rpt/Rr6me001.htm.*

Perks, R. 2004. Rewriting the Rules, 3rd Annual Edition. The Bush Administration's Assault on the Environment. April. Washington, D.C.: Natural Resources Defense Council. Available at *http://www.nrdc.org/legislation/rollbacks/rr2004.pdf.*

Petersen, C.A. 1996. Technical Basis for Classification of Low-Activity Waste Fractions from Hanford Site Tanks. WHC-SD-WM-TI-699, Rev. 2. September. Richland, Wash.: Westinghouse Hanford Company.

Philipose, K. 500-Year Concrete for a Waste Repository. 1988. Presented at the Symposium on Waste Management '88, University of Arizona, Tucson, Ariz.

Philipose, K., R.F. Feldman, and J.J. Beaudoin. 1990a. Qualifying Concrete for a Radioactive Waste Repository—Durability Assessment from Measurement of Ionic Ingress. Presented at the Symposium on Waste Management '90, March, Tucson, Ariz.

Philipose, K., R.F. Feldman, and J.J. Beaudoin. 1990b. Use of Blast Furnace Slag in Making Durable Concrete for Waste Management Repositories. Presented at the Air and Waste Products Management Meeting, May, Pittsburgh, Pa.

Philipose, R.F., R.F. Feldman, and J.J. Beaudoin. 1991. Durability Predictions from Rate of Diffusion Testing of Normal Portland Cement, Fly Ash, and Slag Concrete. SP-126-18. CANMET/ACI International Conference, Montreal.

Philipose, K., J.J. Beaudoin, and R.F. Feldman. 1992. Degradation of Normal Portland and Slag Concrete Under Load due to Reinforcement Corrosion. Atomic Energy of Canada Limited. AECL Report, AECL-10490. August.

PNNL (Pacific Northwest National Laboratory). 2005a. Hanford Site Groundwater Monitoring for Fiscal Year 2004. M.J. Hartman, L.F. Morasch, and W.D. Webber, eds. PNNL-15070. March. Richland, Wash.: Pacific Northwest National Laboratory.

PNNL. 2005b. Hanford Site Environmental Report for Calendar Year 2004. T.M. Poston, R.W. Hanf, and R.L. Dirkes, eds. PNNL-15222. Richland, Wash.: Pacific Northwest National Laboratory.

Purohit, J. 2005. The Second Stage Tank 5 Interim Sludge Removal Strategy. CBU-LTS-2005-00272. Rev. 0. Memorandum from J. Purohit to J.E. Herbert. November 18. Aiken, S.C.: Westinghouse Savannah River Company.

Quigley, K. 2005. Plans for Cleaning and Closure of Idaho Tank Farm. Presentation to the Committee on the Management of Certain Tank Waste Streams Stored in Tanks at Three Department of Energy Sites, July 26, Idaho Falls, Idaho.

Quintero, R., 2005a. E-mail communication from R. Quintero (SRS) to B. Pastina with attachment, Retrieval Cost by Tank. November 17.

Quintero, R. 2005b. E-mail communication from R. Quintero (SRS) to B. Pastina, Re: Hanford Responses, with attachments, Single-Shell Tank Retrieval Evaluation Form for Tank 241-C-202 and Single-Shell Tank Retrieval Evaluation Form for Tank 241-C-203. September 16.

Quintero, R. 2006. E-mail communication from Roger Quintero (SRS) to B. Pastina, Re: A few extra questions. January 18.

Raymond, K.N. 1990. Biomimetic metal encapsulation. Coordination Chemistry Reviews. 105:135.

Reboul, S.H., and P.J. Hill. 2005. Characteristics and Understanding of SRS Tank Farm Waste. Presentation to the Committee on the Management of Certain Radioactive Tank Waste Streams Stored in Tanks at Three Department of Energy Sites, April 14, Augusta, Ga.

Reddick, G.W., Jr. 2004. Campaign Report for the Retrieval of Waste Heel from Tank 241-C-106. RPP-19919, Rev. 0. Richland, Wash.: CH2M Hill Hanford Group, Inc.

Reynolds D.A. 2004. Retrieval Pumps-Recent Experience. Hanford/ Savannah River Technical Exchange, January 13-15, Richland, Wash.

Roberson, J. 2004. Testimony by Jesse Roberson, Assistant Secretary for Energy for Environmental Management, before the U.S. Senate Subcommittee on Strategic Forces, Committee on Armed Services on Waste Incidental to Reprocessing, February 25. Transcript of statement available at *http://lgraham.senate.gov/index.cfm?mode=speechpage&id=222465*.

Rosenberger, K.H., B.C. Rogers, and R.K. Cauthen. 2005. Saltstone Performance Objective Demonstration Document (U) CBU-PIT-2005-00146, Revision 0. June. Aiken, S.C.: Westinghouse Savannah River Company.

Ross, S. 2005. Savannah River Site Meeting Performance Objectives for On-site Disposition of Tank Waste. Presentation to the Committee on Management of Certain Radioactive Waste Streams Stored in Tanks at Three Department of Energy Sites, April 14, Augusta, Ga.

Ross, S. 2006a. E-mail communication from S. Ross (SRNL) to M. Lowenthal, Re: Question concerning F Tank Farm residual inventory. January 5.

Ross, S. 2006b. E-mail communication from S. Ross (SRNL) to M. Lowenthal, Re: SRS F Tank Farm residual inventory. February 6.

Saldivar, E. 2002. Savannah River Site Waste Removal Program Past, Present, and Future. WM'02 Conference, February 24-28, Tuscan, Ariz.

Sams, T.L. 2004. Basis for Exception to the Hanford Federal Facility Agreement and Consent Order Waste Retrieval Criteria for Single-Shell Tank 241-C-106. June. Richland, Wash.: DOE Office of River Protection.

Sams, T.L. 2005. The Certain Tank Waste Committee Discussion of WMA Cleanup and Closure/Tank Closure Technologies. Presentation to the Committee on the Management of Certain Tank Waste Streams Stored in Tanks at Three Department of Energy Sites, July 22, Richland, Wash.

SCDHEC (South Carolina Department of Health and Environmental Control). 2004a. Proper Closeout of Wastewater Treatment Facilities 61-82. State Register 28(9), effective September 24.

SCDHEC. 2004b. Hazardous Waste Management Regulations 61-79. State Register 28(9), effective September 24.

SCDHEC. 2005a. Letter from Mr. David E. Wilson, Jr., South Carolina Department of Health and Environmental Control to Mr. Matthew Duchesne, U.S. Department of Energy, Office of Environmental Management, Re: South Carolina Department of Environmental Control Comments Regarding the Draft Section 3116 Determination for Closure of Tank 19 and Tank 18 at the Savannah River Site. Attachment: SCDHEC Comments Draft Section 3116 Determination for Closure of Tank 19 and Tank 18 at the Savannah River Site. November 21.

SCDHEC. 2005b. Letter from Mr. David E. Wilson, Jr., South Carolina Department of Health and Environmental Control to Mr. Randall Kaltreider, U.S. Department of Energy, Office of Environmental Management, Re: Enclosed Written Comments Regarding the Draft 3116 Determination for Salt Waste Disposal at the Savannah River Site. Attachment: SCDHEC Comments on Draft Section 3116 Determination Salt Waste Disposal Savannah River Site. May 12.

Schaus, P.S. 2005. Waste Management Area Integration Study—Waste Management Area C. RPP-PLAN 25062. Rev 0. July 28. Richland, Wa: DOE Office of River Protection.

Schepens, R. 2005. ORP Opening Remarks. Presentation to the Committee on the Management of Certain Tank Waste Streams Stored in Tanks at Three Department of Energy Sites, July 22, Richland, Wash.

Smith, B. 2004. What the DOE Knows It Doesn't Know About Grout: Serious Doubts Remain About the Durability of Concrete Proposed to Immobilize High-Level Nuclear Waste in the Tank Farms at the Savannah River Site and Other DOE Sites. October 18. Takoma Park, Md.: Institute for Energy and Environmental Research. Available at *http://www.ieer.org/reports/srs/grout.pdf*.

Spears, T.J. 2005. Salt Waste Processing Facility (SWPF) Project Line Item 05-D-405. Project Status Review for the National Academy of Sciences. Presentation to the Committee on the Management of Certain Radioactive Waste Streams Stored in Tanks at Three Department of Energy Sites, May 5, Augusta, Ga.

SRS FFA (Savannah River Site Federal Facility Agreement). 1993. Federal Facility Agreement for the Savannah River Site. The United States Environmental Protection Agency Region IV and the United States Department of Energy and the South Carolina Department of Health and Environmental Control. Federal Facility Agreement under Section 120 of CERCLA and Sections 3008(h) and 6001 of RCRA. Effective Date August 16. Administrative Docket No: 89-05-FF. WSRC-OS-94-42. Available at *http://www.srs.gov/general/programs/ soil/ffa/ffaa.pdf*.

Steiger, M.D., and M.C. Swenson. 2005. Calcined Waste Storage at the Idaho Nuclear and Technology Engineering Center. INEEL/EXT-98-00455. Rev. 2. January. Idaho Falls, Idaho: Idaho National Engineering and Environmental Laboratory.

Terhune, C., and J. Kasper. Salt Waste Processing Facility (SWPF) Project Technical and Schedule Review Waste Management Committee. Presentation to the SRS Citizens Advisory Board. December 13, 2005.

Thompson, M.C., M.A. Norato, G.F. Kessinger, M.C. Thompson, M.A. Norato, G.F. Kessinger, R.A. Pierce, T.S. Rudisill, and J.D. Johnson. 2002. Demonstration of the UREX Solvent Extraction Process with Dresden Reactor Fuel Solution. WSRC-TR-2002-00444, Revision 0. Aiken, S.C.: Westinghouse Savannah River Company.

Trever, K. 2005. E-mail communication from K. Trever, INL Oversight Office, State of Idaho to B. Pastina, Re: Table Revisions, attachment, NAS Table X.

Trever, K. 2006. Letter from Ms. Kathleen Trever, INL Oversight Office, State of Idaho to Mr. Keith Lockie, Idaho Operations Office, U.S. Department of Energy and Mr. Xiaosong Yin, Office of Nuclear Material Safety and Safeguards, U.S. Nuclear Regulatory Commission, Re: State of Idaho Document Review Report on §3116 Determination for INL Tank Farm. January 17. Idaho Falls, Idaho.

Triay, I. 2005. DOE Perspective on National Research Council Study of Certain Tank Wastes. Presentation to the Nuclear and Radiation Studies Board, September 12, Washington, D.C.

Unger, A.J.A., B. Faybishenko, G.S. Bodvarsson, and A.M. Simmons. 2004. Simulating Infiltration Tests in Fractured Basalt at the Box Canyon Site, Idaho. Vadose Zone Journal 3(February):75-89.

USNRC (U. S. Nuclear Regulatory Commission). 1989. Recent Safety-Related Incidents at Large Irradiators. Information Notice No. 89-82. December 7. Washington, D.C.: U.S. Nuclear Regulatory Commission.

USNRC. 1995. Final Branch Technical Position on Concentration Averaging and Encapsulation, Revision in Part to Waste Classification Technical Position. Letter by Malcolm R. Knapp to Commission Licensees, January 17. Washington, D.C.: U.S. Nuclear Regulatory Commission.

USNRC. 1997a. U.S. Nuclear Regulatory Commission Review of the Department of Energy at Savannah River High-Level Waste Tank Closure Methodology. Washington, D.C.: U.S. Nuclear Regulatory Commission.

USNRC. 1997b. NUREG-1200, Standard Review Plan for the Review of a License Application for a Low-Level Radioactive Waste Facility. Washington, D.C.: U.S. Nuclear Regulatory Commission.

USNRC. 1999. Classification of Savannah River Residual Tank Waste as Incidental. SECY-99-284. December 15. Washington, D.C: U.S. Nuclear Regulatory Commission.

USNRC. 2002. Decommissioning Criteria for the West Valley Demonstration Project (M-32) at the West Valley Site; Final Policy Statement. Federal Register 67(22): 5003-5012 February 1. Available at *http://www.epa.gov/fedrgstr/EPA-IMPACT/2002/February/Day-01/i2373.htm.*

USNRC. 2003. Letter from Lawrence E. Kokajko, Acting Branch Chief, Environmental Performance and Assessment Branch, Division of Waste Management, Office of Nuclear Material Safety and Safeguards, U.S. Nuclear Regulatory Commission to Joel T. Case, Director, INTEC Waste Programs, U.S. Department of Energy, Idaho Operations Office, Re: NRC Review of Idaho Nuclear Technology and Engineering Center Draft Waste-Incidental-to-Reprocessing Determination for Tank Farm Residuals—Conclusions and Recommendations, June 18.

USNRC. 2005. Draft interim concentration averaging guidance for waste determinations. Federal Register 70(241):74846-74850. December 16.

USNRC. 2006. Request for additional information on the Draft Section 3116 Determination for Idaho Nuclear and Engineering Center Tank Farm Facility. DOE/NE-ID-11226. January 10. Washington, D.C.: U.S. Nuclear Regulatory Commission.

Utsunomiya, S., L-M. Wang, M. Douglas, S.B. Clark and R.C. Ewing. 2003. The effect of ionizing radiation on uranophane. American Mineralogist 88:159-166.

Vesco, D.P., B.L. Burks, S.D. Van Hoesen, and D.J. Kington. 2001. Lessons Learned and Final Report for the Houdini Vehicle Remote Operations at the Oak Ridge National Laboratory. American Nuclear Society Paper 083. Available at http://www.tanks. *org/ttgdoc/f083.pdf.*

Wagoner, J.D. 1995. Letter from J.D. Wagoner, Manager, Department of Energy, Richland Operations Office to J.T. Conway, Chairman, Defense Nuclear Facilities Safety Board, December 16, forwarding U.S. Department of Energy, Richland Operations Office (RL) Complete Historical Tank Layering Models—Update of Defense Nuclear Facilities Safety Board (DNFSB) Recommendation 93-5, Commitment 1.16. 95-CHD-090. Richland, Wash.: DOE.

Washington State, Departments of Ecology and Health. 2000. Draft Environmental Impact Statement for the Commercial Low-Level Waste Disposal Site; Richland, Wash. 00-05-010. August. Available at *http://www.ecy.wa.gov/biblio/0005010.html.*

Welch, T.D. 2001. Tank Waste Transport, Pipeline Plugging, and the Prospects for Reducing the Risk of Waste Transfers. ORNL/TM-2001/157. August. Oak Ridge, Tenn.: Oak Ridge National Laboratory. Available at *http://www.tanks.org/ttgdoc/ornl-tm-2001-157.pdf.*

West, W.L. 1980. Tank 16 Demonstration Water Wash and Chemical Cleaning Results. DPSP-80-17-23. December 16. Aiken, S.C.: E.I. du Pont de Nemours and Company.

Wiegman, S. 2004. Office of River Protection. Presentation to the National Academies Committee on Risk-Based Approaches for Disposition of Transuranic and High-Level Radioactive Waste, March 9, Richland, Wash.

WSRC (Westinghouse Savannah River Company). 1997. Composite Analysis for E-Area Vault and Saltstone Disposal Facilities, Rev 0, WSRC-RP-97-311. Aiken, S.C.: Westinghouse Savannah River Company.

WSRC. 2000. The Maintenance Program for the E-Area Low-Level Waste Facility and Saltstone Performance Assessments and the Composite Analysis, Draft. SWD-SWE-2000-00053. April 6. Aiken, S.C.: Westinghouse Savannah River Company.

WSRC. 2004a. Saltstone Processing Facility Vault Radionuclide Inventory. CBU-PIT-2004-00019, Rev 0. Aiken, S.C.: Westinghouse Savannah River Company.

WSRC. 2004b. Savannah River Site Groundwater Protection Program. WSRC-TR-2004-00152. Aiken, S.C.: Westinghouse Savannah River Company.

WSRC. 2005. Savannah River Site Environmental Report for 2004. A.R. Mamatey, ed. WSRC-TR-2005-00005. Aiken, S.C.: Westinghouse Savannah River Company.

Appendixes

Appendix A

Biographical Sketches of Committee Members

FRANK L. PARKER (NAE), *chair*, is a distinguished professor of civil and environmental engineering at Vanderbilt University. Dr. Parker's research interests include hazardous chemical and radioactive waste disposal policy, risk analysis of hazardous and radioactive waste disposal, thermal pollution, and water resources engineering. He served in the U.S. Army in a variety of engineering positions and worked for the U.S. Bureau of Reclamation and the Rockland Light and Power Company as a civil and water resources engineer. After graduating from Harvard, Dr. Parker worked for a consulting hydraulic engineering firm and then went to Oak Ridge National Laboratory, where he became head of Radioactive Waste Disposal Research. He also served as head of Radioactive Waste Disposal Research for the International Atomic Energy Agency. In recent years, he has focused on radioactive and hazardous chemical waste problems, with increasing attention to the policy questions associated with these problems at both national and international levels. Dr. Parker was elected to the National Academy of Engineering in 1988 for world leadership in the development of basic information required for the safe disposal of high-level radioactive wastes. Dr. Parker has served on several National Academies committees and boards, including the Board on Radioactive Waste Management, which he chaired from 1984 to 1991.

HADI ABU-AKEEL (NAE) is president of AMTENG Corp., an independent consulting firm. Dr. Abu-Akeel recently retired from FANUC Robotics NA, Inc., an industrial robotics firm, where he was senior vice president and chief engineer. His main areas of expertise include optimization of robot design including trade-offs of performance, cost, manufacturability, application requirements, and user friendliness; utilization of robotic devices to overcome manufacturing productivity challenges and provide cost-effective manufacturing process alternatives; development and application of microsensors for intelligent robots, robotic assist devices, autonomous robots, and remote presence; and risk assessment, safety, and safeguarding of robot applications. In 1997, he was elected to the National Academy of Engineering for contributions to the design, control, and implementation of industrial robots. Since then, he has served as a member of the National Academies' Panel for Manufacturing Engineering and Mechanical Engineering Peer Committee.

JOHN S. APPLEGATE is associate dean for academic affairs and Walter W. Foskett Professor of Law at Indiana University School of Law-Bloomington. He teaches and writes about environmental law, regulation of hazardous substances, risk, environmental remediation, and the U.S. Department of Energy. Mr. Applegate cochaired the long-term stewardship and accelerated cleanup subcommittees of the Department of Energy's Environmental Management Advisory Board. He was previously the James B. Helmer, Jr., Professor of Law at the University of Cincinnati, College of Law and chaired the Fernald Citizens Advisory Board. He has served as a visiting professor at Vanderbilt University Law School, a judicial clerk to the United States Court of Appeals for the Federal Circuit, and an attorney in private practice. He is the author or coauthor of more than 20 articles and the author or editor of books on risk and environmental law.

HOWIE CHOSET is an associate professor of mechanical engineering and robotics at Carnegie Mellon University where he conducts research in motion planning and design of serpentine mechanisms, coverage path planning for demining and painting, mobile robot sensor-based exploration of unknown spaces, and education with robotics. In 1997, the National Science Foundation awarded Dr. Choset its career award to develop motion planning strategies for arbitrarily shaped objects. In 1999, the Office of Naval Research started supporting Dr. Choset through its Young Investigator Program to develop strategies to search for land and sea mines. In 2002, the Massachusetts Institute of Technology magazine *Technology Review* elected Dr. Choset as one of its top 100 innovators in the world under 35. Dr. Choset directs the undergraduate robotics minor at Carnegie Mellon and teaches an overview course on robotics that uses a series of custom-developed Lego labs to complement the course work. Finally, Dr. Choset is a member of an urban search and rescue response team using robots with the Center for Robot Assisted Search-and-Rescue.

ALLEN G. CROFF retired from Oak Ridge National Laboratory (ORNL) in 2003. While employed at ORNL, Mr. Croff was involved in technical studies and program development focused on waste management and nuclear fuel cycles. Mr. Croff chaired a committee of the National Council on Radiation Protection and Measurements that produced the 2002 report *Risk-Based Classification of Radioactive and Hazardous Chemical Wastes*; he also chaired the Nuclear Energy Agency's Nuclear Development Committee for a decade. Mr. Croff is currently vice-chair of the U.S. Nuclear Regulatory Commission's Advisory Committee on Nuclear Waste and a member of the Nuclear Energy Research Advisory Committee. Mr. Croff is currently serving on the National Academies' Nuclear and Radiation Studies Board and has worked on numerous National Academies committees.

PATRICIA J. CULLIGAN is a professor of civil engineering and engineering mechanics at Columbia University. Her research focuses on applying geoengineering principles to understand and control the migration of contaminants from waste disposal sites. In particular, she studies the behavior of miscible contaminants and nonaqueous-phase liquids in soil and fractured rock and the effectiveness of in situ remediation strategies for the cleanup of waste sites. Her research interests also include the design of land-based disposal sites for waste materials. Dr. Culligan has received numerous awards including the Arthur C. Smith Award for Undergraduate Service (1999) and the National Science Foundation CAREER Award (1999). She is the author or coauthor of more than 50 journal articles, book chapters, and refereed conference papers.

KEN CZERWINSKI is an associate professor in the Chemistry Department at the University of Nevada, Las Vegas, and director of the radiochemistry Ph.D. program. His expertise is in actinide chemistry, focusing on understanding, evaluating, and predicting the chemical forms of actinide elements in differing conditions, with research efforts in speciation of actinides in the environment, actinide separations in the nuclear fuel cycle, and actinide chemical forms in solids. Dr. Czerwinski has been an associate professor in the Nuclear Engineering Department at the Massachusetts Institute of Technology and an associate research scientist for the Institut für Radiochemie Technische Universität München. He has been accorded the Presidential Early Career Award in Science and Engineering.

RACHEL J. DETWILER is senior engineer at Braun Intertec Corporation in Minneapolis, Minnesota. Her areas of expertise are construction forensics, construction troubleshooting, concrete durability, transport properties, microstructure, and test methods for concrete and cement-based materials. Dr. Detwiler served in an advisory role for the initial development of a formulation of grout for the stabilization of radioactive and hazardous waste in underground storage tanks at the Savannah River Site until 1996. She has served as a principal engineer at Construction Technology Laboratories; an assistant professor at the University of Toronto; a post-doctoral research fellow at Norges Tekniske Høgskole, Trondheim, Norway; and a design and materials engineer with ABAM Engineers, Inc. She is a member of the American Society for Testing and Materials and a fellow of the American Concrete Institute, where she has served as chair of Committee 227 on Radioactive and Hazardous Waste Management and as a member of Committee 234 on Silica Fume in Concrete. She has received a Norges Teknisk-Naturvitenskapelige Forskninsgråd Fellowship and the Carlson-Polivka Fellowship, and has published more than 50 technical papers related to concrete microscopy, durability, and testing.

EDWIN E. HERRICKS is professor of environmental biology in the Department of Civil and Environmental Engineering at the University of Illinois at Urbana-Champaign. His areas of expertise include aquatic ecology and stream ecosystem and watershed management, and he has broad experience in the identification, assessment, and restoration of the adverse effects of man's activities on streams, rivers, lakes, and their watersheds. His current research has focused on the development of methods to restore stream habitat, manage wildlife on and around airports, manage stormwater runoff, and develop ecohydrology and ecohydraulics methods. Dr. Herricks has recently served on National Research Council panels dealing with endangered species and the Platte River and evaluation of the U.S. Army Corps of Engineers project process. He has written numerous articles and papers on the broad theme of improving engineering design and environmental decision making. He is a member of the Urban Water Resources Research Council of the American Society of Civil Engineers and is chairman of a task group on receiving system effects from urban runoff.

TISSA H. ILLANGASEKARE is the AMAX Distinguished Chair of Environmental Sciences and Engineering and a professor of civil engineering at the Colorado School of Mines (CSM). He is also the director of the Center for the Experimental Study of Subsurface Environmental Processes (CESEP) located at CSM. His expertise is in mathematical and numerical modeling of flow and transport in porous and fractured media, unsaturated and saturated zone processes, surface-subsurface interaction, snow hydrology, multiphase flow, aquifer remediation, and physical modeling of flow and transport in laboratory test tanks. He is a registered professional engineer and a fellow of the American Geophysical Union. He is the hydrology editor of *Earth Science Review* and serves on the editorial boards of *Water Resource Research*, *Journal of Hydrology*, *Journal of Contaminant Hydrology*, and *Vadose Zone Journal*.

MILTON LEVENSON (NAE) is nationally recognized for his ability to apply creative new insights to major engineering challenges in the nuclear industry and for his organizational and leadership skills. Currently an independent consultant, Mr. Levenson is a chemical engineer with more than 50 years of experience in nuclear energy and related fields. His technical experience includes work related to nuclear safety, fuel cycle, water reactors, advanced reactors, and remote control. His professional experience includes research and operations positions at the Oak Ridge National Laboratory, the Argonne National Laboratory, EPRI (formerly the Electric Power Research Institute), and Bechtel. He was elected to the National Academy of Engineering in 1976. Mr. Levenson is a fellow and past president of the American Nuclear Society, a fellow of the American Institute of Chemical Engineers, and a recipient of the American Institute of Chemical Engineers' Robert E. Wilson Award in Nuclear Chemical Engineering. He is the author of more than 150 publications and presentations and holds three U.S. patents. Mr. Levenson also is a member of the National Academies' Nuclear and Radiation Studies Board and has served on several National Academies committees.

PAUL A. LOCKE is a visiting scholar in the Department of Environmental Health Sciences at the John Hopkins Bloomberg School of Public Health. Dr. Locke has worked extensively on environmental health and policy issues, including radiation protection and radioactive waste disposal, indoor air quality, alternatives to animal testing, and risk assessment. Dr. Locke currently serves on the U.S. Environmental Protection Agency's Clean Air Act Advisory Committee, is a member of the National Academies' Nuclear and Radiation Studies Board, and is a councilor of the National Council on Radiation Protection and Measurements. He is also a member of the editorial board of *Risk Analysis: An International Journal* and is a past councilor of the Society for Risk Analysis. Dr. Locke is a lawyer licensed to practice before the bars of the District of Columbia and the United States Supreme Court.

MICHAEL H. MOBLEY is a private consultant on regulatory radiation-related issues, particularly in the area of commercial low-level waste processing. He is a retired director of the Tennessee Division of Radiological Health and has worked in every aspect of the division's Radiation Control Program. He has represented the State of Tennessee since 1984 as a commissioner for the Southeast Low-Level Radioactive Waste Management Compact Commission. Mr. Mobley is a past chairperson of the Conference of Radiation Control Program Directors, Inc. (CRCPD), has served as that organization's treasurer, and has served on numerous committees and working groups for the CRCPD. He served on the Federal Facilities Committee, which was given the charge by the CRCPD to develop and coordinate information regarding federal facility radiological impact issues. Mr. Mobley received the Gerald S. Parker Award in 1996 for his significant contributions to radiation protection and to the CRCPD. In 2000, he was awarded life member status to the CRCPD (one of four awarded in 35 years). Mr. Mobley has also served as the state liaison officer for Tennessee to the U.S. Nuclear Regulatory Commission.

DIANNE R. NIELSON is the executive director of the Utah Department of Environmental Quality. Her current responsibilities include regulating the Envirocare commercial low-level waste facility and the White Mesa and Ticaboo Uranium Mills, and maintaining state primacy for implementing federal programs. Dr. Nielson is a member of the American Association of Petroleum Geologists and a fellow of the Geological Society of America. She has served as a member of the National Academies' Board on Earth Sciences and Resources and on several National Academies committees. In addition to her expertise in geology, Dr. Nielson also brings a state perspective to the committee.

KEN E. PHILIPOSE is a project manager with the Decommissioning and Waste Management Business Unit at Chalk River Nuclear Laboratories of Atomic Energy of Canada Limited. His current responsibilities include research and development on the storage of cement-grouted fissile high-level liquid waste (in particular molybdenum-99) and decommissioning planning of large, buried carbon steel tanks containing heels of high-level waste. Mr. Philipose has more than 30 years of experience in durable concrete development studies and applications, waste management and decommissioning, design coordination, and project management of nuclear structures and facilities. Mr. Philipose has participated in several international studies concerning material research and development and has authored or coauthored several publications.

ALFRED P. SATTELBERGER is associate laboratory director for the Physical, Biological and Computing Sciences divisions at Argonne National Laboratory. Prior to his appointment at Argonne in 2006, he was a senior laboratory fellow and former director of the Chemistry Division, Office of Science Programs, and the Science and Technology Base

Program Office at Los Alamos National Laboratory (LANL). Dr. Sattelberger's research interests include actinide coordination, organometallic chemistry, technetium chemistry, and metal-metal multiple bonding. He was elected a fellow of the American Association for the Advancement of Science in 2002 in recognition of his scientific contributions to early transition metal and f-element chemistry. Before joining LANL in 1984, Dr. Sattelberger held a faculty appointment in the Chemistry Department at the University of Michigan. He is a former chair of the Inorganic Chemistry Division of the American Chemical Society and serves on the board of directors for the Inorganic Syntheses Corporation and on the editorial advisory board of the *Journal of Coordination Chemistry*. He served as a member of the 1996 general inorganic chemistry Environmental Management Science Program merit review panel. He has also served as a member of several National Academies committees examining radioactive waste management issues at the U.S. Department of Energy.

ANNE E. SMITH is an expert in integrated assessment of environmental and energy problems, specializing in risk management, decision analysis, benefit-cost analysis, and economic modeling. She has applied these techniques to issues such as contaminated site management, nuclear waste management, global climate change, air quality, and food safety. Dr. Smith has experience in assessing societal values for risk changes or environmental benefits. She has developed and reviewed decision support tools for risk-based ranking of contaminated sites and for making risk trade-offs in selecting remediation alternatives. Dr. Smith is a vice president of CRA International in Washington, D.C. Previously, she was a vice president of Decision Focus Incorporated and an economist with the U.S. Environmental Protection Agency. She has served on several National Academies committees examining issues involving risk management within the U.S. Department of Energy's Environmental Management Program.

LESLIE SMITH is the Cominco Chair in Minerals and the Environment at the University of British Columbia in Vancouver. His expertise is in the areas of subsurface hydrology and contaminant transport processes. His current research interests include transport processes in fractured rock masses, hydrologic processes in unsaturated waste rock piles, hydrogeological decision analysis and risk assessment, inverse modeling, and radionuclide transport in watersheds near the Chernobyl Nuclear Power Plant in Ukraine. Dr. Smith has served on several National Academies committees, including the Committee for a Review of the Hanford Site's Environmental Remediation Science and Technology Plan and the Committee to Review Specific Scientific and Technical Safety Issues Related to the Ward Valley, California, Low Level Radioactive Waste Site.

DONALD W. STEEPLES is currently the Dean A. McGee Distinguished Professor of Applied Geophysics, Department of Geology, at the University of Kansas and president of Great Plains Geophysical, Inc. Previously, he held positions at the Kansas Geological Survey. Dr. Steeples is involved in the development and application of noninvasive geophysical techniques, specifically shallow seismic reflection methods applied to environmental and groundwater problems. Dr. Steeples also chairs the geoscience reviews of the Laboratory Director's Advisory Board at the Idaho National Laboratory. He has published more than 100 articles on the application of geophysical methods and is currently an editorial referee for more than 20 scholarly journals. Dr. Steeples has served on several National Academies committees, including one on noninvasive techniques for characterization of the shallow subsurface for environmental engineering applications.

Appendix B

Statement of Task

The objective of this study is to review and evaluate the Department of Energy's (DOE's) plans to manage certain radioactive waste streams stored at its sites as identified below.

The waste streams to be addressed in this study are the streams of waste from reprocessed spent nuclear fuel that:

1) exceed the concentration limits for Class C low-level waste as set out in Section 61.55 of Title 10, Code of Federal Regulations;
2) the Department plans to dispose of on the sites specified below rather than in a repository for spent nuclear fuel and high-level waste; and
3) are stored in tanks at the following sites:

   (A) Savannah River Site, South Carolina.
   (B) Idaho National Engineering and Environmental Laboratory, Idaho.
   (C) Hanford Reservation, Washington.

This study shall evaluate:

1. the state of the Department's understanding of the physical, chemical, and radiological characteristics of the waste referred to above, including an assessment of data uncertainties;
2. any actions additional to those contained in current plans that the Department should consider to ensure that its plans to manage its radioactive waste streams will comply with the performance objectives of Part 61 of Title 10, Code of Federal Regulations;
3. the adequacy of the Department's plans for monitoring disposal sites and the surrounding environment to verify compliance with those performance objectives;
4. existing technology alternatives to the current management plan for the waste streams mentioned above and, for each such alternative, an assessment of the cost, consequences for worker safety, and long-term consequences for environmental and human health;
5. any technology gaps that exist to effect improved efficiency in removal and treatment of waste from the tanks at the Hanford, Savannah River, and Idaho sites; and
6. any other matters that the committee considers appropriate and directly related to the subject matter of the study.

The committee may develop recommendations it considers appropriate and directly related to the subject matter of the study, including:

1. improvements to the scientific and technical basis for managing the waste covered by the study, including the identification of technology alternatives and mitigation of technology gaps; and
2. the best means of monitoring any on-site disposal sites from the waste streams referred to above to include soil, groundwater, and surface water monitoring.

Appendix C

Section 3116, Order 435.1, and Performance Objectives

Table C-1 is a side-by side comparison of excerpts from Section 3116 of the Ronald Reagan National Defense Authorization Act (NDAA) for Fiscal Year 2005 and Department of Energy (DOE) Order 435.1. The performance objectives in Title 10 Code of Federal Regulations Part 61 (10 CFR 61) referred to in both documents are reproduced in Table C-2.

TABLE C-1 Side-by-Side Comparison of Text Relevant to Waste Determinations from Section 3116 of the NDDA and DOE Order 435.1

Topic	NDDA Section 3116	DOE Order 435.1
General definition of waste from the reprocessing of spent nuclear fuel that is not high-level waste	The term ''high-level radioactive waste'' does not include radioactive waste resulting from the reprocessing of spent nuclear fuel that the Secretary of Energy . . . in consultation with the Nuclear Regulatory Commission . . . determines—does not require permanent isolation in a deep geologic repository for spent fuel or high-level radioactive waste	Waste resulting from reprocessing spent nuclear fuel that is determined to be incidental to reprocessing is not high-level waste, and shall be managed under DOE's regulatory authority in accordance with the requirements for transuranic waste or low-level waste, as appropriate. When determining whether spent nuclear fuel reprocessing plant wastes shall be managed as another waste type or as high-level waste, either the citation or evaluation process described below shall be used
Evaluation Process		
Radionuclide removal requirements	Waste has had highly radioactive radionuclides removed to the maximum extent practical	Wastes have been processed, or will be processed, to remove key radionuclides to the maximum extent that is technically and economically practical
Radiation protection performance objectives when managed as low-level waste	In compliance with the performance objectives set out in subpart C of part 61 of title 10, Code of Federal Regulations	Meet safety requirements comparable to the performance objectives set out in 10 CFR Part 61, Subpart C, *Performance Objectives*
Radiation protection performance objectives when managed as greater-than-Class C waste	In compliance with the performance objectives set out in subpart C of part 61 of title 10, Code of Federal Regulations	Managed as transuranic waste (subject to approval by the Administrator of the EPA) as described below or low-level waste as describe above
Radiation protection performance objectives when managed as transuranic waste		Will be incorporated in a solid physical form and meet alternative requirements for waste classification and characteristics, as DOE may authorize

continued

TABLE C-1 Continued

Topic	NDDA Section 3116	DOE Order 435.1
Acceptable disposal destinations	Applies only to material disposed of on-site in South Carolina and Idaho otherwise covered by this section that is not transported from the state	Not restricted
Regulatory oversight for waste disposal when managed as low-level waste	Pursuant to a state-approved closure plan or state-issued permit, authority for the approval or issuance of which is conferred on the State outside of this section	Are to be managed, pursuant to DOE's authority under the *Atomic Energy Act of 1954*, as amended, and in accordance with the provisions of Chapter IV of this Manual, provided the waste will be incorporated in a solid physical form at a concentration that does not exceed the applicable concentration limits for Class C low-level waste as set out in 10 CFR 61.55, *Waste Classification*; or will meet alternative requirements for waste classification and characterization as DOE may authorize
Regulatory oversight for waste disposal when managed as greater-than-Class C waste (Section 3116) or as transuranic waste (DOE Order 435.1)	Pursuant to a atate-approved closure plan or atate issued permit, authority for the approval or issuance of which is conferred on the atate outside of this section; and pursuant to plans developed by the Secretary in consultation with the Commission	Are managed pursuant to DOE's authority under the *Atomic Energy Act of 1954*, as amended, in accordance with the provisions of Chapter III of this Manual, as appropriate
U.S. Nuclear Regulatory Commission role in the disposal plan	The [Nuclear Regulatory] Commission shall, in coordination with the covered atate, monitor disposal actions taken by the Department of Energy . . . for the purpose of assessing compliance with the performance objectives set out in subpart C of part 61 of title 10, Code of Federal Regulations	No provision
Citation Process		
Criteria for determining reprocessing waste to not be high-level waste	No provision	Waste incidental to reprocessing by citation includes spent nuclear fuel reprocessing plant wastes that meet the description included in the Notice of Proposed Rulemaking (34 FR 8712) for proposed Appendix D, 10 CFR Part 50, Paragraphs 6 and 7. These radioactive wastes are the result of reprocessing plant operations, such as, but not limited to: contaminated job wastes including laboratory items such as clothing, tools, and equipment

PERFORMANCE OBJECTIVES

Table C-2 contains excerpts from Title 10 Code of Federal Regulations Part 61 (10 CFR 61) Licensing Requirements for Land Disposal of Radioactive Waste, Subpart C: Performance Objectives.

TABLE C-2 Performance Objectives in 10 CFR 61, Subpart C

Section 61.40 General requirement	"Land disposal facilities must be sited, designed, operated, closed, and controlled after closure so that reasonable assurance exists that exposures to humans are within the limits established in the performance objectives in Section 61.41 through 61.44"
Section 61.41 Protection of the general population from releases of radioactivity	"Concentrations of radioactive material which may be released to the general environment in groundwater, surface water, air, soil, plants, or animals must not result in an annual dose exceeding an equivalent of 25 mrem to the whole body, 75 mrem to the thyroid, or 25 mrem to any other organ to any member of the public. Reasonable effort should be made to maintain releases of radioactivity in effluents to the general environment as low as reasonably achievable (ALARA)"
Section 61.42 Protection of individuals from inadvertent intrusion	"Design, operation, and closure of the land disposal facility must ensure protection of any individual inadvertently intruding into the disposal site and occupying the site or contacting the waste form at any time after active institutional controls over the disposal site are removed"
Section 61.43 Protection of individuals during operations	"Operations at the land disposal facility must be conducted in compliance with the standards for radiation protection set out in part 20 of this chapter [10 CFR 20], except for releases of radioactivity in effluents from the land disposal facility, which shall be governed by Section 61.41 of this part. Every reasonable effort shall be made to maintain radiation exposures ALARA"
Section 61.44 Stability of the disposal site after closure	"The disposal facility must be sited, designed, used, operated, and closed to achieve long-term stability of the disposal site and to eliminate to the extent practicable the need for ongoing active maintenance of the disposal site following closure so that only surveillance, monitoring, or minor custodial care are required"
Compliance time frame	"Even though a time of compliance is not mentioned in 10 CFR Part 61, a period of 10,000 years has been recommended by Nuclear Regulatory Commission staff in its guidance on performance assessments, but this recommendation has not been approved by the Commission, so it does not constitute official agency policy"

AIR AND WATER PERFORMANCE OBJECTIVES

Table C-3 contains the low-activity waste disposal performance objectives for air and water protection based upon DOE Order 435.1, the National Primary Drinking Water Regulations, 40 CFR 141, and the National Emission Standards for Hazardous Air Pollutants (NESHAP; 40 CFR 61H and 40 CFR 61Q) as adapted from the *Hanford Immobilized Low-Activity Waste Performance Assessment: 2001 Version* (Mann et al., 2001).

TABLE C-3 Radiological Performance Objectives

Protection of Groundwater Resources [a, b, c]	
Alpha emitters	
^{226}Ra plus ^{228}Ra	5 pCi/L
All others (total)	15 pCi/L
Beta and photon emitters	4 mrem in a year
Protection of Surface Water Resources [a]	
Alpha emitters	
^{226}Ra plus ^{228}Ra	0.3 pCi/L
All others (total)	15 pCi/L
Beta and photon emitters	1 mrem in a year[d]
Protection of Air Resource [a, e, f]	
Radon (flux through surface)	20 pCi m^{-2} s^{-1}
All other radionuclides	10 mrem in a year

[a] Evaluated for 1,000 and 10,000 years, but calculated to the time of peak or 10,000 years, whichever is longer.
[b] Evaluated at the point of maximum exposure, but no closer than 100 m (328 feet) from the disposal facility (DOE O 435.1; DOE M 435.1).
[c] Main driver is the National Primary Drinking Water Regulations (40 CFR 141).
[d] This is the limit used at Hanford to meet Washington State regulation (WAC 173-201A) and minimize reporting requirements. The EPA drinking water standard is 4 mrem in a year.
[e] Evaluated at the disposal facility.
[f] Main driver is NESHAP (40 CFR 61H and 40 CFR 61Q).

Appendix D

Information-Gathering Meetings

Following is a list of presentations received by the committee during its information-gathering meetings, which were open to the public and included opportunities for public comment.

INFORMATION-GATHERING MEETINGS

Meeting 1: March 7-8, 2005, Washington, D.C.

Background on Congressional Request to Do the Study, Congressman John Spratt, D., South Carolina, member House Armed Services Committee and ranking member House Budget Committee, and Mike Lieberman, Legislative Assistant to Congressman John Spratt

Environmental Management, Keeping Our Commitments: Proven to Deliver, Paul Golan, Principal Deputy Assistant Secretary for Environmental Management, U.S. Department of Energy (DOE-EM)

Department of Energy Tank Wastes, Ken Picha, Engineer, DOE-EM

NRC's [U.S. Nuclear Regulatory Commission's] Role in Waste Determinations, Larry Camper, Director, Division of Waste Management and Environmental Protection, Office of Nuclear Materials Safety and Safeguards, NRC

NRC's Previous WIR [Waste Incidental to Reprocessing] Reviews, Scott Flanders, Deputy Director, Division of Waste Management and Environmental Protection, NRC

States' Perspectives, Mike Wilson, Nuclear Waste Program, Washington State Department of Ecology, and Kathleen Trever, Manager, Idaho National Laboratory Oversight Program

Environmental Public Interest Group, Tom Cochran and Geoff Fettus, Natural Resources Defense Council

Meeting 2: April 13-15, 2005, Savannah River Site (SRS), Augusta, Georgia

Tour of F Tank Farm, H Tank Farm, Saltstone Facility, and Pump Test Tank at TNX

Characteristics and Understanding of SRS Tank Farm Waste, Scott Reboul and Pete Hill, Westinghouse Savannah River Company (WSRC)

Tank Waste Removal Processes, Doug Hintze, Director, Waste Disposition Programs Division, U.S. Department of Energy Savannah River Operations Office (DOE-SR)

Savannah River Site Removed Waste Treatment Overview, Terrel J. Spears, Director, Salt Processing Division, DOE-SR

Savannah River Site Meeting Performance Objectives for On-site Disposition of Tank Waste, Sherri Ross, Engineer, Programs Division, DOE-SR

Savannah River Site Concentration Averaging, Challenges, and Factors of Safety for On-site Disposition of Tank Waste, Sherri Ross, Engineer, Programs Division, DOE-SR

Monitoring Activities, James Heffner, WSRC

NRC's Previous SRS WIR Review, Anna Bradford, Senior Project Manager, Division of Waste Management and Environmental Protection, Office of Nuclear Material Safety and Safeguards, NRC

NRC's Technical Review of Tank Closure at SRS, David Esh, Senior Systems Performance Analyst, Division of Waste Management and Environmental Protection, Office of Nuclear Material Safety and Safeguards, NRC

South Carolina Department of Health and Environmental Control (SCDHEC), David Wilson and Shelly Sherrit, SCDHEC

Waste Removal and Treatment Technology, Tom Caldwell, Program Integration and Technology, Closure Business Unit, WSRC

Strontium-Actinide Separations, David T. Hobbs, Advisory Scientist, Waste Treatment Technology, Savannah River National Laboratory (SRNL)

Waste Treatment Technology for On-site Dispositioned Streams: Caustic-Side Solvent Extraction, Harry D. Harmon, Development Manager, Tank Focus Area Salt Processing Project Research and Development Program, Pacific Northwest National Laboratory

Tank Closure Grouts, Christine A. Langton, Advisory Scientist, SRNL

Waste Disposition Heel Removal, Noel F. Chapman, Engineering Manager, Tank Closure Projects, WSRC

Meeting 3: May 5-6, 2005, SRS, Augusta, Georgia

High-Level Waste System Analysis, Mark Mahoney, Program Integration and Technology, Closure Business Unit, WSRC

Tank Space Overview, Mark Mahoney, Program Integration and Technology, Closure Business Unit, WSRC

Safety Case, Doug Hintze, Director, Waste Disposition Programs Division, DOE-SR

Removal of Heels—Bases for Decisions and Methods of Testing, Tom Caldwell, DOE-SR

Monitoring Activities, James Heffner and Daniel Wells, WSRC

Performance Assessment Results in Waste Determination, Elmer Wilhite, James Cook, SRNL

Grout Waste Form—Mixing, Encapsulation, Durability, and Performance, Christine A. Langton, SRNL, and Tom Caldwell, DOE-SR

SRS Waste Tank Sampling Programs, Peter J. Hill, Scott H. Reboul, and Bruce A. Martin, WSRC

Salt Waste Processing Facility (SWPF) Project Line Item 05-D-405, Project Status Review, Terrel J. Spears, Federal Project Director, Salt Processing Division, DOE-SR

Meeting 5A: July 21-22, 2005, Hanford Site, Richland, Washington

The National Academies: River Protection Project Overview, Roy Schepens, Manager, U.S. Department of Energy Office of River Protection (DOE-ORP)

The National Academies: Opening Remarks, Roy Schepens, Manager, DOE-ORP

Understanding of Hanford Tank Wastes, Jim Honeyman, CH2M Hill Hanford Group, Inc.

The National Academies: The Certain Tank Waste Committee [Committee on Management of Certain Radioactive Waste Streams Stored in Tanks at Three DOE Sites] Use of Characterization Information in Determining Tank Waste Disposition, Don Wodrich, YAHSGS LLC

Immobilized Low-Activity Waste (ILAW) Properties and Disposal in Integration Disposal Facility, Dr. Frederick M. Mann, CH2M Hill Hanford Group, Inc.

Waste Retrieval Technologies, Ryan A. Dodd, Vice President, Closure Operations C Farm Project, CH2M Hill Hanford Group, Inc.

The National Academies: The Certain Tank Waste Committee Tank Waste Residuals, William Hewitt, YAHSGS LLC, and Terry Sams, CH2M Hill Hanford Group, Inc.

The Certain Tank Waste Committee Discussion of WMA [Waste Management Area] Cleanup and Closure/Tank Closure Technologies, Terry L. Sams, CH2M Hill Hanford Group, Inc.

Pre- and Post-closure Monitoring, Moses Jaraysi, CH2M Hill Hanford Group, Inc.

The National Academies: The Certain Tank Waste Committee Radionuclide Removal Technologies and Law and Secondary Waste, Bill Hamel and Billie Mauss, DOE-ORP

Scientific and Technological Challenges, Jim Honeyman, CH2M Hill Hanford Group, Inc.

Historical Perspectives on Current Challenges by Hanford Retired Technical Experts, John L. Swanson, Harry Babad, and Robert C. Roal

Remarks to the National Academies Committee on Management of Certain Radioactive Waste Streams Stored in Tanks at Three DOE Sites (The Certain Tank Wastes Committee), Nick Ceto, Program Manager, Hanford/INL Project Office, U.S. Environmental Protection Agency (USEPA), Region 10, Richland, Washington

National Academy of Sciences Hanford Public Meeting Washington State Department of Ecology Comments, Jane Hedges and Suzanne Dahl, Nuclear Waste Program

Oregon Department of Energy, Ken Niles and Dirk Dunning, Nuclear Safety Division

Yakama Nation, Russell Jim, Director, Environmental and Waste Management Program and Ray Givens, Attorney, Yakama Nation

Nez Perce Tribe, Gabriel Bohnee, Director, Environmental Restoration and Waste Management

Interpreted Extent of Subsurface Contamination Resulting from the 241-BX-102 Tank Leak, Stan Sobczyk, Environmental Restoration and Waste Management, Nez Perce Tribe

Hanford Advisory Board, Todd Martin, Chair

Meeting 5B: Idaho National Laboratory July 25-26, 2005, Idaho Falls, Idaho

Overview of Reprocessing, Waste Generation, Tank Farm, and Calcine Storage, Keith Lockie, U.S. Department of Energy, Idaho Operations Office (DOE-ID)

Overview of Tank Cleaning Experience, Keith Quigley, Project Manager, CH2M-WG Idaho

Tank Farm Contaminated Soils and Groundwater, Lorie S. Cahn, Tank Farm Soils Technical Lead, CH2M-WG Idaho

INL Opening Remarks, John Kotec, DOE-ID Deputy Manager

Origin and Characteristics of Idaho Tank Waste, Arlin Olson, Idaho National Laboratory, Battelle Energy Alliance

Plans for Cleaning and Closure of Idaho Tank Farm

- *Overview of Tank Closure Scope and Objectives,* Keith Locke, DOE-ID
- *Development of Tank Cleaning and Grouting Approaches —Tank Cleaning Experiences to Date,* Keith Quigley, CH2M-WG Idaho
- *Analytical Results of Tank Cleaning to Date,* Nick Stanisich, Portage Environmental

Remote Jet Grouting of Residual Waste Heels, Joe Faldowski, Senior Project Manager, AEA Technology Engineering Services, Inc.

Calcine Characteristics and Plans for Calcine Retrieval and Disposal, Mike Patterson, CH2M-WG Idaho

Safety of Closure Activities: Idaho Tank Farm Closure Performance Assessment, Dave Thorne and Nick Stanisich, Portage Environmental

Historical Perspectives on Current Challenges by INL Retired Technical Experts, Ernie Nieschmidt, John Commander, Bill Echo, Richard Green, and Dean Maindiloff

NRC's [U.S. Nuclear Regulatory Commission's] Incidental Waste Reviews at INEEL [INL], David Esh, Division of Waste Management and Environmental Protection, Office of Nuclear Material Safety and Safeguards, NRC

INL Citizens Advisory Board, David Kipping, Chair

Shoshone-Bannock Tribal Nations, Willie Preacher

State of Idaho, Kathleen Trever, Manager, Idaho National Laboratory Oversight Program

Appendix E

Interim Report Summary and Follow-up

This appendix presents the summary of the committee's interim report (NRC, 2005a) and an overview of developments since that report. In this final report, the committee stands by the findings and recommendations presented in its interim report, and elaborates on some of them.

SUMMARY OF THE COMMITTEE'S INTERIM REPORT

The full text of the interim report is available on-line, free of charge, at *http://www.nap.edu/catalog/11415.html.*

Summary

In the Ronald Reagan National Defense Authorization Act of 2005 (Section 3146 of Public Law 108-375), Congress directed the Department of Energy (DOE) to request a study from the National Academies that evaluates DOE's plans for managing certain radioactive wastes stored in tanks at its sites in Idaho, South Carolina, and Washington.[1] The wastes addressed in this study are from reprocessing of spent nuclear fuel, exceed certain concentration limits,[2] and are planned for disposal at the sites mentioned above.

Congress asked the National Academies[3] for an interim and a final report addressing this task. According to the Defense Authorization Act, the interim report "shall address any additional actions the Department should consider to ensure that the Department's plans for the Savannah River Site, including plans for grouting the tanks, will comply with the performance objectives [of 10 CFR 61[4]] in a more effective manner" (Section 3146 (e)(A)). This document fulfills the interim report request.

Congress requested this study at the same time another provision of the same law (Section 3116) provided the basis for DOE, in consultation with the U.S. Nuclear Regulatory Commission (USNRC), to determine that tank wastes at the South Carolina and Idaho sites meeting certain listed criteria are not high-level waste (HLW).[5] Such wastes may then be disposed of on-site.

TECHNICAL BACKGROUND

The Savannah River Site has 51 underground tanks that are used for storing 138,000 cubic meters (36.4 million gallons) of hazardous and radioactive waste from chemical processing of spent nuclear fuel and related operations.[6] Tank construction and characteristics vary, but the typical tank is a large cylindrical carbon steel and reinforced concrete structure buried at a shallow depth (1 to 3 meters below the surface). The tanks' sizes range from about 2,800 cubic meters (m^3) to 4,900 m^3 (750,000 to 1.3 million gallons). The largest tanks are approximately 26 meters (85 feet) in diameter and 11 meters (35 feet) from the inner tank floor to the center of a domed ceiling. Most of the tanks are equipped with dense networks of vertical and horizontal cooling pipes, referred to as cooling coils (see Figure S-1). These cooling coils are used to remove heat produced by radioactive decay in the waste.

Twenty-seven of the tanks have a full secondary containment (i.e., a tank inside another tank) and are considered "compliant

---

[1]The full statement of task can be found in Appendix A.

[2]These limits define the maximum radionuclide concentrations for Class C low-level waste for radioactive waste disposal facilities regulated by the U.S. Nuclear Regulatory Commission. The limits are found in Part 61, Title 10 of the Code of Federal Regulations (10 CFR 61) titled "Licensing Requirements for Land Disposal of Radioactive Waste." For the purpose of this study, the committee interprets this concentration criterion to apply to the waste streams stored in tanks prior to waste processing.

[3]The National Academies appointed a committee to carry out this study. Biographical sketches of committee members can be found in Appendix C.

[4]The performance objectives of 10 CFR 61 can be found in Appendix A and contain four mandates: (1) protect the general population from releases of radioactivity, (2) protect individuals from inadvertent intrusion, (3) protect individuals during operations, and (4) provide stability of the site after closure. Regulatory guides use a time period of 10,000 years for the performance period.

[5]The term "high-level waste" is used in this report according to its legal definition in the U.S. Code, Title 42, Chapter 108, Nuclear Waste Policy, Section 10101 (see page 13, footnote 8). There is no particular radioactivity concentration or dose limit associated with this definition.

[6]Reprocessing operations at the Savannah River Site started in 1953 and continue on a reduced scale to this day. A map of the site can be found in Appendix E.

FIGURE S-1 Photograph of the interior of a Type I tank (Tank 4) prior to receipt of wastes.
SOURCE: Caldwell (2005a).

tanks" under the site's Federal Facility Agreement,[7] which regulates storage and disposal of hazardous waste at the site. The remaining tanks do not have complete secondary containment and are considered noncompliant. Visual inspections and conductivity probes in the tanks and in the annuli of the tanks have shown that about half of the noncompliant tanks have leaked in the past (although the leaks were confined to the tank's annulus in all but one case).

Although the composition of waste in each tank varies, the tanks generally contain a bottom layer of a peanut-butter-like deposit of insoluble solids (referred to as sludge), a layer of crystalline solids (the saltcake), and a salt solution (the supernate). The term "salt waste" is sometimes used to refer to saltcake and supernate. Although the sludge represents less than 10 percent of the volume, it contains about half of the radioactivity in the waste tanks,[8] mainly from insoluble actinides and strontium salts. The other half of the radioactivity is mostly in the supernate, where the soluble radionuclides, mainly cesium-137, are in solution. A fraction of the soluble radionuclides is also trapped as liquid in the interstices of the saltcake.

DOE has argued that it is impractical to dismantle and remove the tanks after the waste has been retrieved because of the exposures incurred by workers from radioactive residues and because of the overall prohibitive costs of exhuming such large structures. The committee has not seen analyses to support this claim. For each tank, the general plan is to retrieve the bulk of the waste, clean up the tank to the "maximum extent practical,"[9] and close the tank in place, according to milestones agreed to in the site's Federal Facility Agreement. Because of practical limitations on waste retrieval, "emptied" tanks will still contain variable amounts of the radioactive waste (the "heel"), depending on the success of the retrieval and cleanup process.

DOE plans to close emptied tanks by placing layers of engineered grout to encapsulate and stabilize the tank heel and a controlled low-strength material to provide structural support against tank collapse and act as a physical barrier that inhibits the flow of water through the residual waste. Tanks that do not have a concrete roof would have a high-strength layer of grout that would serve as an intruder barrier. An engineered cover to retard infiltration to the tanks after closure is also under consideration.

DOE's plan to manage the bulk of the waste retrieved from the tanks is to separate the radioactive from the nonradioactive components, the latter of which make up most of the waste volume. This processing generates two waste streams: (1) a high-activity waste stream, which will be immobilized and disposed off-site in a high-

[7]This is an agreement among DOE, the Environmental Protection Agency, and the South Carolina Department of Health and Environmental Control and contains the plan for tank closure.

[8]The radionuclides of concern for this study are short-lived but highly radioactive isotopes, such as strontium-90 and cesium-137 and their decay products; long-lived (>30 years) radionuclides such as uranium and plutonium isotopes; and especially long-lived and highly mobile radioisotopes, such as iodine-129, technetium-99, tin-126, selenium-79, and neptunium-237.

[9]One of the criteria that DOE must use according to Section 3116 of the Defense Authorization Act to determine whether waste is not HLW and can be disposed as low-level waste (LLW) is if this waste has had highly radioactive radionuclides removed to the "maximum extent practical." DOE is authorized to make this determination in consultation with the USNRC at the Savannah River and Idaho sites.

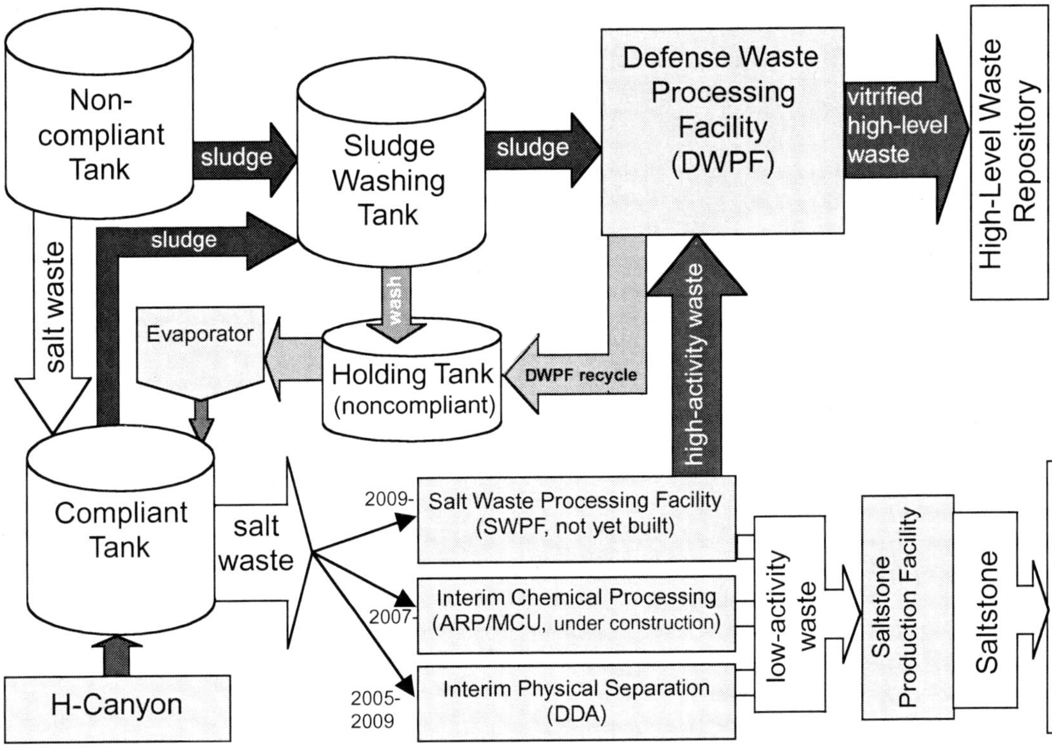

FIGURE S-2 Waste flows in the Savannah River Site waste management plans. Note that the sizes do not necessarily scale with the sizes of the waste flows.

level waste repository,[10] and (2) a low-activity waste stream, which is to be disposed on-site.

At the Savannah River Site, DOE already retrieves sludge and then processes and immobilizes it in glass at its Defense Waste Processing Facility (DWPF). These operations generate as a secondary product a relatively low-activity liquid waste, referred to as the DWPF recycle stream, which is returned to the HLW tanks. To separate highly radioactive constituents of the salt waste, DOE proposes to utilize three different processes[11] that will be available at different times and have different capabilities. Two low-capacity processes are expected to be available sooner and are referred to as "interim" processing by DOE. These are the deliquification, dissolution, and adjustment (DDA) process, which could begin immediately upon approval of the waste determination by the Secretary of Energy in accordance with Section 3116 of the 2005 National Defense Authorization Act, consultation with the USNRC, and permitting by the state of South Carolina; and the actinide removal, modular caustic-side solvent extraction process (ARP/MCU), which is expected to begin operations in 2007. A high-capacity chemical processing facility, called the Salt Waste Processing Facility, is scheduled to be available in 2009 and could be supplemented by the ARP, if needed.

DOE indicated to the committee that the Savannah River Site is facing a "tank space crisis" because of net waste inputs from current waste processing and waste removal operations. To alleviate the tank space crisis, DOE is proposing to begin processing salt waste using DDA as soon as possible (see Figure S-2). The low-activity waste streams from these three processes will have varying concentrations of radioactivity and will be mixed with cementitious material to form "saltstone" and disposed on-site as a monolith in near-surface concrete vaults.

FINDINGS AND RECOMMENDATIONS

Although DOE, its regulators, and others worked with the committee to provide the information needed for this study, some data were not available (not yet collected, not yet generated, or not yet made public), and some plans had not yet been formulated or finalized when this report was written.[12] Appendix B describes the main documents to which the committee had access and the missing

[10]The high-activity waste stream is outside the scope of this report, which focuses solely on waste disposed on-site.

[11]DOE refers to this as a two-phase, three-step approach. The committee has not adopted this way of describing the approach because it suggests that all wastes undergo each process, which is inconsistent with DOE's plan.

[12]The information-gathering phase for the interim report lasted from March through June 2005. Under the Federal Advisory Committee Act Amendments of 1997 (Public Law 105-153), any document provided to the committee from outside of the National Academies must be made available to the public, unless the document is exempt from disclosure under the Freedom of Information Act (Public Law 89-554) and its amendments. As a result, the committee could not accept any document that was undergoing security review, internal scientific review, or legal and policy review and was therefore not ready for public release.

pieces of information to assess DOE plans for compliance with the performance objectives set forth in 10 CFR 61.

Therefore, the committee was unable to evaluate fully what, if any, actions are needed for DOE to comply with these performance objectives. However, the committee was able to evaluate factors that reduce risk and recommends actions to (1) reduce the waste left on-site and (2) increase DOE's understanding of the long-term performance of waste forms and other barriers to the release of radionuclides. These actions will increase confidence in DOE's ability to comply with the performance objectives in general and conform with the requirement to take actions to make releases of radioactivity to the environment as low as reasonably achievable (ALARA), with economic and social considerations taken into account. Findings and recommendations address four major issues: (1) near-term and long-term risks; (2) the tank space crisis; (3) Class C limits and performance objectives; and (4) research and development needs. The following findings and recommendations are based on the information available to the committee at the time of writing this interim report and may be extended in the committee's the final report.

Near-Term versus Long-Term Risks

Finding 1a: By far the greatest reductions in near-term probability and quantity of radionuclide and hazardous chemical releases to the environment are achieved by bulk removal and immobilization of liquid, salt, and sludge from the noncompliant high-level waste tanks. The tank heels that remain after bulk removal contain a smaller quantity of waste that is less mobile and constitutes a much lower near-term probability of release.

Finding 1b: The Savannah River Site Federal Facility Agreement has schedules for waste removal from and closure of the noncompliant tanks. For some tanks, the tank-closure step immediately follows the waste-removal step, making them appear to be coupled. This coupling could limit the time available for tank-waste removal and consequently could determine how much waste can be removed to "the maximum extent practical." A decoupled schedule is already planned for a limited number of tanks, as shown in Appendix F. Decoupling allows the consideration of a wider set of options for removing and/or immobilizing residual waste (especially for tanks that have significant obstructions that complicate waste removal), which could reduce long-term risks.

Recommendation 1: DOE should decouple tank waste removal and tank closure actions on a case-by-case basis where there are indications that near-term (5-10 year) techniques could become available to remove tank heels more effectively, safely, or at a lower cost. In evaluating schedules for each tank, DOE should consider the risks from postponing tank closure compared with the risk reductions that could be achieved if the postponement improves heel removal. Although the committee believes that postponing tank closure need not extend the closure dates of the tank farms, DOE should work with the State of South Carolina to revise the schedule for closure of a limited number of the tanks that contain significant heels, if necessary.

The committee agrees with DOE's and South Carolina's overall approach to cleanup at the Savannah River Site: bulk removal of the waste containing the majority of the mobile radionuclides is the highest priority to reduce release of radioactive materials to the environment in the near term. The noncompliant tanks, about half of which have a history of leakage, demand attention first, but nearly all of the tanks are beyond their design lifetimes.

Filling a tank with grout is, from a practical point of view, an irreversible action, although it is conceivable to open a tank and excavate the grout if absolutely necessary. Moreover, postponing closure of some tanks for several years would appear to have essentially no effect on near- or long-term risk. The current approach of coupling cleanup and closure schedules forecloses options that may become available in the near future (e.g., using alternative technologies to reduce the radioactive heel [source] and/or using other types of immobilizing material to fill the tank).

DOE should decouple cleanup and closure schedules, keep as many options open as practical, and regularly assess technology developments and alternatives to reduce long-term risks presented by the tank heels. DOE should make additional investments in research and development to enhance tank waste retrieval (reducing the source term), improve residual waste immobilization (stabilizing the source term), or reduce the ingress of water once the tanks are closed (protect the source term), as stated in Recommendation 4. In some cases, tank closure need not be delayed, such as in tanks that have small heels (i.e., as small as the heels in Tanks 16, 17, and 20) and/or low concentrations of radionuclides, or if risks specific to the tank require early closure (i.e., as soon as waste removal is completed). Conversely, delaying closure may be warranted for tanks with large heels or high concentrations of radionuclides. This approach need not necessarily affect the final closure date of the tank farm, which will occur later than 2022, the milestone for closure of the noncompliant tanks. If new technologies become available in the near future (i.e., 5-10 years), it may be possible to clean up and/or close tanks faster (possibly leaving less waste behind), thus meeting the final milestone for the tank farms.

As DOE considers delaying closure for some tanks, it has to evaluate the advantages and disadvantages from both a risk and a cost perspective. If DOE can relax other constraints on tank waste removal, such as the tank space problem, delaying tank closure could free up funds planned for closure activities, and those funds could be devoted to enhancing waste removal, waste processing, and confidence in the near- and long-term performance of the waste immobilization and tank fill materials. Similarly, research and development require funds, but if they are successful they could result in lower costs and increased safety overall (see Finding and Recommendation 4).

Tank Space Crisis

Finding 2a: The lack of compliant tank space does appear to be a major problem because of continuing waste inputs and the anticipated future needs for space to support site operations and tank cleanup. As presently operated, sludge waste processing results in a net addition of waste to the compliant tanks. Salt waste processing will also require storage volume in compliant tanks for batch preparation and other operations.

Finding 2b: DOE plans to use the deliquification, dissolution, and adjustment process to free up space in compliant tanks.

While DOE analyses so far suggest that the wastes from this process would meet the performance objectives in 10 CFR 61, it achieves less radionuclide separation than other planned processes. While waste from the DDA process represents only 8 percent of the volume of low-activity waste to be generated during salt waste processing, it contains 80-90 percent of the radioactivity that is projected to be sent to the Saltstone Disposal Vaults.

Recommendation 2: DOE and other involved parties should consider options other than DDA to alleviate the impending crisis in usable storage in compliant tanks. Options include actions that (1) reduce waste inputs to the tanks, such as redirecting the DWPF recycle stream for disposition in the Saltstone Facility; and (2) actions that free up usable volume in compliant tanks, such as using noncompliant tanks not known to have leaked for emergency storage volume.

Waste retrieval, processing, and tank cleaning operations continuously add secondary wastes to the tanks; in addition, space in compliant tanks is needed to prepare feeds for the high-level and salt waste processing facilities. Moreover, DOE is maintaining the equivalent of a full tank capacity—4,900 m^3 (1.3 million gallons)—in empty compliant space for emergency purposes at all times. Hence the "tank space crisis."

DOE plans to address the tank space problem in the short term by implementing the DDA process. This process uses physical rather than chemical means to accomplish cesium separation (i.e., draining interstitial liquid present in the saltcake and then dissolving the saltcake and grouting it into saltstone (see Figure S-2).[13] The saltstone from this process is expected to contain cesium concentrations that are two orders of magnitude higher than the waste from the chemical processes that eventually will be used in the Salt Waste Processing Facility (albeit still considerably lower than Class C limits). Even these higher levels of cesium may not cause projected doses from the Saltstone Vaults to exceed dose limits, although as noted earlier, details underlying a performance assessment for DDA saltstone were not available for committee examination. However, this raises the following question: Does this process remove radionuclides to the maximum extent practical?

The tank space crisis forces DOE to engage in increasingly complex operations to ensure that there is sufficient space to continue waste processing. Hence, the tank space crisis may increase the possibility of accidental worker exposure to radiation, the chance of operational accidents, and the chance of waste leakage during transfers. In its recommendation, the committee suggests alternative options to DDA to mitigate the tank space crisis.

Class C Limits and Performance Objectives

Finding 3: The future site-specific risks posed by wastes disposed of on-site is the primary issue of concern in this study. Such risks are determined by the radionuclide and chemical quantities and concentrations, their conditioning, their interactions with the environment, and their bioavailability, not by the relationship of radionuclide concentrations to generic limits such as those for Class C low-level waste. The National Defense Authorization Act Section 3116 requires the use of the performance objectives in 10 CFR 61 to limit and minimize these risks.

Recommendation 3: When deciding what wastes may be disposed of on-site, DOE and other involved parties should ensure that discussions focus on how radionuclide and chemical quantities and concentrations, their conditioning, their interactions with the environment, and their bioavailability affect site-specific risk.

The Class C limits are not a criterion for acceptability of on-site disposal of tank wastes from reprocessing of spent nuclear fuel under the present law but are sometimes discussed as if they were. The Class C limits were developed for a diverse commercial sector to establish limits on what is generally acceptable for near-surface disposal, based in part on assumptions about the overall set of wastes destined for disposal. According to Section 3116, comparison of radionuclide concentrations in waste to Class C limits is relevant to waste disposition decisions *only* procedurally, in that DOE must develop its disposal plans in consultation with USNRC.

Rather than Class C limits, site-specific risk assessments are the bases for determining whether the facility meets the performance objectives in the regulations. These risks depend on radionuclide quantities and concentrations, their conditioning, and their interactions with the environment.[14] The performance objectives and waste acceptance criteria constrain the overall quantity of radioactive material that can be disposed in a facility.[15]

Acceptable radionuclide concentrations (and/or inventories) and distributions should be determined as a result of a properly constituted and implemented risk assessment[16] that takes into account measured and/or projected radionuclide concentrations, spatial variability of the concentrations, and attendant uncertainties. Such a risk assessment was not available at the time of report writing (see Appendix B).

Congress recognized the importance of the performance objectives for evaluating site-specific near-surface disposal of waste in Section 3116 of the 2005 National Defense Authorization Act by explicitly including these objectives as the basis for determining whether waste is HLW instead of relying on the radionuclide concentrations that define the upper boundary of Class C waste. All substantive technical criteria that DOE's determination must meet (e.g., performance objectives, remove highly radioactive radionuclides to the maximum extent practicable) apply irrespective of whether a waste is less than or greater than Class C.

[13]DOE plans to send what it has identified as the least radioactive salt wastes from the tanks through the DDA process.

[14]Regulatory guides for 10 CFR 61 state that 10,000 years is an appropriate time frame for the performance assessments.

[15]Waste acceptance criteria take into account broader considerations than performance objectives, such as waste "processibility" (i.e., compatibility of waste and secondary products with the chemical and physical processes prior to disposal) and other site-specific requirements.

[16]A recent National Research Council report *Risk and Decisions about Disposition of Transuranic and High-Level Radioactive Waste* describes a framework for decision-making processes in the presence of risk and uncertainties (NRC, 2005).

Research and Development Needs

Finding 4: Focused research and development could help DOE reduce the amount, improve the immobilization, and test some of the assumptions used in performance assessment of tank waste to be disposed of at the Savannah River Site. These actions could reduce the risks to humans and the environment and improve confidence in DOE's risk estimates. These research and development activities could also increase DOE's ability to demonstrate compliance with the performance objectives in 10 CFR 61.

Recommendation 4: DOE should fund research and development efforts focused on providing deployable results within 5-10 years on the following topics: (1) in-tank and downstream processing consequences of chemical tank-cleaning options, (2) technologies to assist in tank-waste removal, including robotic devices, and (3) studies of the projected near- and long-term performance of tank-fill materials such as grout.

To reduce long-term risks to the site and test the assumptions in the performance assessment, the committee recommends that DOE perform focused research and development to enhance tank waste retrieval and residual waste immobilization. Tank waste retrieval could be enhanced using better mechanical or chemical tools. Tank waste retrieval is currently performed using hydraulic technologies (i.e., water jets) and, to a certain extent, robotic devices and chemical cleaning agents (i.e., oxalic acid). The committee believes that additional research and development on mechanical tools, including but not limited to robotic devices and chemical cleaning could reduce the tank heels, especially in tanks with cooling coils. DOE should further evaluate the effectiveness of residual waste immobilization by conducting durability studies of grout (and alternative fill materials).

These activities may increase confidence in DOE's management plans or may cause DOE to revise some of the assumptions used in the performance assessment. Testing assumptions and improving DOE's knowledge base might increase its ability to comply with the performance objectives specified in the law. Research and development activities should be limited to those technologies that are promising and at a near-deployment stage (i.e., they could provide results within 5 to 10 years, in time to be implemented during the tank closure process). All noncompliant tanks are scheduled to be closed by 2022. A technology developed in the next 5-10 years could be deployed in time to address the most challenging tanks (i.e., those with cooling coils).

The committee believes that a nonradioactive test bed for retrieval technologies that can be adapted to simulate a variety of tank situations (i.e., recalcitrant heels, cooling coils, debris) should be maintained. The Pump Test Tank, a partial Type IV tank mockup at the mostly decommissioned TNX facility used for testing and equipment before deployment, and similar test beds at other sites, are candidates for this role. The Hanford Site also has a mockup of a single-shell tank used for similar purposes. The committee will further address the need for experimental retrieval facilities in its final report.

FUTURE PLANS FOR THE STUDY

The committee's full task is to review and evaluate DOE's plans to manage radioactive waste streams from reprocessed spent fuel that exceed the Class C concentration limits and are planned for on-site disposal at the Savannah River Site, the Idaho National Engineering and Environmental Laboratory, and the Hanford Reservation. Congress requested assessments of the following: DOE's knowledge of the characteristics of the wastes; additional actions DOE should take in managing these wastes to comply with the performance objectives; monitoring plans; existing technologies and technology gaps for waste management; and any other matters that the committee considers appropriate and directly relevant. For its interim report, the committee was charged to examine whether DOE's plans to manage its radioactive waste streams at the Savannah River Site will comply with the performance objectives of 10 CFR 61.

Compliance with the performance objectives depends upon the amount of radioactive material left onsite, the manner in which it is immobilized, its interaction with the environment, and its interaction with ecological and human receptors. As noted above, some critical data, analyses, and plans were not available when this report was written: the performance assessment for closed tanks; plans for residual waste characterization; plans for tank annuli and tank-system piping; support for assumptions, estimated levels of conservatisms, and sensitivity analyses for performance assessment calculations; and long-term monitoring plans are examples of the missing information. In this interim report, the committee has fulfilled the charge to the extent possible by focusing mainly on the amount of waste left in the tanks and in the Saltstone Vaults at the Savannah River Site. The committee has made findings and recommendations on four major issues:

1. near-term and long-term risks in the context of tank waste removal and the schedule for tank closure;
2. the tank space crisis and options to alleviate the crisis;
3. the roles of the Class C limits and the performance objectives in determining whether on-site disposal is acceptable; and
4. research and development needs, particularly in-tank and downstream consequences of chemical cleaning options, technologies to assist in tank waste removal, including robotic devices, and studies of the projected near- and long-term performance of tank fill materials, such as grout.

The committee is still examining the interactions of the tanks and the saltstone with the surrounding environment; the role of environmental monitoring; the role of the point of compliance in meeting the performance objectives; and the role of modeling in the performance assessment. These topics are relevant to all three sites and will be addressed in the final report, along with the rest of the statement of task. For a substantive analysis, the information described above will be needed at all sites. In addition, because the wastes and the site conditions differ, the topics investigated in this report will also be examined at the Hanford and Idaho sites. These investigations at other sites will have an impact on the committee's views on the Savannah River Site. Hanford will likely offer the committee the greatest challenge because it is the oldest site, has many tanks that have leaked, and has the most complicated wastes because of the various management practices and several chemical processes that generated the wastes, including the earliest processing technologies. The committee may also extend the comments on the Savannah River Site found in this report as additional information on this site becomes available during the period of this study.

DEVELOPMENTS SINCE THE COMMITTEE'S INTERIM REPORT

The committee has received direct and indirect feedback from DOE and the South Carolina Department of Health and Environmental Control (SCDHEC) on the findings and recommendations in its interim report. Inez Triay, U.S. Department of Energy Office of Environmental Management's (DOE-EM's) chief operating officer, informed the Nuclear and Radiation Studies Board on September 12, 2005, that while it agrees in principle with the findings and recommendations, DOE disagrees with some of the details. DOE (Triay, 2005)

- Reiterated its commitment to the schedule for closing tanks;
- Said that the committee misunderstood the tank space problem and the solutions the committee proposed were irrelevant or unworkable;
- Expressed that it had no concerns about the recommendation concerning the class C limits, except to note that the committee could be most helpful by sticking to the National Academies' strengths, which are science and research; and
- Asked for more specific guidance on what research is needed.

SCDHEC representatives in interviews with reporters reiterated SCDHEC's commitment to the schedule for closing tanks and disagreed with the committee's conclusion that delaying filling of tanks with grout would be beneficial from the perspective of risk.

Postponement of Tank Grouting

As noted in its final report, the committee remains convinced not only that postponing tank filling after tank cleanout should be kept as an option, but that DOE is already effectively doing this with some of its tanks at Savannah River Site. Some tanks are scheduled for waste removal years before they are scheduled for closure. This is likely to become an even more important option as DOE continues cleanup and encounters tanks from which waste retrieval promises to be more difficult.

Tank Space Crisis

Concerning the tank space crisis at SRS, the committee's concerns have continued to increase. Quoting from the committee's interim report:

> One committee concern is what will happen with salt waste processing if the [Salt Waste Processing Facility, SWPF] or the interim chemical processing cannot be brought into operation on schedule. The committee did not review the engineering readiness of the salt waste processing, but the schedule to bring the facilities on-line (ARP/MCU by 2007 and the high-capacity SWPF by 2009) and operating to specifications (i.e., processing waste at the expected throughput and meeting the waste acceptance criteria) is ambitious.
>
> Based on DOE's prior experience with developing and initiating operations at major waste processing facilities, it is prudent to plan for the possibility that salt waste will not be removed from the tanks at the planned pace. In other words, DOE needs a contingency plan for tank space. More generally, the committee cautions that in a schedule-driven system there is the danger that wastes could be sent through the process that is currently available rather than the one that is most suited to the wastes. The committee recognizes, of course, that there are other considerations (e.g., safety, risk, and cost) involved in such decisions. The committee here offers some suggestions to reduce waste inputs to tanks and to free up compliant tank space.

The committee suggested alternative options for the Defense Waste Processing Facility recycle stream. On this point, the situation was changing during the committee's information gathering for the interim report, and the committee did not have the most current information when the interim report was released (see Sidebar E-1). By that time, DOE had already moved its concentrated Defense Waste Processing Facility recycle waste from compliant tanks to a noncompliant tank with no history of leakage. There also appeared to be some confusion about the committee's suggestion that noncompliant (but nonleaking) tank space be used for emergency reserve. Some understood the committee to say that DOE should transfer waste into noncompliant tanks. Although this practice has been used on a temporary basis by DOE with approval from SCDHEC and could potentially be used to alleviate short-term space crises,[17] this was not the committee's suggestion. Instead, the committee suggested keeping the emergency reserve storage, which is empty tank space, in noncompliant tanks that have no history of leakage. Waste would be transferred into the reserve only in an emergency, such as a major leak discovered in another tank.

Since the interim report was released, the main point the committee was making—that DOE and its regulators must think creatively to solve the tank space crisis—has become more salient. DOE has encountered additional significant delays in bringing the Salt Waste Processing Facility (SWPF) on-line. The facility is being redesigned due to seismic concerns raised by the Defense Nuclear Facilities Safety Board (Allison, 2005). DOE now expects a 26-month delay in the start of operations, from August 1, 2009, to September 30, 2011 (Terhune and Kasper, 2005). Although DOE has not stated the specific consequences of the delay, it previously emphasized that the schedule was crucial, and

[17]In essence, this is what DOE has done with its concentrated DWPF recycle stream, which was stored in compliant tank space but now is in a noncomplaint (Type IV) tank.

SIDEBAR E-1
The DWPF Recycle and Tank Space at SRS

The Defense Waste Processing Facility (DWPF) receives tank waste and processes and immobilizes that waste in glass logs. Some 5,680 m^3 (1.5 million gallons) per year of waste called DWPF recycle is sent back to the tank farms. DWPF recycle is a combination of several liquid waste streams from the DWPF, including overheads[a] from the Melter Off-gas System and the Sludge Receipt and Adjustment Tank and the Slurry Mix Evaporator. The latter waste streams are collected in the Slurry Mix Evaporator Condensate Tank (SMECT). These "mildly contaminated" waste streams are alkalized with sodium hydroxide (1 molar) and corrosion-inhibited with sodium nitrite (1 molar) before being sent to noncompliant (Type IV) tanks: Tanks 21 and 22. They are then sent to the 2H evaporator system, which comprises the feed tank (Tank 43, Type III), the 2H evaporator, and the concentrate receipt (drop) tank (Tank 38, Type III). The drop tank takes the evaporator bottoms. The evaporator overheads are sent to the Effluent Treatment Facility.

The evaporator bottoms sent to Tank 38 are essentially all concentrated supernate (liquor), with high concentrations of sodium hydroxide. Until recently (November 2004), the liquor was cycled back to the feed tank and run through the evaporator repeatedly. As a result, sodium hydroxide built up in the 2H evaporator system and the volume reductions achieved by evaporation worsened.

In November 2004, DOE concluded it could, and had to, transfer the roughly 3,000 m^3 (800,000 gallons) of liquor from the 2H evaporator system to another tank for storage to allow new DWPF recycle to be sent to the 2H evaporator. The concentrated liquor was stored temporarily in Tank 49, a Type III (compliant) tank, from November 2004 until April 2005 when it was transferred to Tank 24, a virtually empty Type IV (noncompliant) tank. Tank 49 had previously been emptied in preparation for its role as the DDA settling tank and, therefore, was available for temporary storage of the 2H evaporator liquor. Before the waste transfer to a noncompliant tank with no history of leakage (Tank 24) was approved, DOE expressed some concern that the waste would continue to occupy compliant tank space needed for salt waste operations, but that potential problem never became a reality.

Considering the capacity in Tanks 21 and 22 and the 2H evaporator system, DOE projects that DWPF recycle will not have storage problems for several years to come. In the future, DOE plans to use the 2 molar DWPF recycle to adjust the sodium molarity of the dissolved saltcake (7.5-8 molar) down to the range that is best for saltstone feed (5-6 molar). The sodium molarity of the concentrated DWPF recycle stream in Tank 24 is 11. DOE notes that gibbsite and aluminosilicates (primarily cancrinite) will form if DWPF recycle or the liquor is mixed with other tank wastes, but these solids are expected to be resuspended easily in the liquids, rather than forming the agglomerated masses that have proven to be problems in the 2H evaporator and in tanks that have zeolite in the form of ion-exchange media.

DOE said that the Savannah River Site has no way to get the DWPF recycle stream directly to the Saltstone Production Facility (Triay, 2005). Upon further discussion with DOE, the statement was clarified to mean that currently there is no way to bypass the tank farm entirely. In fact, an option explored by Mahoney and d'Entremont (2004) is to send a portion of the DWPF recycle stream to Tank 50, which is the feed tank for the Saltstone Production Facility. DOE did not select this option. In short, DOE can send the DWPF recycle stream to saltstone. However, without a new transfer line, DOE cannot avoid neutralizing the waste stream because it has to go through the tank farm to get to the saltstone facility.

[a] Overheads are the vapors arising from waste in a waste evaporator. Once condensed, they constitute another liquid waste stream.

this new delay can only exacerbate the tank space problems unless DOE (1) decreases the rate of waste additions from sludge processing and canyon operations; (2) increases the amount of waste sent through interim processing, including the deliqufication, dissolution, and adjustment (DDA) process, or (3) finds alternative storage options. In its interim report, the committee recommended that DOE consider options other than DDA to alleviate the impending crisis in usable storage in compliant tanks. DOE is examining what alternatives it has that will allow for continued full-capacity operation of the DWPF without increasing the radioactivity in saltstone above the amount the state has already agreed to (3-5MCi). DOE hopes to put forward a new strategy for salt processing and tank space in January 2006.

Class C Limits

In December 2005, the USNRC issued a Draft Interim Concentration Averaging Guidance for Waste Determinations (FR 74846 v. 70, n. 241, Dec. 16, 2005) that would allow averaging concentrations of residual waste in tanks over the volume of the grout in which the waste is mixed or that is needed to stabilize the waste. However, it would not allow averaging over the volume of the overlying grout because there is neither substantial mixing nor encapsulation. In essence, the committee's recommendation that discussion focus on risk assessments rather than concentration limits or averaging is consistent with the USNRC's draft interim guidance and prior Branch Technical Position (USNRC, 1995), which say that the determining factor for averaging is the impact on risk.

Appendix F

Waste Retrieval Status

TABLE F-1 Status of Tank Waste Retrieval Operations at the Three Sites

Site	Tank Identifier	Nominal Tank Capacity (gallons)	Volume of the heel (gallons) (percentage of initial volume)	Radioactivity in the heel (percentage of initial radioactivity)	Residual Waste Retrieval Technology Used	Comments
Savannah River Site (51 tanks)	Tank 16[a]	1.06 million	9.5 kg solids[a]	830 Ci (30.7 TBq)	Water washing, chemical cleaning	Completed to the limit of technologies. Most of the radioactivity is due to insoluble strontium-90 inventory. The tank annulus requires additional cleaning.
	Tank 17[b]	1.3 million	2,200 gallons sludge	2,400 Ci (89 TBq)	Water washing, Sluicing and pumping	Completed (closed)
	Tank 20[b]	1.3 million	1,000 gallons sludge	500 Ci (18.5 TBq)	Sluicing and pumping	Completed (closed)
	Tank 18[b]	1.3 million	4,300 gallons wet solids	27,600 Ci (1.02 PBq)	Sluicing and pumping	Completed. Most of the radioactivity is due to cesium-137 and barium-137 trapped in residual zeolites (46% of the heel volume and ~88% of the total curies). Strontium-90 and yttrium-90 make up another ~10% of the radioactivity
	Tank 19[b]	1.3 million	15,100 gallons wet solids	96,000 Ci (3.6 PBq)	Sluicing and pumping	Most of the radioactivity is due to cesium-137 and barium-137 trapped in residual zeolites (66% of the heel volume and ~99% of the total radioactivity)
Hanford	C-106[c]	530,000	2,768 gallons (370 cubic feet)	136,700 Ci (5.06 PBq)	Modified sluicing, oxalic acid dissolution	Completed to the limit of technologies. Most of the radioactivity is due to insoluble strontium-90 inventory
	C-203[d]	55,000	138 gallons (18.5 cubic feet)	36 Ci (1.3 TBq)	Vacuum retrieval system	Completed
	C-202[e]	55,000	147 gallons	Results not available	Vacuum retrieval system	Completed
	S-102[f]	758,000	321,000 gallons (in progress)	Results not available	Modified sluicing, saltcake dissolution	Retrieval still under way

continued

TABLE F-1 Continued

Site	Tank Identifier	Nominal Tank Capacity (gallons)	Volume of the heel (gallons) (percentage of initial volume)	Radioactivity in the heel (percentage of initial radioactivity)	Residual Waste Retrieval Technology Used	Comments
Hanford	S-112[f]	758,000	23,000 gallons	Results not available	Modified sluicing, saltcake dissolution, Salt Mantis	Retrieval complete to the limit of technology for modified sluicing and saltcake dissolution; additional technology (Salt Mantis) deployed, retrieval still in progress
Idaho[g]	WM-180	300,000	7,600 gallons liquids[g] 542 kg solids	1,047 Ci (38.7 TBq)	Pumping and water washing	
	WM-181	300,000	7,300 gallons liquids[g] 246 kg solids	475 Ci (17.6 TBq)	Pumping and water washing	
	WM-182	300,000	6500 gallons liquids[g] 1238 kg solids	2,394 Ci (88.6 TBq)	Pumping and water washing	
	WM-183	300,000	8000 gallons liquids[g] 702 kg solids	1,363 Ci (50.4 TBq)	Pumping and water washing	
	WM-184	300,000	3100 gallons liquids[g] 558 kg solids	1,077 Ci (39.8 TBq)	Pumping and water washing	
	WM-185	300,000	5800 gallons liquids[g] 720 kg solids	1,391 Ci + 3,850 Ci in the sandpad (194 TBq)	Pumping and water washing	
	WM-186	300,000	6,600 gallons liquids[g] 334 kg solids	646 Ci (23.9 TBq)	Pumping and water washing	
	WM-103	30,000	19 kg solids	37 Ci (1.4 TBq)	Pumping and water washing	Conservative estimates of solids based on a biological film layer at the bottom of the tank
	WM-104	30,000	19 kg solids	37 Ci (1.4 TBq)	Pumping and water washing	Conservative estimates of solids based on a biological film layer at the bottom of the tank
	WM-105	30,000	19 kg solids	37 Ci (1.4 TBq)	Pumping and water washing	Conservative estimates of solids based on a biological film layer at the bottom of the tank
	WM-106	30,000	19 kg solids	37 Ci (1.4 TBq)	Pumping and water washing	Conservative estimates of solids based on a biological film layer at the bottom of the tank

NOTE: Only tanks that had most waste retrieved at the time of writing (December 2005) are shown; other tanks may be in process. The table has been fact-checked by the three sites.

[a] Fowler, 1981.
[b] DOE-SRS, 2005a.
[c] Hewitt, and Sams, 2005.
[d] Quintero, 2005b.
[e] Quintero, 2005b.
[f] Dodd, 2005.
[g] DOE-ID, 2005a.
[h] A volume of flush water is left in the tanks after the last wash cycle, to allow for sampling, keep any remaining solids in a state to allow further removal during the grouting phase, and allow enough volume to permit restart of the transfer jets during tank grouting. During the tank grouting phase, the transfer jets will be operated to remove the remaining liquid and whatever solid particles come with it.

TABLE F-2 Summary of Main Waste Retrieval Tools in DOE's Toolbox

Mobilization or Collection Technique	Tool Considered or Developed	Tool Tested	Tool Deployed
"Wet" technologies			
Mixing or sluicing technologies			
Low pressure (<1000 psi)			Slurry pumps (several SRS tanks); waterbrush (SRS Tank 17); Flygt Mixer (SRS Tanks 17 and 19); bladed agitators (several SRS tanks—small processing vessels); Advanced Design Mixer Pump (SRS Tank 18); Hanford C-103 Sluicer (Hanford Tank C-103); Hanford C-106 Sluicer (Hanford Tank C-106);washball or directional nozzle wash system (INL tanks)
Moderate pressure (1,000 < psi < 3,000)			Borehole miner (several ORNL tanks); water mouse (SRS Tank 17)
High-pressure (>3,000 psi)			Hydrolaser/hydrolance (SRS Tank 19); Salt Mantis (Hanford S-112); Confined Sluicing End Effector (CSEE; ORNL); bilateral sluicers (SRS Type I Tanks)
Pulsating mixing devices		AEA Technology Power Fluidics (cold tested at Hanford) Russian Pulsating Mixer Pump (PMP) (tested at PNNL and at ORNL)	
Chemical cleaning			
Using sluicing			Modified sluicing and acid dissolution (Hanford tank C-106)
Using mixers			Oxalic acid (SRS Tanks 16 and 24)
Dry or semidry technologies			
Vacuum			Vacuum retrieval system (Hanford C-202, C-203, C-201)
Scarifier or grinder	Rail or pneumatic wheel-based systems used in the mining industry (never tested or deployed in DOE tanks); dry retrieval system (considered for Hanford Tank C-104 but never deployed)		Burnishing tool (deployed at West Valley); scarifier (deployed at ORNL using both the Houdini™ and the Modified Light Duty Utility Arm (MLDUA)
Mechanical conveyance systems			
Deployment devices or delivery tools			
Simple mast	Delphinus (never deployed)		Vacuum retrieval system (Hanford Tanks C-202, C-203, C-201), Mast Tool Delivery System (West Valley Tank 8D-1)
Multijoined arm	SRS Crawler (designed for deployment in SRS Tank 19 but never deployed), VAC TRAX (never deployed), Pit Hog (never deployed), ESG/LATA Trac-Pump (never deployed)	EMMA (tested for use at Fernald but never deployed), ReTRIEVR (tested for use at Fernald but never deployed), Tarzan (partially built for use at West Valley but never completed)	Light Duty Utility Arm (LDUA) and MLDUA (Oak Ridge)

continued

TABLE F-2 Continued

Mobilization or Collection Technique	Tool Considered or Developed	Tool Tested	Tool Deployed
In-tank vehicle		ARD (never deployed in a tank but used in SRS B-Area solvent tanks)	Houdini™ (deployed at ORNL), Scarab-3 (deployed at ORNL)
Combined systems			
Vacuum plus in-tank vehicle		Mobile retrieval system (tested at the Hanford Cold Test Facility)	
Scarifier plus in-tank vehicle	Grinding mechanism used in combination with the Delphinus arm		Scarifier deployed via Houdini™ (ORNL)
Vacuum plus surface system			Waste Dislodging and Conveyance System (deployed at ORNL)

NOTES: Some tools have been considered or developed in a laboratory setting. Some tools have been tested in a cold test facility. Some tools have been deployed in actual waste tanks (location in parenthesis). Deployment implies previous development and cold testing as well. Status indicates if and where device was deployed or tested. INL = Idaho National Laboratory; ORNL = Oak Ridge National Laboratory; PNNL = Pacific Northwest National Laboratory; SRS = Savannah River Site.

SOURCES: Davis, 1998; Bogen et al., 1999; DOE-TFA, 2000a; Bamberger et al., 2001; Burks, 2005; DOE-SRS, 2005a.

Appendix G

Tank Waste Retrieval Techniques and Experience at West Valley and Oak Ridge

This appendix provides more detailed discussions of tank waste retrieval techniques mentioned in Chapter III. It also elaborates on the experience with tank cleanup at the West Valley Demonstration Project in New York and the Oak Ridge National Laboratory in Tennessee, mentioned in Chapter III.

HYDRAULIC TECHNIQUES

By far the most popular techniques for mobilizing residual wastes involve the use of pressurized water. As with bulk sludge retrieval, the water is usually formed by reusing supernatant or some other internal recycle stream to reduce the amount of makeup water introduced into the system. One class of hydraulic devices is moved around the tank to wash radioactive material from the walls and top the tank bottom. Devices in this class include a rotating flusher nozzle using low-pressure water (<1,000 pounds per square inch [psi]) much as in a dishwasher, or a manually directed jet of moderate-pressure water (1,000 to 3,000 psi), which has been used to remove the more recalcitrant deposits in smaller areas. While the jet can remove materials that are unaffected by the rotating nozzle, it must be directed manually (through remote controls) at the area of concern. Hence, this technique is labor intensive. Moreover, because of the introduction of fluids, these techniques must be used carefully in tanks that are known or potential leakers to prevent additional release of radionuclides to the environment.

A second class of hydraulic device uses low-pressure water from sluicing jets to mobilize the solids present on the bottom of a tank after bulk waste retrieval or to wash the internal tank surfaces and collect the material around a transfer pump, a process similar to using a water hose to mobilize and corral debris on a driveway.

Each site has developed its version of hydraulic techniques for tank cleaning. The Savannah River Site has used rotary spray jets (Tank 16) and a water monitor to wash the internal surfaces of the tanks, a moderate-pressure water jet called the "water mouse" to break up solid deposits on the bottom of the tank, and a sluicing jet known as the "waterbrush." It has also used a high-pressure jet (10,000 to 30,000 psi) called the Hydrolaser/hydrolance in Tank 19 to break up a 42-inch high by 30-inch diameter mound of zeolite. This technique is very effective for hard materials within a couple of meters of the nozzle but ineffective beyond this range. It should be noted that the solids resulting from breaking up zeolite deposits can be difficult to mobilize, collect, and transfer because they settle very quickly.

The Hanford Site has used a manually directed jet to wash internal tank surfaces and a sluicing technique that involves two directional nozzles. This system was used in Tanks C-106, and in Tank S-112. For Tank S-112 a new tool, called the Salt Mantis (also known as the Hydrolaser), involving a high-pressure water lance is being tested with promising results. The Salt Mantis can fit down a 10-inch riser and then unfold to form a large cross. The Mantis has two tires, one on each cross-beam of the cross, that are each hydraulically operated to manipulate the Mantis inside the tank. The long member of the cross provides stability when the water lance is working. The Mantis puts out a low-volume (6 gallons per minute [gpm]) of high-pressure water (variable between 2,000 and 35,000 psi) to break up the waste, which is then removed by sluicing. In a 10-hour demonstration test in Tank S-112, the Salt Mantis uncovered 30 percent of the tank bottom. One of the advantages of this tool is that is has a "low head" (i.e., does not provide a significant driving force to cause leaks). The Salt Mantis works best with approximately 1 foot of water that its high pressure jets can agitate the standing water and use it to break up the waste.

The Idaho site has used a device called the "washball" (see Figure G-1) in combination with directional nozzles that direct a relatively low-pressure (less than 100 psi) stream of water onto a vertical slice of the internal tank surfaces (walls, roof) above the waste to wash radioactive material onto the bottom of the tank. Figure G-2 shows the jets from the directional nozzles on the tank walls.

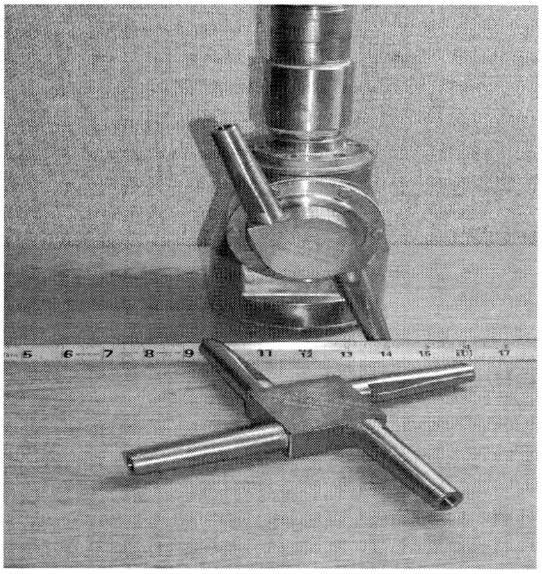

FIGURE G-1 Tank cleaning washball used at the Idaho site. SOURCE: Lockie et al., 2005.

FIGURE G-2 High-pressure water jet used to clean the tank walls at the Idaho site. Note the horizontal cooling coils on the walls. SOURCE: Lockie, 2005b.

VACUUM TECHNIQUES

For situations in which using significant amounts of water is not acceptable—for example, in tanks that have already leaked or are at risk of leaking—vacuum technologies in combination with small amounts of fluids are used. The Hanford Site has developed a vacuum retrieval device consisting of a vacuum head, a vacuum pump, a slurry vessel, and slurry transfer pumps. Air and water are added to assist in transferring the waste to another tank pneumatically. This device operates much like a steam carpet cleaner where the water is injected and almost immediately removed. This prevents any significant accumulation of water and thus reduces the potential for leaking water to the environment. Additionally, small amounts of high-pressure water can be introduced via a scarifier to dislodge waste and help suspend heavy particles so that they can be removed.[1]

The vacuum retrieval system has been used successfully in completing waste retrieval from Tanks C-202 and C-203 at Hanford. The heel volumes in Tank C-202 and C-203 are estimated at 19.6 and 18.5 cubic feet, respectively, in both cases meeting the Hanford Federal Facility Agreement and Consent Order retrieval criteria (limits of technology and less than 30 cubic feet). Hanford plans to use vacuum retrieval to retrieve residual waste and may use it for bulk waste retrieval in tanks where leakage becomes a problem.

[1]A scarifier is a tool that can be used to break up and loosen hard surfaces such as concrete walls or road pavements. It is usually employed to treat surfaces for deactivation and decommissioning purposes.

MECHANICAL TECHNIQUES

Mechanical retrieval devices have been used for two purposes. First, they have been used to break up solid deposits. Such devices use remotely controlled grinding or scraping tools. The resulting particles can then be mobilized and collected for removal by a transfer pump using techniques described above. Such techniques can be very effective but the deposit must be accessible to the device and its use is labor intensive (Figure G-3).

The second purpose for using mechanical retrieval devices is to move wastes without having to introduce water into the tank. Examples of such devices are pushers that move waste to a transfer pump or move debris out of the way so the waste is amenable to hydraulic techniques or hooks or pincer claws that remove debris from the tank.

DEPLOYMENT TECHNOLOGIES

All physical techniques for retrieving residual wastes in tanks require some type of deployment technology of which there are three:

1. A jointed arm or mast that is installed through the tank riser and maneuvered to access various portions of the inside of a tank by human control;
2. A tethered vehicle that is normally connected via wires inserted through a tank riser or potentially electronically to a human controller; and
3. An autonomous robotic device programmed to adapt to the requirements at hand with minimal human inter-

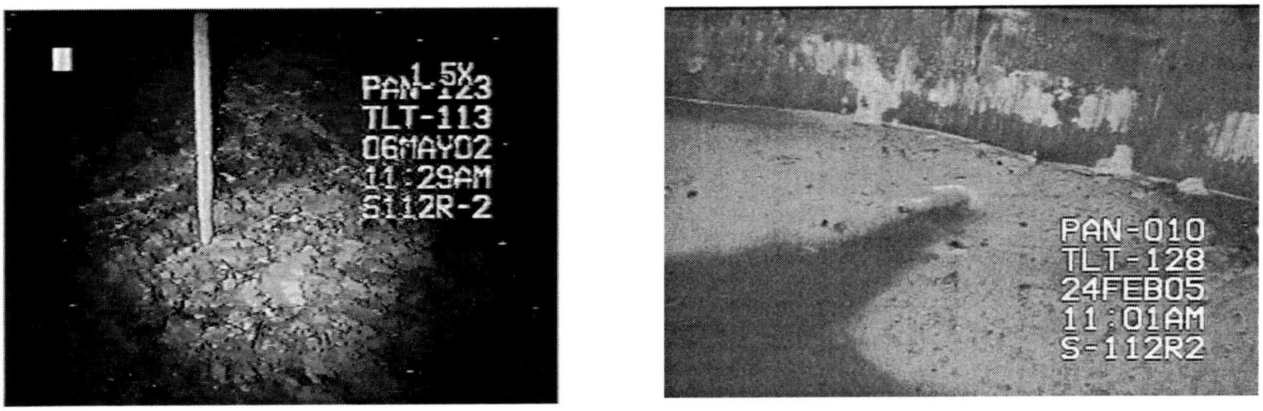

FIGURE G-3 Residual waste on the bottom of Tank S-112 after modified sluicing and saltcake dissolution. SOURCE: Barton, 2005a.

vention. Such a device may be powered by on-board rechargeable battery packs and an in-tank recharging station or by electrical cables inserted through a riser.

To date the dominant deployment technology for tools used to retrieve residual wastes has been mechanical arms. The most popular type of arm used at Department of Energy Sites (DOE) sites is a relatively simple mast inserted through a riser from which an arm containing one or two joints projects and some type of retrieval device extends. This technology has been used to deploy washing technologies such as the wash ball, water jets (see Figure G-3), vacuum devices, and mechanical devices and waste mobilization and collection technologies such as sluicing jets and water brooms. Mechanical arm technology has two main advantages: (1) it is a demonstrated and available technology, and (2) it can reach virtually any part of a less impeded (i.e., without cooling coils) tank. The disadvantages of this technology are that the lifting capacity of the arm across the necessary distances in underground tanks is limited, it requires detailed manual control by a human, and its operation can be impeded severely by some internal tank structures. Moreover use of these technologies for deploying tools is more time consuming than simply mixing and pumping because operating a remote arm in a cluttered environment can be tedious and demands concentration so operators have to be rotated periodically. However, the advantage of articulated arms is to access waste that cannot be accessed easily using centralized mixing systems.

Tethered vehicles have been used at the Oak Ridge National Laboratory, Savannah River, and Hanford Sites to deploy various tools to the bottom of underground tanks containing radioactive waste (Burks, 2005). The size of the tanks in which these devices have been tested or deployed varies from 5.7 – 644 m^3 (1,500-170,000 gallons; Oak Ridge) to about 3,785 m^3 (1 million gallons, Hanford and Savannah River). These are best envisioned as small tracked vehicles (crawlers). Such vehicles have deployed tools to wash tank internal surfaces (e.g., water jet), hydraulically or mechanically break up solid deposits (e.g., water mouse, scarifier), hydraulically mobilize and collect wastes (e.g., the "water brush" used at the Savannah River Site, or mechanically mobilize and collect waste by acting as a mini-bulldozer (see Figure G-4).

Tethered devices have two main advantages: (1) the ability to access tank surfaces that are inaccessible to mechanical arms, and (2) the ability to deploy relatively heavy tools to distant tank locations that are beyond the capabilities of mechanical arms. The disadvantages of in-tank tethered devices are similar to those for articulated arms: the need for detailed manual control by skilled humans, the need to rotate operators, the complications of maintaining the on-board electronics and sensors, and maneuvering their tethers in tanks having internal structures such as cooling coils. Deployment of autonomous retrieval technologies is beyond the current state of the art and is discussed further in the section on advanced technologies.

FIGURE G-4 Houdini™ in-tank crawler. SOURCE: RedZone Robotics.

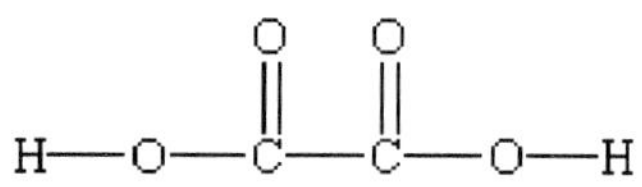

Oxalic acid, $H_2C_2O_4$

FIGURE G-5 Chemical composition of oxalic acid.

CHEMICAL TECHNOLOGIES FOR RESIDUAL WASTE RETRIEVAL

When physical technologies for retrieving residual waste have not been deemed sufficient, DOE has employed chemical technologies to remove remaining tank wastes. Tests performed in the late 1970s and early 1980s (Bradley and Hill, 1977; Hill, 1978; West, 1980) proved oxalic acid solutions to be the most effective reagent for dissolving sludge while not corroding the carbon steel tank. Oxalic acid ($H_2C_2O_4$; see Figure G-5) is a reactive chelating agent and a weak acid ($pK_{a1} = 1.2$; $pK_{a2} = 4.2$).

Oxalic acid solutions dissolve some radioactive materials that are found in recalcitrant forms or difficult-to-access locations. Most of the resulting solution is readily retrievable by pumping. Oxalic acid treatment has been used in Tank 16 at the Savannah River Site and in Tank C-106 at Hanford. In both instances, oxalic acid allowed further retrieval of residual waste. Oxalic acid treatment was also applied to Tank 24 at the Savannah River Site which contained 37.9 m^3 (10,000 gallons) of zeolite (used as an ion exchanger to separate cesium-137), but it proved ineffective in this application.

Previous Savannah River Site experience indicates that using oxalic acid can be effective at removing some residual wastes from tanks. However, Savannah River Site staff expressed reluctance to use oxalic acid extensively for the final stage of cleaning due to concerns about criticality, downstream processing, and cost.

Criticality Concerns Using Oxalic Acid

Since its interim report, the committee has received additional information from Savannah River Site staff about criticality concerns. The following is based on the information provided by the site (DOE-SRS, 2005e). According to the Savannah River Site staff, sludge heels after bulk waste removal may contain enough fissile mass,[2] on average, to cause a criticality event. Savannah River Site staff could not share with the committee the exact amounts of fissile mass for reasons of homeland security. However, the staff told the committee that some tanks have uranium enriched to 60 to 70 percent. Safe storage in waste is achieved by diluents and neutron poisons. Bulk removal and water washing do not significantly affect the relative ratios of diluents and poisons to fissile material, but dissolution of a sludge heel with oxalic acid may cause the diluents and poisons to separate from the fissile material. The following sections describe information on solubility available in the literature, simulated waste dissolution tests, and actual waste dissolution tests that Savannah River Site staff provided in response to the committee's request for the rationale on criticality concerns.

Karraker (1998) reviewed the available solubility data for uranium, plutonium, iron, and manganese in oxalic acid solutions, and concluded that the neutron poisons are considerably more soluble than plutonium, which could result in separation of some plutonium from the poison (Karraker, 1998). The specific chemical compound for each of the sludge components affects the relative rate of dissolution. Although solubility differences exist, the dissolution rates could affect the observed behavior with real waste. Relative reaction rates for the poisons are discussed in Adu-Wusu (2003). Actual rates relative to uranium and plutonium are not available.

One waste dissolution test with Tank 16 sludge at the Savannah River Site indicates that some portion of the plutonium could be highly insoluble, but sample sizes were too small to be conclusive (Bradley and Hill, 1977). A follow-up analysis for the extent of insoluble plutonium showed that about 30 percent of the plutonium remained insoluble after 99 percent of the bulk sludge was dissolved including the primary neutron poisons (Hill, 1978). Hobbs (2003) conducted laboratory oxalic acid dissolution testing on sludge from Savannah River Site Tank 8 to evaluate the downstream impacts. The sludge contained high levels of sodium oxalate although the test data were inconclusive regarding any increased leaching of plutonium and uranium.

Although not directly applicable to Savannah River Site waste, an example of laboratory tests with Hanford waste show that 5 to 20 percent of the total alpha-emitting isotopes are dissolved with oxalic acid. Up to 50 percent of the iron and 80 percent of the manganese dissolved in the same sample (Bechtold et al., 2003). The data from samples of waste remaining after each acid wash cycle during Tank 16 heel dissolution with oxalic acid provide enough information to identify a relative change in plutonium content with poisons. These tests showed that plutonium appeared to decrease compared to iron relative to the initial sludge composition, but remained more or less constant with each subsequent acid strike (Bradley and Hill, 1977; West, 1980, Table 6).

As DOE recognizes, the available data are inconclusive with regard to the potential for criticality during oxalic acid dissolution because the bulk of the material, including the fissile material, may be removed before the ratio of fissile material to poison falls below safe storage limits. Savannah River Site staff told the committee that each tank heel needs

[2]Fissile mass is made of isotopes having a high probability of undergoing nuclear fission when struck by neutrons and, in the right quantities and configurations, can sustain a nuclear chain reaction.

to be evaluated before attempting dissolution with oxalic acid to determine whether any criticality potential exists and how to mitigate the risk. The committee has not seen any additional calculations regarding the criticality issue and cannot determine whether the concern is real, the committee recommends that DOE investigates the matter further (see Chapter IX).

Downstream Problems Using Oxalic Acid

According to DOE, there are additional downstream complications when using oxalic acid due to the large amounts of acid needed to neutralize the waste and to dissolve the residual sludge (e.g., approximately 276 m^3 [73,000 gallons] of 4 weight percent for Tank 16). The process flowsheet for the use of oxalic acid in tank cleaning is estimated to add about 26,000 to 38,000 kg of sodium oxalate per 19 m^3 (5,000 gallons) of sludge residual removed. These oxalates eventually are processed in the Savannah River Site evaporator system and will create some operational problems based on past experience. For example, in the 2H evaporator it was found that the crystalline structure of the sodium aluminum silicate (which was formed by mixing high-silica streams with high-aluminum streams) has an affinity for uranium but not for any of the neutron poisons, so there was a selective concentration of uranium in the evaporator that raised some criticality issues. If it is added to the sludge before washing, oxalic acid goes to the evaporator and may precipitate in the feed tank. If oxalic acid is added to the salt tank it may precipitate and form a hard salt layer that would require a lot of water to remove.

Other problems cited as part of downstream concerns with oxalic acid are the following:

1. *Foaming during evaporation.* Savannah River Site staff stated that the addition of oxalic acid has been observed to cause foaming during evaporation of recycle streams to reduce volume. Such foaming compromises the function of the evaporator.
2. *DWPF off-gas flammability.* To prevent the possibility of combustion or explosions in Defense Waste Processing Facility (DWPF) off-gas processing equipment, the amount of organic material fed to the DWPF is limited. Substantial use of oxalic acid could lead to these limits being exceeded.

Melter performance in the presence of oxalates does not seem to be a concern at the site. The melter can tolerate considerable amounts of oxalate in the feed (spent oxalic acid wash from up to 10 to 15 tanks).

Cost Of Using Oxalic Acid

DOE's estimate for oxalic acid washing in 1999 was $1,050,000 per tank, including disposal costs at the Savannah River Site. DOE concluded that, for the F Area Tank Farm, oxalic acid washing of the 10 remaining tanks would add approximately $10,500,000. DOE expects results for individual tanks in the H Area Tank Farm to be similar in terms of additional costs (USNRC, 1999).[3] A 2005 study estimates the cost of using oxalic acid to be $15,000,000 each for Tanks 18 and 19, attributing additional costs to the need to retrofit the two tanks completely (remove old equipment and purchase and install new equipment, including a chemical addition system)[4] (Gilbreath, 2005). Cost estimates would be lower for other tanks if the oxalic acid addition were performed immediately after waste residual removal and before the tank is isolated for closure. According to Savannah River Site staff, additional costs are due to the nuclear criticality safety and evaluation reports that DOE would have to do on a tank-by-tank basis to show the effect of oxalic acid on the neutron poisons.

DOE Plans to Use Oxalic Acid in the Future

DOE has indicated it is planning to use oxalic acid as a final cleaning step at the Savannah River Site on a tank-by-tank basis, taking into account the factors listed above (DOE-SRS, 2005d). The site is interested in chemical cleaning technologies. A team was formed in December 2005 at the Savannah River Site to evaluate the use of nitric acid and its application to sludge removal in Savannah River Site waste tanks. However, the committee believes that the site is lacking a strong impetus for research and development at this time (DOE-SRS, 2005e). In Chapter IX, the committee provides findings and recommendations on research and development needs in chemical cleaning.

Hanford is not planning to use oxalic acid in the future. According to DOE, oxalic acid was only modestly successful in breaking the hard waste residual in Tank C-106 and was used in conjunction with other hydraulic retrieval technology. Barton reports that three acid batches and one sluicing operation on tank C-106 removed more than two-thirds of the waste. The last three oxalic acid batches did not react further with the waste (based on pH readings). Water sluicing was able to mobilize part of this residual waste and move it toward the transfer pump (Barton, 2005b).

Hanford's tank wastes were formed by many different chemical processes (Bismuth Phosphate Process, REDOX

[3]This cost estimate is based on the major assumption that oxalic acid would be added directly after the bulk and residual waste removal activities had been declared complete. This figure included the purchase and installation of the chemical addition system, the disposal of oxalic acid waste, and the safety bases analyses required at the time. The existing waste removal equipment (slurry pumps, transfer pumps, spray wash devices, etc.) was assumed to be functional and to accommodate the use of oxalic acid.

[4]This cost estimate includes removal of existing equipment (old waste removal pumps, jets, etc., that occupy risers that are necessary for access), purchase and installation of new pumps and other equipment, and upgrades to the design safety analyses.

Process, PUREX Process) and waste separations techniques, each contributing a different mix of chemicals to the tanks. As a result, any chemical technique used for hard heel removal has to be tailored to the chemical characteristics (origin) of the heel in each tank. The chemical must (1) break up the hard heel matrix to enable removal by sluicing, (2) not add substantial new chemical bulk to be removed in pretreatment, and (3) not put the actinides and strontium-90 in soluble forms (which would defeat solids-liquids separations in the ultrafiltration system) to avoid the same criticality issues identified by Savannah River Site staff.

In the case of the oxalic acid used on Tank C-106, oxalate solids were formed that now must be removed by ultrafiltration in the Waste Treatment Plant. The ultrafiltration system is a bottleneck in the Waste Treatment Plant due to the filters being used for several operations (e.g., sludge washing, oxidative leaching). Systemically adding chemical burden to the ultrafiltration system could impact the feed rates to the Waste Treatment Plant high-level waste melters and low-level waste melters. Dissolution of additional strontium and transuranic waste could also result in more wastes that require additional pretreatment to ensure that immobilized low-activity waste glass meets regulatory limits and ALARA (as low as reasonably achievable) package handling requirements. These additional pretreatment steps will slow the overall production of the waste treatment plant, resulting in additional years of operation of the combined system.

Accordingly, Hanford's waste retrieval staff is more interested in mechanical and high-pressure hydraulic techniques such as the plow blade on the Mobile Retrieval System (MRS) and the high-pressure water lance on the Salt Mantis. Both robotic techniques allow DOE to break up hard heel materials to perform more complete retrievals without further complicating the tank farm and Waste Treatment Plant chemistry (Hewitt, 2005a). Idaho is not planning to use oxalic acid because its tanks do not have sludge to remove.

TECHNOLOGY ADVANCES FOR MECHANICAL RESIDUAL WASTE RETRIEVAL

Pulsating Mixer Pump Technology

DOE is considering the use of the Power Fluidic Technology™ developed by AEA Technology and a similar Russian Pulsating Mixer Pump Technique for bulk waste retrieval at the Hanford and the Savannah River Sites (Murray, 2005). This technology involves using two or more nozzles in a tank to establish a "back-and-forth" motion of the sludge and a vacuum induced by fluid flow through a "jet pump system," as shown in Figure G-6.

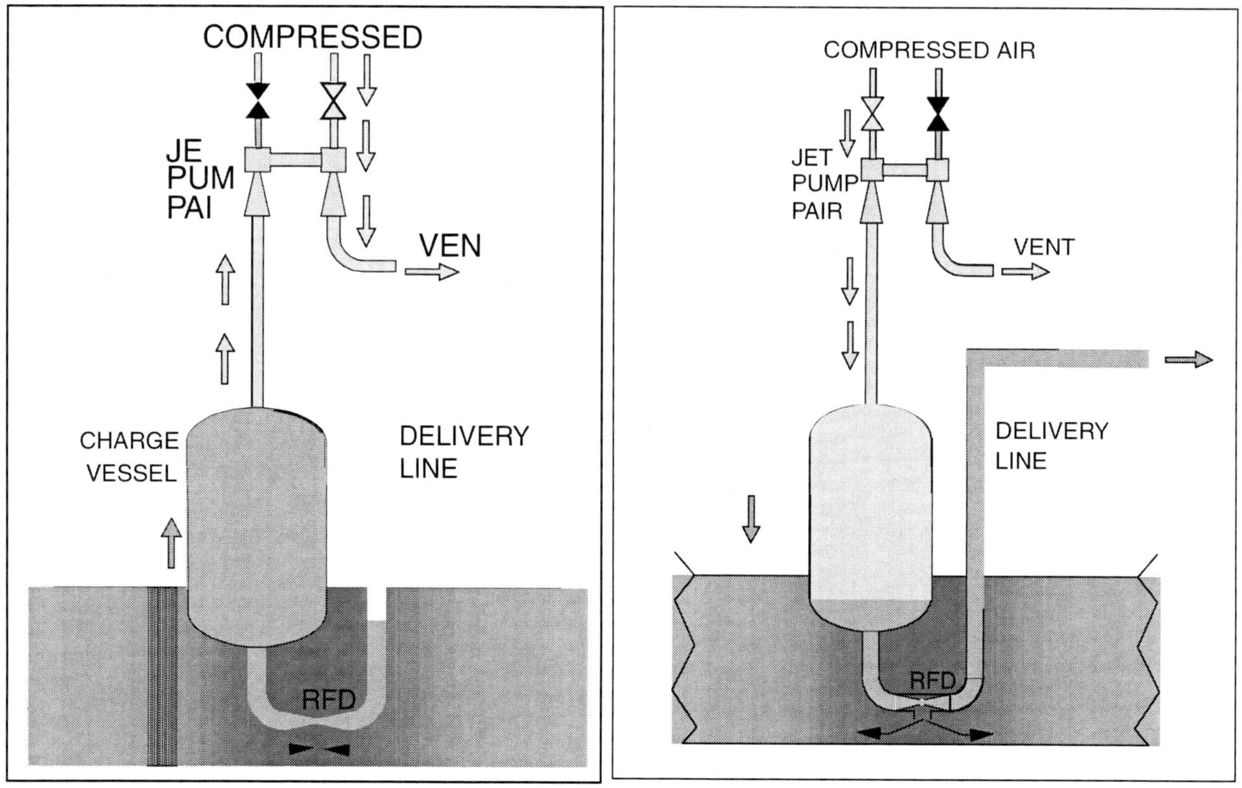

FIGURE G-6 Key components of a pulse jet system. SOURCE: AEATES, 2005a.

This technology does not have any mechanical devices inside the tank and is claimed to use water more efficiently. Fluidic technology can be used for bulk waste as well as residual waste retrieval and for mixing the grout with waste residuals that cannot be further removed. However, its effectiveness on waste in tanks with vertical cooling coils is uncertain as with other technologies.

Hanford tested a prototype using this technology in a full-scale cold test facility, and determined that the complex air and fluid controls and air handling systems would result in significant maintenance and ALARA concerns for operations in the single-shell tank farm system. No further deployment of fluidic technology for waste retrieval from single-shell tanks is planned. Use of fluid jets for mixing grout with waste residuals is still under consideration.

Hanford Mobile Retrieval System

The Mobile Retrieval System consists of a vacuum retrieval device mounted on a mechanical arm (mast) combined with a remotely operated in-tank vehicle (ITV) (see Figure G-7).

The ITV is equipped with a plow blade and a water jet used to move waste within the range of the vacuum device. The plow is used to move or break up the waste. The vacuum system uses a small amount of water to mobilize the waste as it is sucked up by the vacuum. The waste is collected in the slurry vessel where it is then pumped from the tank. The articulated mast is fitted with a vacuum head, vacuum pump, slurry vessel, and slurry transfer pump (the same type of vacuum retrieval apparatus that was successfully used to remove sludge from Tanks C-203 and C-202).

The Mobile Retrieval System (MRS) removes waste from a single-shell tank by using raw water to mobilize the sludge in the tank and remove it with a vacuum device. Recycled supernatant is then used to transport the waste to the double-shell receiving tank. The recycle loop may include skid-mounted equipment to dewater the recycle stream to reduce water usage. As a result of being composed of devices already used in other tank retrievals, the MRS represents an incremental improvement over previous retrieval technology. Like the Salt Mantis, the MRS has a low head since there is little or no standing water in the tank. Therefore, these tools are appropriate for tanks that have a high risk of leakage. The mobile retrieval system will be deployed for the first time in Tank C-101.

Delivery Tools

Essentially all of the techniques for retrieving residual wastes in tanks described in this chapter must be deployed

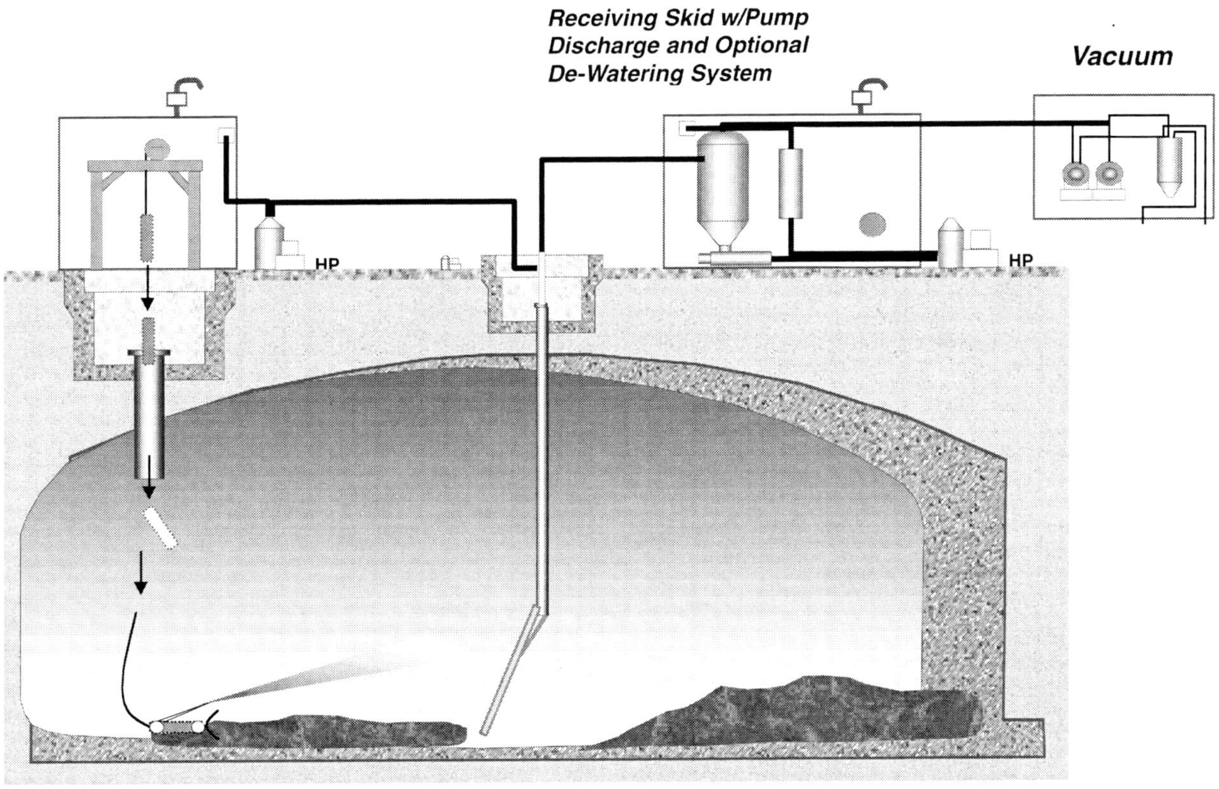

FIGURE G-7 Mobile Retrieval System. SOURCE: Gasper: 2005.

by delivery tools, such as multijoined mechanical arms and in-tank tracked and tethered vehicles. DOE has developed advanced mechanical arms having multiple joints and improved guidance capabilities. The most prominent example of this is the Light-Duty Utility Arm (LDUA) and its modified version (MLDUA). In September 1996, the LDUA's stereo viewing systems and gripping capacities were demonstrated successfully in Hanford Tank C-106. The LDUA was deployed at the Idaho site for approximately two months in early 1999 to inspect and sample Tank WM-188, and later in the year to inspect and sample Tanks WM-182 and WM-183. The MLDUA was used at Oak Ridge National Laboratory (ORNL) from June 1997 through September 2000 to retrieve waste from two tanks with 161 m^3 (42,500 gallon) capacity and five tanks with 644 m^3 (170,000 gallon) capacity (Glassell et al., 2001). These demonstrations yielded mixed success: The LDUA used at Idaho was able to deploy multiple tools within its reachable workspace but had significant downtime due to hardware failures. Since this demonstration, DOE sites have used simpler mechanical arms for residual waste retrieval as described previously. It is not known whether the need to retrieve wastes from more complicated tanks will lead to further development of the LDUA to improve its reliability. The MLDUA used at ORNL had almost no hardware failures over the time (3 years and 4 months) it was deployed (Glassell et al., 2001).

As discussed earlier, several tracked tethered vehicles have been developed in the DOE complex to operate inside a tank in limited amounts (less than 10 cm deep) of sludge or liquid. Some of these vehicles (Houdini™, ESG/LATA) are collapsed to fit through a 61 cm- (24 inch) diameter riser and then expanded to a 1.5 by 1.5 meters working platform; others vehicles are fixed frame and must fit through the riser openings. Some of the crawlers are available commercially (Houdini™, ARD). The committee did not find evidence of any foreseeable dramatic improvements in crawler technology, but incremental engineering improvements are likely to continue (see Chapter IX).

GENERAL WASTE RETRIEVAL ISSUES

Previous sections have alluded to future difficulties that can be expected when retrieving waste from DOE's "complicated" tanks. The following sections describe some of the important complications that are anticipated.

Recalcitrant Waste Deposits

Waste may be encrusted on internal tank surfaces or structures in a semidry form, or it can agglomerate in chemical forms that resist physical removal technologies. These recalcitrant waste deposits must be mobilized before they can be collected and removed from the tanks.

One of the most common types of recalcitrant waste deposit is composed of inorganic zeolite ion-exchange material that has reacted with tank chemical contents to yield sodium and calcium aluminosilicates that form a solid deposit (see Figure G-8). Zeolites were used to separate

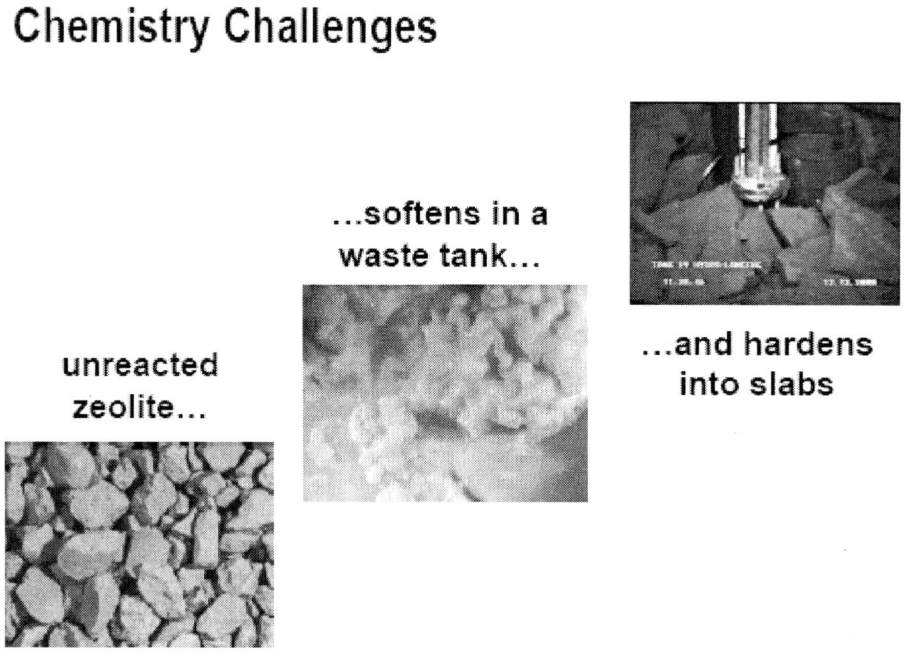

FIGURE G-8 Zeolite deposits can be difficult to remove because the fine particles formed when the slabs of zeolite are broken up by waste retrieval tools settle very quickly at the bottom of the tank. SOURCE: Caldwell, 2005a.

cesium-137 from the waste and are potentially important because they contain significant amounts of this radionuclide. The Savannah River Site had 11 tanks containing zeolite deposits, of which 2 (Tanks 18 and 19) that do not contain cooling coils have recently been cleaned. Eight of the remaining nine zeolite tanks have coils (see Appendix F).

Zeolite compounds are recalcitrant for two reasons. First, they are difficult to dissolve. Based on experience at the Savannah River Site and Hanford, chemical cleaning using oxalic acid is moderately successful at best. Savannah River is also considering using nitric or hydrofluoric acid to dissolve the zeolite particles but these two acids are very corrosive on carbon steel tanks. Second, although the deposits can be disaggregated into sand-like particles using mechanical techniques, the particles are difficult to keep in suspension for recovery with transfer pumps. Savannah River Site has considered the use of in-tank vehicles (such as the mini-bulldozer used in Hanford) to corral the zeolites, but anything with a tether makes deployment in tanks with coils very challenging. Savannah River Site staff has requested the help of DOE's Office of Cleanup Technologies (EM-21) for research and development on zeolite removal.

Hanford Site staff noticed hardened saltcake deposits in Tanks S-112 and S-102. Waste retrieval in these two tanks is being pursued using high-pressure devices such as the Hydrolaser/hydrolance (in S-112) and the Salt Mantis (in S-102).

Waste Accessibility

To remediate a tank, the waste has to be accessible to waste removal tools. One of the main challenges of waste removal is the number of physical obstacles to the waste and other issues related to tank design that complicate waste retrieval operations. The following is a list of waste retrieval challenges due to tank design features or physical obstacles that impede access to the waste or due to tank integrity concerns that limit the choice of waste retrieval tools. This list was developed by the Tanks Focus Area in 2002 for the Savannah River Site, but many of the challenges apply to the Hanford Site and a few to Idaho National Laboratory as well.

Waste retrieval challenges identified by the Tanks Focus Area (adapted from Saldivar, 2002) include the following:

- Horizontal cooling coils (Savannah River Site and Idaho National Laboratory)
- Vertical cooling coils (Savannah River Site only)
- Tank integrity limiting the choice of retrieval tools (Hanford Site)
- High-level Waste environment on tank top and in surrounding area
- Tank bottoms located 45 to 50 feet below ground surface
- Contamination containment for potentially leaking equipment
- Nonsymmetrical riser positions
- Confined spaces (Savannah River Site only)
- Limited openings into the primary tank and annulus space—no larger than 24 inches in diameter (Savannah River Site only; see Figure G-9)
- Removal of waste from the ventilation duct at the bottom of the annulus space (Savannah River Site only)
- High radiation rates in tank and at riser openings
- Tank top loading is limited
- All transfers out of the tanks are from one riser location
- Tank support columns produce shadowing effects (all sites)

Of these, the most difficult to overcome is likely to be the vertical cooling coils in Savannah River Site tanks, which severely impede the ability to maneuver water jets, mechanical arms, and in-tank vehicles to access the waste.

The Hanford tanks, although they do not have cooling coils, have some internal obstructions such as the air-lift circulators used to stir and suspend the sludge in high-heat tanks. The designs vary, with some welded to the floor with "guy wire" supports and others suspended from risers at the top of the tanks. A single tank can contain as many as 22 air-lift circulators. The following tanks have air-lift circulators installed:

- 15 single-shell tanks in SX Farm
- 6 single-shell tanks in A Farm
- 4 single-shell tanks in AX Farm
- 2 double-shell tanks in AY Farm
- 2 double-shell tanks in AZ Farm

According to DOE, the circulators in Tank AZ-101 (see Figure G-10) are probably the most complex in the Hanford tank farms.

In-Tank Debris

Most underground storage tanks, especially at the Hanford and Savannah River Site contain objects labeled as debris. These can include failed pumps, tapes for measuring waste depth, instrument trees, sluicers, hoses, and miscellaneous items. Some items have been left in place (e.g., suspended from the top of the tank); while others (such as waste surface measuring tapes) have been dropped into the tanks. In some cases, miscellaneous materials were added to the tanks (generally during the 1960s and 1970s) as a way to dispose of them. Such debris can interfere with bulk or residual waste retrieval by impeding the flow of water necessary to mobilize and collect wastes at the location of a transfer pump. Some of the in-tank debris at the Hanford Site includes the following:

- Six single-shell tanks that were past leakers have as much as 95,000 kg of diatomaceous earth per tank that was added to absorb liquids;

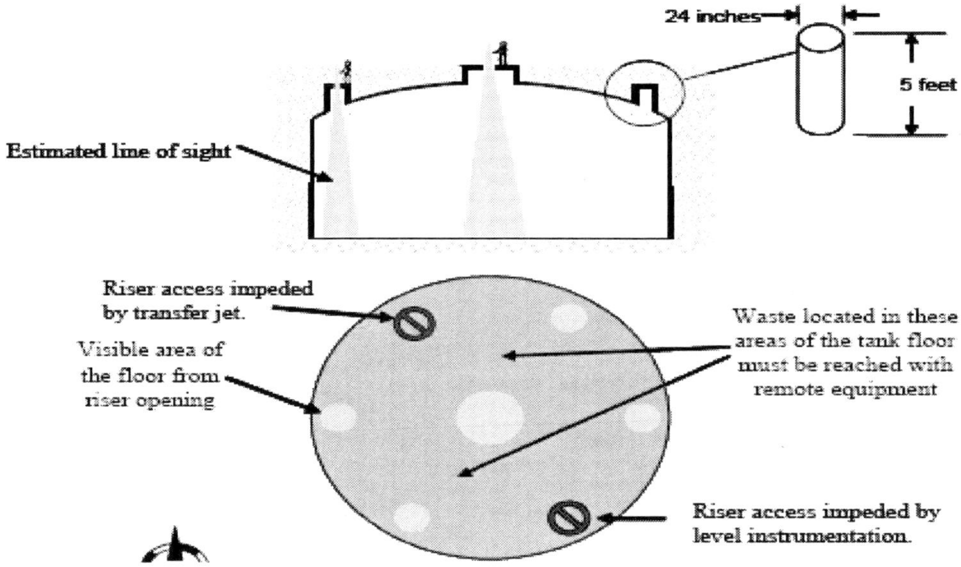

FIGURE G-9 Diagram of Tank 19 access area for heel removal equipment. SOURCE: DOE-SRS, 2005a.

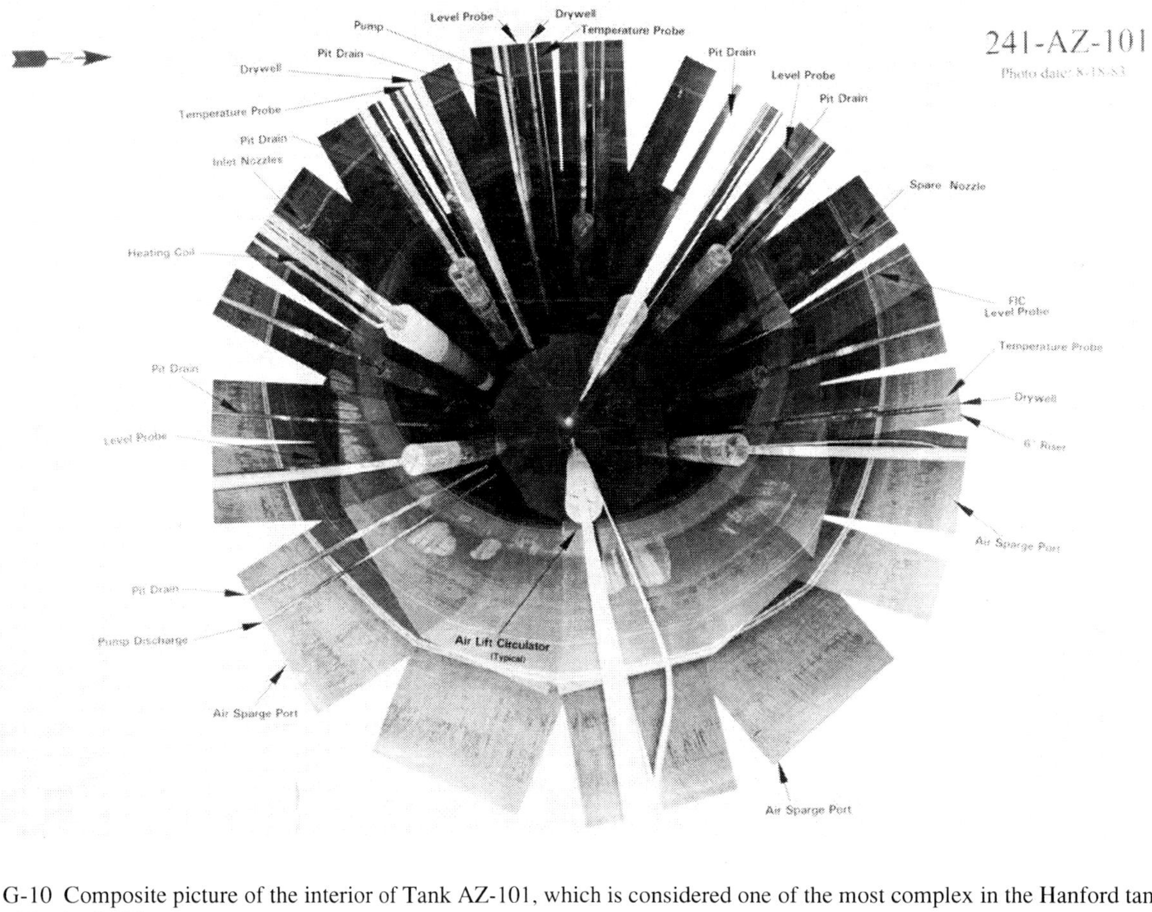

FIGURE G-10 Composite picture of the interior of Tank AZ-101, which is considered one of the most complex in the Hanford tank farms. SOURCE: Hewitt, 2005a.

FIGURE G-11 Steel tape grappling device developed at the Savannah River Site. SOURCE: DOE-SRS, 2005g.

- At least one single-shell tank contains approximately 57,000 kg of Portland cement, which was added to bind liquids;
- One single-shell tank contains 7 m^3 of organic ion-exchange resin that was disposed of in the tank;
- One single-shell tank contains 16 plastic bottles (3 inches in diameter x 54 inches long) containing uranium solutions that were disposed of in the tanks;[5] and
- One single-shell tank contains six cask loads of experimental fuel elements, shroud tubes, and samarium ceramic balls that were disposed of in the tank.

To address the issue of in-tank debris, DOE has developed techniques (e.g., grappling devices to retrieve tape measures (see Figure G-11) that can remove some of the debris or move the debris to different locations to allow the underlying waste to be retrieved. Neither the Savannah River Site nor the Hanford Site identifies in-tank debris as a major obstacle to waste retrieval (DOE-RL, 2005; DOE-SRS, 2005d). If heel retrieval systems that use vehicles or mobile in-tank systems are deployed, debris may become a problem even if these systems are capable of removing this debris or moving it around the tanks.

Residual Waste in Pipelines and Ancillary Equipment

The large underground storage tanks that are the primary focus of retrieval are interconnected by myriad pipelines and ancillary equipment, such as small underground storage tanks, pumps; valve boxes used to transfer tank wastes among the large tanks, and other hardware used in waste retrieval operations (see Figure G-12). As a consequence, to varying degrees, these pipelines and ancillary equipment contain tank waste that will likely have to be removed to the maximum extent practical. The closure of pipelines and ancillary equipment has been considered part of the draft Section 3116 waste determination at the Idaho National Laboratory (INL; DOE-ID, 2005a). At the Savannah River Site pipelines and ancillary equipment will be addressed in the tank farm closure determination. Hanford staff recently produced a study on plugged and abandoned pipelines (Lambert, 2005).

As part of routine practice, the majority of pipelines and ancillary equipment is flushed with nonradioactive water after use so the remaining waste is expected to contain amounts of radionuclides that are small relative to what is being left in the tanks. As an example, at the Idaho National Laboratory (which has the most compact array of tanks) there are about 2 miles of pipelines associated with the tanks containing an estimated 30 Ci in about 15 kg of waste.

There are challenges associated with the pipelines, foremost among which is the fact that a few of the pipelines plugged during use so that they could not be flushed to remove most of the residual radionuclides. This is especially the case at Hanford where an estimated 100 pipelines are plugged; they contain about 900 liters (232 gallons) of waste (Lambert, 2005). A lesser problem is that records of a limited number of pipelines may never have existed or have been lost during the intervening decades. Nondestructive methods

[5]According to DOE, criticality does not appear to be an issue for this type of debris (Boomer et al., 1993; Hewitt, 2005b).

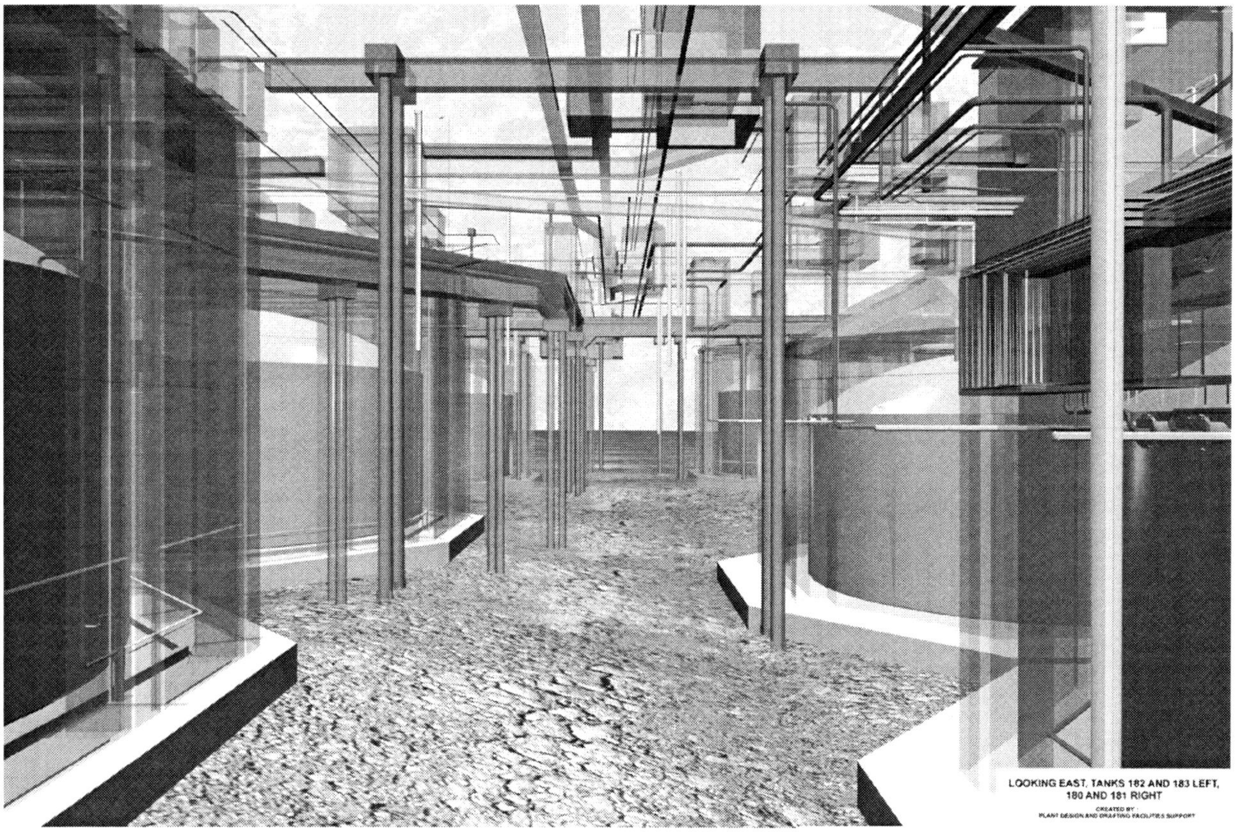

FIGURE G-12 Computer-generated image of underground pipelines connecting the tanks at Idaho National Laboratory. Similar underground structures are present at the other two sites. SOURCE: Lockie et al., 2005.

such as ground-penetrating radar may be used to locate the pipelines (NRC, 2000e; 2001b).

As a result of the relatively small amount of radioactive material believed to be in the pipelines and ancillary equipment, and because some of this system is still in use, DOE has accorded characterization and remediation planning a low priority and a formal decision process has not yet been initiated. Information obtained by the committee as a result of site visits and conference calls indicates that DOE is inclined to propose that most pipelines and ancillary equipment be flushed and grouted in situ while considering exhumation of plugged pipelines where necessary (Harbour et. al. 2004; Schaus, 2005). The Savannah River and Hanford Sites have both demonstrated grouting of pipelines (Harbour, 2000). Savannah River Site staff has demonstrated this ability on "non-tank farm" underground lines that contain relatively low (if any) amounts of contamination. Additional vents had to be installed on some of these lines to obtain proper flow of grout (this is impractical in the tank farms because of the relative depth of transfer lines, radiation rates to install vents, and difficulty in even reaching some of these locations due to obstructions). Harbour (2005) demonstrated that with proper grout flow characteristics and "vent and delivery hose in one," grouting of lines over long distances is possible. This has not been demonstrated practically in Savannah River Site tank farms to date. Experience with pipeline plugging and unplugging from Russian Nuclear Defense Material production sites has also been published (Florida International University, 2001). Other technology reports on pipeline locating, unplugging, and cutting technologies were published as Innovative Technologies Summary Reports within DOE's former Office of Science and Technology (ITSR, 1999a, 1999b, 1999c, 1999d, 1999e). The Tanks Focus Area also sponsored several projects on waste transfer line plugging prevention and unplugging methods involving chemical cleaning, pressure cycling, or vibration (e.g., DOE-TFA, 2000b, Welch, 2001).

Leaks to the Environment

To varying degrees, all three DOE sites have inadvertently released wastes contained in the tanks into the environment. At Hanford the releases are substantial and primarily the result of leakage from the single-shell tanks. At the Idaho site, leaks to the environment were primarily the result of spills during tank transfers between reprocessing facilities and of tank and valve failures within the tank farms. At the Savannah River Site, the only leak to the environment is due to a crack in Tank 16 that leaked waste into the annulus pan. Such releases raise two issues: (1) the degree of retrieval from the tanks to be required given the context provided by the amount of radionuclides in the immediately surrounding environment, and (2) application of, removal to the "maximum extent practical" to leaked radionuclides.

Regarding the first issue, the context provided by leaked radionuclides is one factor to be considered in determining whether radionuclides have been removed to the maximum extent practical. This issue is discussed further in Chapter X. Regarding the second issue, DOE has stated that it intends to address the remediation of radioactive material leaked from tanks using a Comprehensive Environmental Response, Compensation, and Liability Act (CERCLA) process.

SITE-SPECIFIC ISSUES

In addition to the general issues, a number of site-specific issues may limit the extent to which wastes can be retrieved, as discussed in the following section.

Savannah River Site

Retrieval at the Savannah River Site is favored by the fact that the integrity of the tanks is good with only one known minor leak to the environment (from Tank 16). High retrieval efficiencies obtained in Tanks 16, 17, and 20 may not be matched in the future because of site-specific conditions, as discussed below.

Waste Retrieval from Tanks with Vertical Cooling Coils

Among the 51 tanks, 43 contain a "forest" of vertical cooling coils (see Figure G-13). The coils in these tanks can impede conventional bulk retrieval technologies by creating numerous hydraulic "shadowed" zones where waste can settle, as well as providing numerous additional surfaces to which waste deposits can adhere. The coils also impede physical residual retrieval technologies because they hinder the use of tools deployed on mechanical arms or tethered devices.

Chemical mobilization of waste residuals was successfully applied in a tank with cooling coils at the Savannah River Site (Tank 16). Tank 16 contained at least 265 m^3 (70,000 gallons) of sludge that were sent directly to this tank from H Canyon during the late 1950s and early 1960s. Minimal salt formed on the "cool" areas of the tank and where the supernate flowed into the annulus pan (after repeated evaporation via the annulus ventilation system, the supernate eventually dried and left behind a salt residue). Tank 16 was subjected to chemical cleaning with oxalic acid. DOE reports that bulk waste removal and spray water washing removed 97.98 percent of radioactivity, and oxalic acid wash and rinse removed 99.98 percent of radioactivity in the tank (DOE-SRS, 2002).

As previously mentioned, DOE is in the initial stages of bulk retrieval of waste from a tank containing sludge and cooling coils (Tank 5). This effort should provide valuable information related to the effectiveness of available bulk retrieval technologies from a"complicated" tank and submersible mixer pumps. The increased effort required to retrieve sludge and the complications imposed by the vertical cooling coils make it more likely that additional retrieval measures will be required for such tanks to obtain an acceptably small heel. Such measures may include the need to use chemical cleaning methods or new physical methods for retrieving tank residuals.

As DOE acknowledges, Tank 11 will be particularly challenging to clean because it has 76 m^3 (20,000 gallons) of high-activity sludge and vertical cooling coils. DOE is considering multiple retrieval tools for this tank. One of them is the Power Fluidics™ AEA Technology, which could also be used in combination with oxalic acid (DOE- SRS, 2005d).

The committee asked DOE whether it was possible to cut or bend the coils to improve accessibility to the waste. The committee agrees with DOE that cutting the coils is not a practical solution because the coil debris would have to be disposed in situ due to the limited riser accessibility. There are about 4 miles of coils per tank and cutting them or bending them would entail high doses to workers. Similarly, increasing the size of the risers to remove the coils would also involve high workers doses. Finally, because the coils are attached to the tank top and bottom, any force on the coils to cut or bend them would impose a stress on the tank structure. It is worth noting that Savannah River Site staff has bent (albeit very slightly) the coils in Tank 5 to introduce some of the waste removal equipment.

Retrieval of Waste in Tank Annuli

While only a very small amount of tank waste has leaked to the environment at the Savannah River Site, larger amounts have leaked from the primary containment into the annulus. Four Savannah River Site tanks are known to have waste accumulation in the annulus: Tank 9, Tank 14 (both with about 10-12 inches of salt waste in the annulus), and Tanks 10 and 16 (both with about 2 inches of salt waste in the annulus).

A photo of such waste deposits is shown in Figure G-14). It is not yet resolved whether such waste deposits will require retrieval, in part because the amount and composition of the waste in the annuli is not fully characterized.

Savannah River Site staff estimate that there are about 400,000 Ci of cesium in all the annuli (value extrapolated for all tanks based on inches of deposits observed). Most of the deposits in the annuli are believed to be salt waste from

FIGURE G-13 Tank 4 Interior Photograph dated 1954. SOURCE: Caldwell, 2005b.

FIGURE G-14 Salt deposits in the annulus of Tank 16 at the Savannah River Site. SOURCE: Mahoney, 2005.

supernatant seepage though small leaks. If waste is to be removed from the annulus, the general approach would be to add hot water and possibly steam to the annular space, allow the water to sit for roughly two weeks to dissolve residual salt waste, and then transfer the resulting salt solution to the primary waste tank. A combination of sampling and visual inspection would then be used to monitor and verify the progress of the cleaning effort (Badheka and Elraheb, 2004, p. 5).

While these methods should work on highly soluble salt deposits, they did not work when Tank 16 underwent annulus cleaning (DOE-SRS, 2005g, and references therein). During the leak investigation, the outside of the tank's primary wall was cleaned using a sand-blasting technique. The silicates in the sand reacted with sodium and aluminum in the waste and formed sodium aluminum silicates which did not dissolve with hot water and agitation. DOE is considering using steam jets, agitation, and/or chemical cleaning to mobilize this waste in the annulus of Tank 16.

If waste deposits do not dissolve with hot water and steam, technology to retrieve such deposits is, at best, developmental, or, at worst, nonexistent. Development of a practical technology to clean the annuli is complicated because of very limited access to the annuli surrounding the tanks and very poor access to the ventilation ducts. Given that the annulus is a confined area with one less barrier to protect the environment from a potential release, the issue of waste retrieval needs serious consideration. DOE describes techniques for annulus cleaning in its "Heel Removal and Annulus Cleaning Technology Development Suspension Plan" (Cantrell, 2004), but the site acknowledges that techniques for annulus cleaning "need further development." (DOE-SRS, 2005g).

Hanford

Retrieving waste from Hanford tanks could be easier than from Savannah River Site tanks for two reasons. First, Hanford tanks do not contain cooling coils. Heat was removed by evaporation and in some of the hottest tanks, the waste boiled off water at the rate of 114 to 151 liters per minute (30 to 40 gallons per minute). That water was condensed and returned to the tank for cooling. Unimpeded access to the tanks means that delivery tools for mobilizing, collecting, and removing residual waste can be brought to bear with much greater effect. In some cases however, the Hanford single-shell tanks contain failed equipment, thermocouples, metal tapes, and some equipment with air-lift circulation devices that may prevent unobstructed access to the waste. Second, the bottom of Hanford tanks is downwardly concave (like a dish), which makes it much easier to collect the residual waste in a small area to be pumped from the tank.

The foregoing notwithstanding, the Hanford Site does face two major waste retrieval issues. First, retrieval techniques must be compatible with corroding carbon steel single-shell tanks. According to DOE, 67 of the 149 single-shell tanks are known or suspected to have leaked as much as 3,785 m^3 (1 million gallons) of waste and between 0.45 million and 1.8 million Ci to the subsurface environment beneath the tanks (see Chapter II). The problem of leaks has been managed so far by transferring liquid waste from the single-shell tanks to the double-shell tanks and leaving in the single-shell tanks just saltcake and sludge in the single-shell tanks. However, as mentioned previously, waste mobilization technologies involve the introduction of water or recycled liquids to mobilize residual waste. Reintroducing fluids into these tanks has the potential to release waste to the environment as at Savannah River. The second issue is that Hanford wastes are the result of multiple reprocessing technologies (i.e., Bismuth Phosphate, REDOX, PUREX) and chemical separations processes that resulted in significant tank-to-tank differences in waste chemistry, which must be factored into the retrieval techniques used. High phosphate wastes, in particular, create potential filtration and pipe plugging issues. Furthermore, as Hanford progresses in its waste processing campaign, it may face tank space challenges similar to those at Savannah River.

DOE has developed retrieval devices that use small amounts of water (see vacuum retrieval and Mobile Retrieval System earlier) in the hope of preventing leaks or reducing the amount of waste leaked to the environment. DOE has also developed devices such as the high-resolution resistivity leak detectors that can detect water entering the unsaturated soil beneath the tanks and allow DOE to cease retrieval before large amounts of waste are released (Barton, 2005c). However, the success of these devices in practice remains to be determined.

Idaho National Laboratory

The Idaho site has 11 underground tanks containing liquid sodium-bearing waste and 6 bins of calcined high-level waste solids.

Contaminated Sand Pads

Nine of the eleven tanks at the Idaho National Laboratory were built on a layer of commercial-grade sand that is 6 inches in depth at the circumference of the tank and 2 inches in depth at the tank center, which serves to cushion the tank bottom from an underlying concrete slab. This volume contains approximately 41,400 kg of sand (assumed density, 1.75 g/cc). The sand pads beneath two of the nine tanks (WM-185 and WM-187) contain radioactive material due to accidental releases into the surrounding vaults in March 1962 (Latchum et al., 1962). Before and after these releases, water from precipitation, spring runoff, and irrigation leaked into the vault areas and provided the mechanism to flush some radionuclides from the sand. The residual

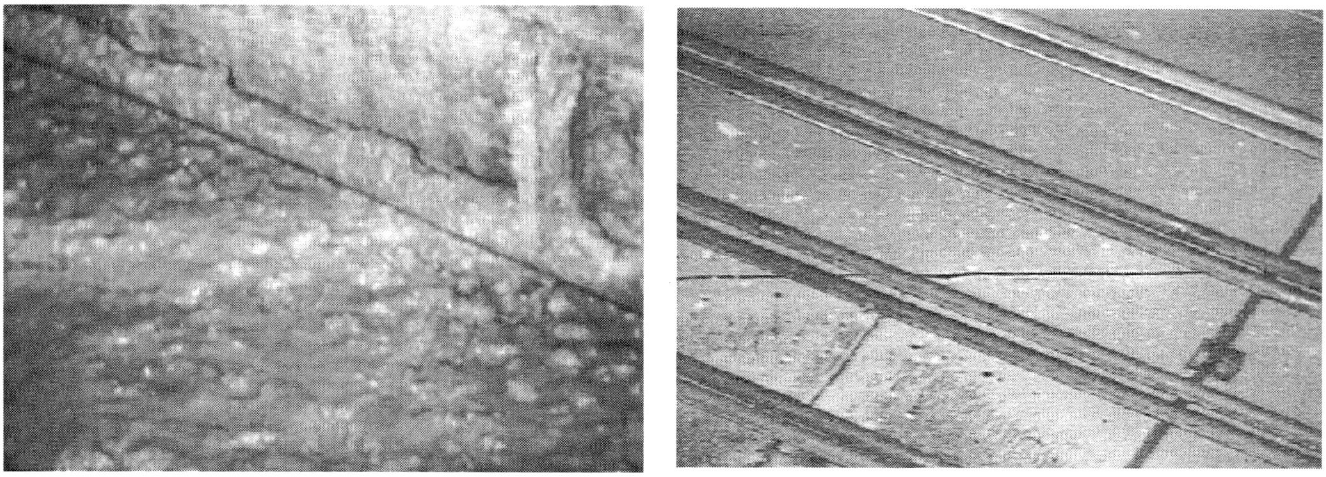

FIGURE G-15 Interior of the Idaho National Laboratory Tank WM-183 before (*left*) and after (*right*) waste retrieval. SOURCE: DOE-ID, 2005a, Figures 11 and 12, pp. 29-30.

inventory predicted for 2012 is based on 38 such "flushing" events when water infiltrated the vault from leaks in the tank or vault roof and was then jetted out of tank vaults.

DOE estimates that the total radioactivity at closure is approximately 21,993 Ci (814 TBq), assuming that the four tanks remaining to be cleaned contain the same number of curies as Tank WM-182, which is the tank with the greatest residual contamination among those that have been cleaned (see Figure G-15). The sand pads contain approximately 30 percent of the activity at closure (6,598 Ci, 244 TBq), and the tanks contain approximately 70 percent of the residual curies (15,395 Ci, 570 TBq) in the fine layer of solids that is distributed unevenly over the bottom of the 300,000-gallon tanks. The piping contains about 0.12 percent of the activity (approximately 30 Ci, 1.1 TBq). Both the U.S. Nuclear Regulatory Commission (USNRC) and the State of Idaho commented that there is significant uncertainty about the inventory of the sand pads and recommended that DOE measure or better estimate the contaminated sand pad radionuclide inventories (Trever, 2006; USNRC, 2006). Neither DOE nor the committee has identified practical means to measure the inventory or to remove additional radioactivity from the sand pads.

Retrieval of Calcined Waste from Bins

Retrieval of calcine waste at the Idaho National Laboratory is very different from tank waste retrieval. As described in Chapter II, most reprocessing waste at the Idaho site was converted to a fine granular powder (calcine) by high-temperature heating. The calcine was transferred pneumatically to groups of stainless steel silos (bin sets) for storage pending final disposition. DOE's plans call for the calcine to be retrieved from the bins, packaged, and shipped off-site to a high-level waste repository by December 31, 2035 (DOE-ID, 2002). However, these plans depend on the opening of a high-level waste geologic repository and on meeting the waste acceptance criteria for such a repository (potentially at Yucca Mountain, Nevada, but no license has been issued to date).

DOE proposes that the calcine be retrieved by reversing the process by which it was emplaced: vacuuming it from the bins through the risers. The committee viewed a video tape of simulated calcine retrieval by vacuuming which showed that the simulant was readily retrievable by this means: any caked calcine was readily dislodged by physical contact with the vacuum head, and the calcine could be transported pneumatically over significant distances (650 feet with vertical elevation changes) to processing facilities (Patterson, 2005). A cold demonstration in September 2005 by AEA Technology confirmed that solid deposits of the simulated calcine could be broken up using light mechanical impact with the retrieval nozzle. AEA Technology also simulated the retrieval of calcine exposed to various humidity levels up to 100 percent with similar success (AEATES, 2005b).[6] According to DOE, vacuum retrieval will remove more than 99 percent of the calcine (DOE-ID, 2002). Work in progress on calcine retrieval in cooperation with AEA Technology includes optimizing the tip of the vacuuming nozzle to mobilize compacted calcine deposits as well as optimizing the control systems to minimize worker fatigue during vacuuming.

However, whether the actual calcine will behave like the simulant after residing in the bins is for several decades

[6]Tests with simulated calcine containing 100 percent humidity are overestimating the amount of water in the bins because the temperatures inside the bins exceed the boiling temperature of water.

uncertain. Information related to the effectiveness of this bulk retrieval technique, techniques for retrieval of residual calcine, and disposition of the emptied bins has not yet been developed. Moreover, the geometry of the bins and the configuration of the risers in the bins may also decrease the effectiveness of the vacuuming method (see Chapter II).

Waste retrieval experience for Silo 3 at DOE's Fernald site shows how retrieval of calcine can be challenging. Silo 3 is a concrete domed silo 27 m (80 feet) in diameter and about 10 m (33 feet) above ground level. Five man-ways on the dome of the silo have an internal diameter of approximately 51 cm (20 inches). Silo 3 contained an estimated 3,890 m^3 of finely powdered metal oxides from uranium recovery operations. Raffinate streams from solvent extraction were processed for storage through calcination. The predominant radionuclide of concern identified within the material is thorium-230, a radionuclide produced from the natural decay of uranium-238. The material is classified as 11.e(2) by-product material under the Atomic Energy Act of 1954, as amended.[7] Silo 3 contents consist of two-thirds of dry, loose, fine powder located in the upper portion of the silo. The remaining third is compacted powder located toward the middle and bottom of the silo. Miscellaneous debris can also be found within the silo. Silo 3 was built in 1952 and was in service until 1957.

Calcine retrieval from Silo 3 began in March 2005 and, as of January 2006 was approximately 90 percent complete. Calcine is retrieved via manual pneumatic (vacuum) retrieval in combination with mechanical retrieval operated from an opening in the side of the silo (Figure G-16).[8] The levels of radiation are sufficiently low to allow workers to manually operate the vacuum wand through the man-ways on the top of the silo. The calcine was initially vacuumed out until pneumatic retrieval was no longer effective in removing loose material due either to the limits of wand reach or to the compaction of material. At that point, an opening was created on the eastern wall of the silo for at-grade access by a mechanical retrieval tool (the Excavator). The Excavator was used to move the material to a conveyor immediately adjacent to the silo opening. A mockup demonstration of wall cutting was performed on an empty silo at the site (Silo 4). Chemical stabilization and binding reagents are added to the calcine during packaging to reduce metal mobility and dispersability (Fluor Fernald, 2003). All of the waste from Silo 3 will be shipped to the Envirocare low-level waste disposal in Utah.

FIGURE G-16 Pneumatic (vacuum) retrieval from the top of Silo 3 at the Fernald facility in Ohio (*left*) and mechanical retrieval tool (the Excavator) operating from an opening in the wall of the silo (*right*). The wall of the Silo is visible at the right border of the frame. SOURCES: Left: www.nukeworker.com; right: Beckman, 2006.

This example shows how vacuuming dry material after several years of storage (in this case more than 49 years) can be challenging. Compared to the Idaho bins, Fernald's Silo 3 is a structure of a simpler design (cylindrical rather than annular like some of the Idaho bins); the calcine can be accessed directly by workers via five direct access ports rather than one riser per bin, often placed at an angle; and the material has a lower radioactivity content so operations do not have to be completely remotely controlled. Yet according to DOE, "material removal from the bottom of the [Fernald] silo presents challenges" (DOE-OFO, 2006). In addition, Idaho will have to manage the added risk from fine respirable powders that are intensely radioactive.

A previous National Research Council (NRC, 1999b, p. 22) committee observed that, given the little characterization information about the Idaho calcine (see Chapter II),

> it would be difficult to conclude that there would be no problem with pneumatic retrieval. Indeed the committee believes that there will be problems but that they can probably be

[7]Section 11.e(2) of the Atomic Energy Act defines by-product material as "tailings or wastes produced as a result of the extraction or concentration of uranium or thorium from any ore processed primarily for its source material content."

[8]Initially a robotic arm (the ReTRIEVR by Framatome) was considered for calcine retrieval in Silo 3, but in 2001 the site changed the cleanup plan and selected manual vacuuming and creating an opening on the side of the silo to operate the Excavator (Fluor Fernald, 2003).

handled. However, this eventually might require mechanical operations to aid particle flow and more elaborate retrieval methods (e.g., a manipulator arm) than simple pneumatic transfer.

This example also highlights the importance of sharing lessons learned among DOE sites because the experience gathered with Fernald's calcine may be of use at the time of Idaho's calcine bin retrieval.

Previous studies (NRC, 1999b; CRESP, 2005) have concluded that the calcine in the bins can be safely stored for hundreds of years because of the absence of water in the bins, the stability of the calcine, and the dry environment. As a consequence, addressing the disposition of the wastes contained therein has justifiably been accorded a lower priority in the face of more pressing issues at the Idaho National Laboratory and other DOE sites. The committee identifies the calcine retrieval and bin disposition issue as an issue for further examination (see Chapter XI).

Tank Cleanup at the West Valley Demonstration Project

As mentioned in Chapter III, between 1966 and 1972, the West Valley Demonstration Project (WVDP) plant generated approximately 2,317 m^3 (612,000 gallons) of high-level waste. This waste consisted of 2,271 m^3 (600,000 gallons) of basic plutonium-uranium extraction (PUREX) sludge and supernatant and approximately 45 m^3 (12,000 gallons) of acidic thorium extraction (THOREX) waste. The PUREX waste was stored in Tank 8D-2, while the THOREX waste was stored in Tank 8D-4. The site had two tanks used for secondary waste streams during waste processing. The retrieved wastes were vitrified in a vitrification facility similar to the one at the Savannah River Site and poured into 275 canisters intended for disposal in a repository for high-level waste. The vitrification facility operated between 1996 and 2002.

The waste pretreatment process used zeolite to retain cesium-137 and separate it from the low-activity waste components. Additionally, it used a titanium-coated zeolite to promote adsorbtion of actinides and strontium from washed sludge. This process resulted in 74 m^3 of cesium-137-laden zeolite, which was stored in Tank 8D-1 prior to the vitrification campaign. An in-line zeolite grinder was also included in a pump pit overlying Tank 8D-2 to reduce the size of the zeolite chunks retrieved from Tank 8D-1 prior to its addition into Tank 8D-2 and subsequent transfer to the vitrification facility. During the vitrification campaign, the sludge waste, including the zeolite, was retrieved by suspending the solids with mobilization pumps and transferring the resulting slurry with multistage pumps that draw suction near the bottom of the tank, similar to the pumps used at the Savannah River Site. Five or six of these pumps were installed in the waste tanks to mobilize sludge or zeolite. During the vitrification process, additional solids from Tank 8D-1 (zeolite) or 8D-4 (recycle from the vitrification process) were periodically consolidated into Tank 8D-2 and flushed toward the vitrification facility.

Low-pressure water jets (100 psi) have been used to clean interior surfaces of the two largest tanks. Other waste retrieval techniques have been considered and to some extent deployed. Riser-mounted robotics with limited mobility and tethered robotics have also been evaluated for both mobilization and characterization tasks. Because of the number of obstructions in the tanks, tethered robotics has not been developed for use. Riser-mounted arms and positioning systems have been used extensively for characterizing and locating waste residues in the tanks and, to a more limited degree, to mobilizing residues for retrieval (Hamel et al., 2000; DOE-TFA, 2000a). The riser-mounted positioning system washed residues from the tanks' internal surfaces (Hamel and Damerow, 2001). Chemical cleaning was evaluated but not deployed because of concerns with tank integrity. To date, less than 2 percent of the initial 30 million curies of radioactivity, and less then 1 percent of the initial long-lived, alpha-transuranic radioactivity, remains in the tanks (DOE-WVDP, 2006).

The West Valley site has done considerable work on tank heel characterization (O'Brien et al., 2001). For example, a special camera was deployed within the tank to measure two-dimensional spatial mapping of gamma-emitting radiation in real time; neutron track recorders were deployed in the tanks to measure neutron fluxes from many reactions within the tank down to the levels of cosmic ray-induced background; beta-gamma detectors were employed to establish the concentrations of both beta- and gamma-emitting radionuclides remaining on tank surfaces; and a burnishing sampler was deployed within the tank to collect samples of the waste from the walls, columns, and other vertical and horizontal surfaces for radiochemical analysis. During waste retrieval a "bathtub ring" of dry waste on the walls of Tank 8D-2, corresponding to the top of the waste level during site operations became visible and required characterization to determine the final waste inventory. A number of attempts were made to reduce the inventory of radionuclides on the ring, with limited success. Tanks 8D-1 and 8D-2 were placed in safe surveillance and maintenance mode (called "lay-up" mode) in 2003. The site is now preparing an environmental impact statement (EIS) to evaluate alternatives for decommissioning and long-term stewardship of the WVDP site, on which the State of New York is a cooperating agency. If DOE selects an alternative to close the tanks in place, the associated waste determination would be made under the USNRC's final policy statement for decommissioning criteria for the West Valley Demonstration Project (USNRC, 2002). A record of decision, along with the associated USNRC-reviewed decommissioning plan, is not expected for several years.

Tank Cleanup at the Oak Ridge National Laboratory

The liquid low-level radioactive waste system at Oak Ridge National Laboratory has been used to collect, store, and treat wastes generated from laboratories and processes since the mid-1940s when ORNL was constructed as part of the Manhattan Project. As the plant expanded, new liquid low-level radioactive waste storage tanks were added to the system, and other tanks were taken out of service as needed to support changing laboratory missions and comply with evolving regulatory requirements and environmental and technological standards. The tanks no longer in service, or inactive tanks, have been remediated using a wide range of technologies. Remediation ranged from grouting the tanks in place, either as found or following partial or complete waste removal, to complete tank removal. The technology selection depended on a range of factors, from the potential risk posed by the tank and its contents, to the tank construction, to the location of the tank and potential impacts of the remediation on nearby ORNL facilities or operations.

The smaller inactive tanks were generally of stainless steel construction. Waste removal was typically accomplished using spray nozzles and lances to mobilize settled sludge, and the resulting slurry was pumped from the tanks (DOE-EM 2001a; 2001b).

A borehole-miner extendible-nozzle sluicing system was used to remove waste from the five Old Hydrofracture facility tanks prior to grouting the tanks (Bamberger and Boris, 1999). Fluidic pulse jet mixing was used for consolidation of sludge from the active Bethel Valley Evaporator Complex Tanks, as well as for waste removal from three larger inactive tanks prior to tank grouting (ITSR, 1999f).

The Oak Ridge National Laboratory had approximately 103,000 gallons of sludge stored in nine Gunite and associated tank system (Lewis et al., 2002a and references therein). This sludge varied from thick viscous waste to easily flowable liquid. In addition, the tanks contained small quantities of dried waste that had the consistency of chalk. The Gunite tank sludge (solids) contained approximately 85,000 Ci (3.1 PBq). This radioactivity came from uranium, plutonium, thorium, and other long-lived isotopes, as well as from the high concentrations of cesium-137 and strontium-90, which have relatively short half-lives. The tanks also contained organic materials in trace amounts and other heavy metals. Groundwater had leaked into the tanks, adding approximately 1300 m^3 (345,000 gallons) of wastewater. This water accumulated on top of the sludge in an aqueous supernatant layer. The supernate and the tank walls contained an additional estimated 15,000 Ci (555 TBq).

The Gunite tanks were constructed of a mixture of sand, cement, and water that is sprayed through a nozzle over a steel reinforcing framework. The tank walls were built in three layers consisting of (1) an outer Gunite wall approximately 6 inches thick, (2) an approximately 0.5-inch-thick asphalt or bitumen layer applied to the inside of the Gunite layer to provide a leak barrier, and (3) an inner Gunite wall approximately 2 inches thick that was applied onto steel reinforcing wire mesh. In Tank W-5, remote inspections showed that the interior layer of the wall had deteriorated. Pieces of the inner Gunite wall had fallen to the floor of the tank, exposing the metal mesh underneath.

A retrieval campaign conducted in the early 1980s in the large Gunite tanks had used long-range sluicing jets and conventional transfer pumps. This campaign recovered 90 percent of the original sludge present, leaving in the tanks the material that was harder and more difficult to retrieve. The retrieved waste was transferred to the double-contained Oak Ridge Melton Valley Storage Tanks, where it was consolidated, and will be processed and, packaged for transportation to the Waste Isolation Pilot Plant (WIPP) as transuranic waste.

The Oak Ridge Site acquired significant experience with waste characterization and retrieval. The characterization effort included participating in development of a variety of techniques to sample liquid and sludge from a range of tank configurations. Characterization efforts for the Gunite tanks included inspecting the tank walls using both an in-tank video system and a laser mapping technique to determine if the condition of the walls presented retrieval limitations and assess the current state of the walls. Characterization during in-tank retrieval operations included obtaining tank wall samples by scraping material from the inside of, and core samples from core drilling, the Gunite tanks (W-3 through W-10) to determine the nature and extent of contamination in the Gunite walls.

The retrieval strategy for the Gunite and associated tanks focused on the use of a wide variety of tools and systems working either individually or in concert. Axial-flow propeller mixers (i.e., Flygt mixer) were used for waste retrieval from one tank. The Russian Pulsating Mixer Pump (PMP) was used for Tank Th-4. The confined sluicing method, deployed by a combination of the MLDUA robotic arm and Houdini™ remotely operated vehicle system, was used for the remaining tanks to consolidate waste into one tank. An axial-flow propeller mixer (i.e., Flygt mixer) was used in concert with pulsed-air mixers to resuspend the lighter-weight sludge for transfer as all other waste was being moved to the consolidation tank. The transfer system was then reconfigured to move the heavier solids to the nearby active Bethel Valley Evaporator Storage Tanks. The major elements of this strategy involved the use of the following equipment and systems:

- Modified Light-Duty Utility Arm (Glassell et al., 2001);
- Houdini™ remotely operated vehicle system (Vesco et al., 2001);
- Waste Dislodging and Conveyance System, including the confined sluicing end effector (CSEE) and hose management arm (DOE-TFA, 2000a);

- Flygt mixer (Pacquet and Leshikar, 2001);
- Pulsed-air mixer (Lewis et al., 2002b);
- Russian Pulsating Mixer Pump (Hatchell et al., 2001);
- Waste conditioning system (Emison et al., 2002); and
- Heavy waste retrieval system (ITSR, 2001).

All equipment and systems mentioned above are also described in detail by Lewis et al. (2002a, and references therein) and in DOE's Tanks Focus Area's "Heel Retrieval Technology Guide" (DOE-TFA, 2000a).

Oak Ridge National Laboratory tanks that are grouted in place were grouted with a low-strength flowable fill. This fill material had the advantage of being self-leveling when pumped into the tanks, aiding in complete grouting of tanks with limited access and residual radioactivity. The material can also be removed in the future with commercial demolition equipment should removal of some of the closed tanks become necessary in the future as part of an eventual final closure of the Oak Ridge National Laboratory.

Appendix H

Features of a Good Monitoring Program

Data collection and the interpretation and analysis of those data are inextricably linked. It is this linkage that places specific requirements on the data collection process to produce data needed to support decision making. A well-developed technical literature has supported the development of guidance for monitoring. This guidance is found in federal agency documents; particularly good examples specific to environmental monitoring are provided by the U.S. Environmental Protection Agency (EPA) in its Quality Assurance Program (QAP). The Department of Energy (DOE), as part of an intergovernmental working group, has itself produced uniform federal guidance for quality assurance programs that are applicable to environmental monitoring in the Uniform Federal Policy for Quality Assurance Project Plans (IDQTF, 2005).

As used in this context, a quality assurance project plan (QAPP) for a monitoring program goes far beyond review and record keeping, or even the protocols for handling of samples and continuity of possession associated with quality control. Quality assurance here includes ensuring that the goals of the monitoring program are well defined and clearly articulated, that data collection strategies and methods are well matched to the goals, and that the data collected are in a form that enables them to be used as information for analysis (see Sidebar H-1).

The committee has identified seven features of a good monitoring program that are worth noting. A good monitoring program (1) is goal oriented; (2) has an integrated vision of monitoring for the overall site; (3) seeks the relevant information in the right places; (4) observes the environment (both natural and constructed) and the dynamics that affect processes of interest; (5) provides early warning to enable intervention if necessary; (6) is subjected to review on a regular basis and adapts to changing circumstances; and (7) archives data in a durable and accessible form. Each of these features is discussed below.

GOAL ORIENTED

A critical element of a good monitoring program is the identification and clear statement of the objectives that are used to guide monitoring design. A good monitoring program is designed to answer specific questions that underlie the purpose of the program (site characterization, emissions from operations, and so forth. Furthermore, data from one stage of monitoring can be helpful in designing the next stage, provide a useful historical base and supply information to support other efforts (e.g., performance assessment). As noted elsewhere in this report, there are several regulatory drivers for monitoring disposal sites and surrounding areas at Hanford, Idaho Falls, and Savannah: namely, Comprehensive Environmental Response, Compensation, and Liability Act (CERCLA); Resource Conservation and Recovery Act (RCRA); specific elements of the Federal Facility Agreements; and the performance objectives in 10 CFR 61. These regulatory drivers are considered in formulating the goals of the monitoring programs at the sites.

AN INTEGRATED VISION OF MONITORING FOR THE OVERALL SITE

Despite the varied goals and methods used for monitoring different facilities, every environmental monitoring activity at a site contributes data that can be used to some extent to improve the understanding environmental and engineered systems at the site. If monitoring of disposal facilities is designed in the context of monitoring adjacent facilities and the overall, site-wide monitoring program, it is easier to gain useful information from the monitoring data. Even better, if DOE can work with regulators to develop an integrated vision of monitoring at the site, fitting together the various monitoring goals, the monitoring program can operate more efficiently and maximize the information gained from monitoring data.

SIDEBAR H-1
Quality Assurance Project Plan

As stated in the uniform federal policy for quality assurance project plans (IDQTF, 2005):

> A QAPP is a formal document describing in comprehensive detail the necessary quality assurance (QA), quality control (QC), and other technical activities that must be implemented to ensure that the results of the work performed will satisfy the stated performance criteria. A QAPP presents the steps that should be taken to ensure that environmental data collected are of the correct type and quality required for a specific decision or use. It presents an organized and systematic description of the ways in which QA and QC should be applied to the collection and use of environmental data. A QAPP integrates technical and quality control aspects of a project throughout its life cycle, including planning, implementation, assessment, and corrective actions.
>
> A well-implemented QAPP is well documented and has a clearly defined objective that drives data collection. The type and quality of the data collection are judged based on how well they support the objective or decision.

In conducting any type of analysis system behavior due to various stresses, it is necessary to recognize that the different components of the hydrologic system (surface and subsurface) are coupled and are in dynamic interaction. At a specific site however, these interconnections and the coupling between the different components or units may not be significant. In these situations, some of the units may be relatively isolated. For example, DOE's Savannah River Site has three distinct watersheds, so the surface hydrology is distinct. Even though watersheds that contribute to surface flow are isolated, the groundwater flow within these distinct watersheds may be interconnected depending on the geology and the hydraulic head distribution. Because of possible interconnections, complex subsurface flow patterns (e.g., due to perched water), and limited knowledge of geologic heterogeneity and geochemical behavior, attempts to develop an understanding of a complete, integrated site-wide behavior of the hydrologic regime can prove difficult. In general, it is believed that at most of the large DOE sites the possible pathways of contaminant transport off-site are relatively well understood. An integrated vision for site monitoring requires that program managers hypothesize how the hydrologic regime works even though some of the internal connections may not yet be fully understood. At the same time however, the committee notes that lessons learned about one geohydrologic unit may be useful in understanding another unit, even if the units are isolated from each other.

RELEVANT INFORMATION IN THE RIGHT PLACES

To identify what to look for and where to look, a good monitoring program is connected with a conceptual model of the site and, in the context of this study, the performance assessment program in an iterative approach that improves both activities. Historical monitoring data provide the basis for the performance assessment design (see Chapter VII). Performance assessment can then be used to screen contaminants and locations to focus the monitoring program on the most important contaminants and the likely pathways for contaminant migration. The extent to which the set of contaminants and locations can be narrowed depends on the extent of knowledge and understanding of the environmental surrounding, including the disposal site, which argues for site characterization that is continually improved by new data. Although it is appropriate to concentrate the greatest efforts on the media that are most important to meeting the goals and objectives of the program, the committee believes that a good monitoring program would consider all environmental media. This is because good monitoring is used to detect the unexpected or to provide assurance that a concentration of effort is appropriate.

OBSERVATION OF THE ENVIRONMENT (BOTH NATURAL AND CONSTRUCTED) AND THE DYNAMICS THAT AFFECT PROCESSES OF INTEREST

A good monitoring program is based on current understanding of the characteristics and dynamics of the site environment and how those dynamics are altered by site activities, and it is designed to fill gaps in the data and to investigate anomalies. This understanding can be obtained by the integration of data collected on the site over time from a wide range of monitoring activities, whether for process analysis, surveillance, general environmental monitoring, or compliance. The available background data help identify processes controlling the movement and impact of contaminants. As noted above, a good monitoring program that is connected with the performance assessment program in an iterative approach improves both monitoring effectiveness and the validity of models used to predict contaminant movement. The monitoring program will fill data gaps pertaining to environmental understanding and the dynamics effecting processes of interest. It will also help to refine the baseline understanding over time, from initial site characterization through closure and beyond (if necessary) and provide data to test hypotheses about future environmental conditions and dynamics at a site.

The value of monitoring data is measured not in the amount collected but the utility of the data. In monitoring programs where wastes present a clear threat to the environment and public health, there is a need for programs to produce data that can be processed and used in a timely decision-making process that will support actions to minimize immediate threats to the environment and public health.

PROVIDES EARLY WARNING TO ENABLE INTERVENTION

A buffer zone is the region that lies between a facility and a location at which compliance is assessed. The availability of monitoring points in the buffer zone is related to early detection of the migration of hazards. A buffer zone provides a safety margin that allows contaminant detection and corresponding mitigation to proceed while contaminants remain on-site. To enable timely detection of contaminants, a monitoring program may have to locate observation points well before the point of compliance. Further, the need for buffer zones also requires careful assessment of the possible means of any threat of migration off-site and how that threat may change over time and distance with potential change in wastes in the natural environment.

To be effective in this role as sentinel, the monitoring program has to have a clear path for the use of monitoring data to support identification of the necessary action and a plan to mitigate the problem before contaminants or radioactive wastes can leave the site.

REVIEW ON A REGULAR BASIS TO ADAPT TO CHANGING CIRCUMSTANCES

A good monitoring program has a formal mechanism for regular review relative to goals and to changes in the environment. The committee recognizes that just as conditions in the environment may be expected to change over time, the relationship between the environment and wastes in disposal sites may change. Further, the performance of the monitoring system, the technology available for monitoring, or the goals of the monitoring program may change. As a result, monitoring plans may have to adapt to new conditions (e.g., observation locations may need to be relocated) to provide proper surveillance of wastes and data that can be used for rapid assessment and mitigation response to threats that disposal sites pose to the environment and public health.

DATA FROM DIFFERENT SOURCES ARCHIVED IN A DURABLE AND ACCESSIBLE FORM

A monitoring program observes the environment and collects data, but those data do not inform decisions or understanding unless they are recorded, accessible, and understood. It is particularly important to pay attention to metadata, the detailed description of data collection and analysis methodologies, to ensure that future interpretations of archived data are comprehensive and compatible. A good monitoring program, therefore, stores data in a form that is readily understood and accessed with a full metadata complement. In addition, data collected by a monitoring program not only are products of the program but also essential elements of the overall monitoring effort. Further, the data are a historical record of environmental conditions and behavior essential for the analysis of site dynamics. The historical record provides a context for new data and a basis for review and improvement of the program, and so must endure for a long time (decades and perhaps centuries) if it is to continue to support the long-term requirements of site management.

Appendix I

Performance Assessment Process

The following is taken from the NRC report, *Risk and Decisions about Disposition of Transuranic and High-Level Radioactive Waste* (NRC, 2005b): Performance assessment is a process that "estimates the potential behavior of a system or system component under a given set of conditions. It includes estimates of the effects of uncertainties in data and modeling. In the context of radioactive waste, performance assessment is a systematic method for a repository risk assessment." The key steps of the performance assessment (PA) process are presented in Figure I.1.

It is only meaningful to conduct a performance assessment in the context of a decision, such as Does the existing contamination require active remediation? or Does the proposed plan for waste disposal pose acceptable risk? The first step is identification of the performance objectives or decision criteria that will protect human health and the environment for the site and waste stream under consideration. An array of environmental regulations establish performance objectives to protect the following:

1. The general public
2. The inadvertent intruder
3. Groundwater resources
4. Air resources
5. Surface water resources
6. All ecosystem components associated with these media

Specifically, performance assessments evaluating disposal of low-level radioactive materials are developed to help assess compliance with the performance objectives of (Department of Energy) DOE Order 435.1 and Title 10 Part 61 of the Code of Federal Regulations. Some of these performance objectives are definite (e.g., 25 mrem per year exposure for a member of the public) and others are more fluid (e.g., releases to the environment must be as low as reasonably achievable [ALARA]), but determining compliance with even the definite performance objectives involves scientific judgment and regulatory policy decisions that are critical to the results. As part of the performance assessment, it is necessary to identify the points of compliance (or some surrogate, such as a point of calculation), which are the locations at which long-term risks to public health and the environment are evaluated.

After the performance objectives are identified and the point of compliance has been determined, the next step is the development of a conceptual model for the site and the waste that captures fundamental physical and chemical processes, issues of elemental dynamics, and features of the natural and engineered systems that will affect contaminant movement and concentration over time. With models of contaminant dynamics in the site environment, it is then possible to assess overall system performance in relation to identified objectives. The overall conceptual model for evaluating the impacts of a disposal facility must include models for waste release and contaminant transport in the relevant media.

The conceptual systems model is developed using monitoring information that can include historical data on system behavior as well as site characterization data. In addition to the data, a well-developed conceptual model utilizes the knowledge that people have accumulated concerning the site and environmental phenomena that affect the site. Neuman and Wierenga (2003) note that because each site is unique, general principles always must be supplemented by regional and site-specific data to be useful for conceptualization and modeling of subsurface flow and transport at a site, regardless of purpose. The same authors also point out that deficiencies in the conceptualization are far more detrimental to the predictive ability of a model than a suboptimal set of model input parameters. Systematic examinations of the results of variations in input parameter (i.e., sensitivity studies) can enable decision makers to make use of a performance assessment, even in data-sparse situations if the conceptual model represents the real environment reasonably well. However, no amount of parametric examination compensates for an inaccurate conceptual model.

An National Research Council (2000f) study *Research*

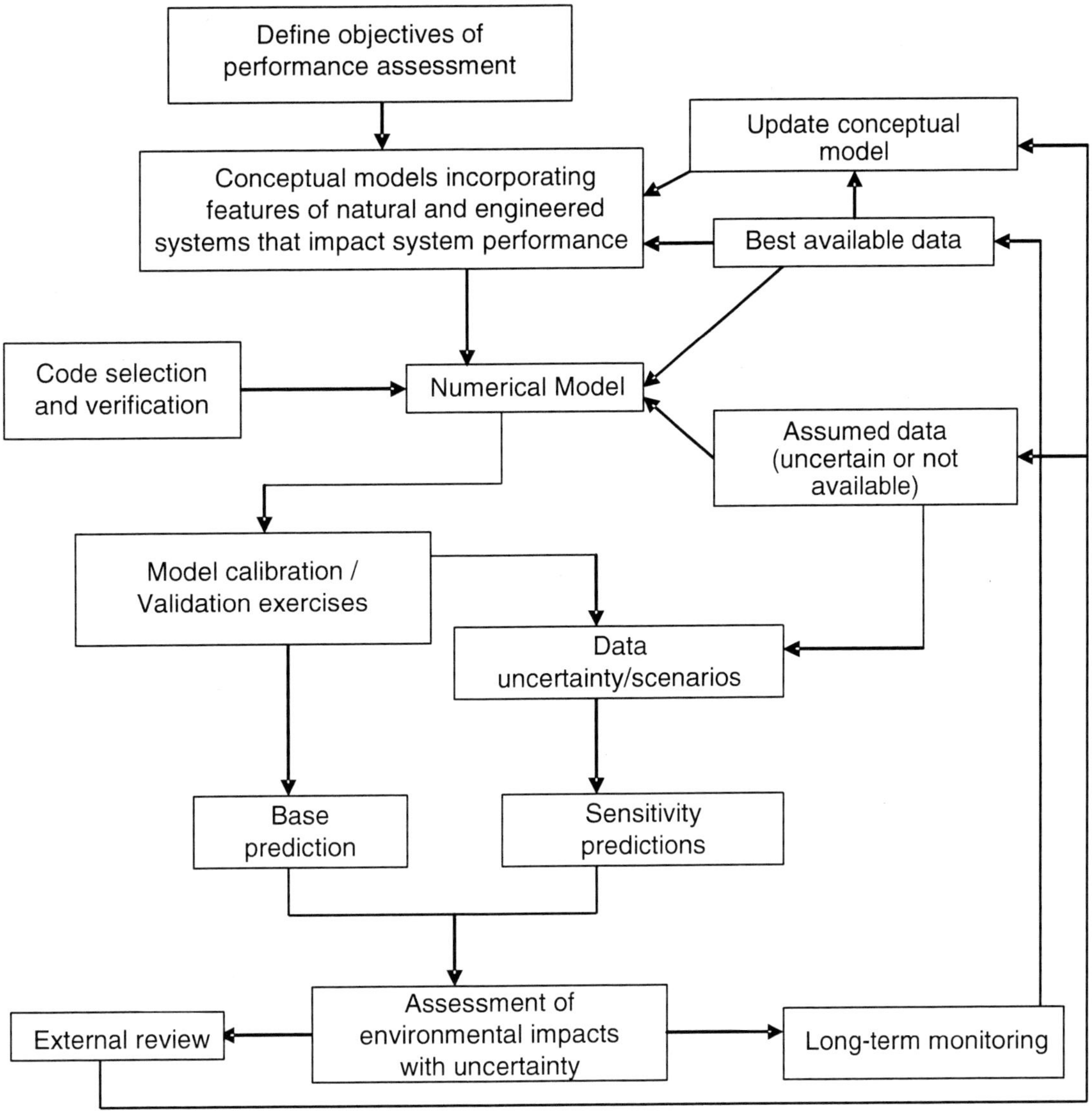

FIGURE I-1 Flow chart diagram illustrating the performance assessment process.

Needs in Subsurface Science for DOE's Environmental Management Science Program recommended research on the development of tools and methodologies for conceptual modeling with an emphasis on heterogeneity, scale, and uncertainty bounds of field experimental data. An NRC (2001e) panel that described the processes through which conceptual models of flow and transport in the fractured vadose zone are developed, concluded that the development of the conceptual model is the most important part of the modeling process.[1]

[1]There is not yet an agreed-upon conceptual model for describing contaminant transport in unsaturated fractured rock.

The next steps are to convert the conceptual model into a set of mathematical equations, referred to as the mathematical model and to solve those equations. The site model is developed by inputting data and boundary conditions specific to the site when the computer code is run. For performance assessment, these mathematical models must be able to reconstruct or predict variations of quantities such as concentrations of a contaminant in air, surface water, or groundwater in both space and time. Computer codes have been developed to calculate results from the governing equations using analytic (exact) solutions or numerical methods to find approximate solutions, depending on the equations and the complexity of the environment to be modeled.

Numerical schemes subdivide time and space, which are continuous, into discrete blocks (the time step and spatial grid) for computation. For the calculation, parameter values do not vary within the time step or within the spatial grid block. This calculation scheme (discretizing continuous variables) inherently produces errors, which are referred to as numerical errors. The magnitudes of these errors are proportional to the spatial grid size and the length of the computational time steps used in the model. In selecting a code, good analysts consider the suitability of the code for the specific application and verify and calibrate the code's ability to simulate conditions in the site environmental system with acceptable prediction accuracy.

Code suitability and acceptable accuracy depend not only on the physical disposal system, but also on the performance objectives. For example, in nuclear waste disposal problems, the simulation time horizons typically are thousands of years. If compliance with the performance objective requires projection of environmental impacts extending thousands of years into the future, a good analyst will establish whether the cumulative errors produced by a model yield meaningful results. It should be noted that the development of a single numerical model to capture all of the processes identified in the conceptual model is unlikely. Actual performance assessment "models" in site-specific analyses typically use a set or ensemble of codes that are linked in a sequential manner. This ensemble may contain both analytically based codes and numerical solutions.

After the selection of the codes(s), the next step in constructing the site model is the development of numerical models that are intended to simulate site-specific conditions. Assumptions about site-specific conditions used in modeling are defined by assumptions about boundary conditions and model input parameters, which are based on site characteristics to the extent practical.

This is a critical phase of modeling because the best available site data must be used in this step of model development to ensure overall model quality and accuracy. Because monitoring data are limited for any site, the same data that are used in conceptual model development are often used in numerical model testing. Additional data may have to be assembled or gathered as required by the specific needs of the code. A characteristic of performance assessment modeling at these sites is the high level of sophistication needed in the code to simulate complex hydrogeochemical processes in the subsurface. This subsurface modeling places a greater demand on the type of data needed by state-of-the-art codes. This demand for specific data resources means that it is reasonable to expect that all of the data required for a model will not be available and the available data will have uncertainties.

The next step in development of the site model is calibration, which involves adjustment of the model parameters to match existing data or observations. The goal of the calibration step is to build confidence in the ability of the code to simulate contaminant movement accurately under site conditions. One expects a calibrated model to be able to simulate the behavior of the system during a period for which both input and output data on system response are available. In groundwater modeling, a rule of thumb is that predications can be made with some confidence for a period equal to the period of the calibrated match (Bredehoeft, 2003). The confidence of the predictions made beyond the length of the calibration period will diminish rapidly. When applied to the disposal of long-lived radioactive waste, the persistence of the hazard creates a tension between the need to meet the performance objectives over a long time horizon and the diminishing confidence in predictions for a time horizon that is orders of magnitude larger than the calibration period. Bredehoeft (2003) states that "the closer we can approach the idea of a long history with which to match the models, even models of nuclear waste facilities, the more confidence we will have in the analysis (and the models, including performance assessment)." He argues that prolonged periods of site monitoring, perhaps as long as 300 to 1,000 years, are required to update codes to calibrate models to gain confidence in long-term predictions before a nuclear waste facility is finally closed.

If the model is "calibrated," a base prediction can be made by simulating the expected behavior of the system for future events. This simulation is a single outcome corresponding to the deterministic dataset used in building the model. This base prediction is not sufficient for performance assessment because uncertainty is inherent in all steps of the process, starting from conceptual model development to scenarios that are assumed in the simulations. According to the NRC (2001e) panel, "It is important to recognize that model predictions require assumptions about future events or scenarios, and are subject to uncertainty." For example, the performance assessments include assumptions about future scenarios that may include the failure of engineered barriers or a change in climate that will alter the frequency and magnitude of floods, droughts, and precipitation. Data limitations and uncertainties associated with the characterization of subsurface heterogeneity also contribute to prediction errors. The reliability of deterministic approaches to conducting flow and transport analysis in complex subsurface systems has been questioned. Methods that use geostatistical techniques to describe the spatial variability and scaling and stochastic analysis of fluid flow and solute transport have become the trend (Neuman and Wierenga, 2003). Stochastic methods enable analysts to explore data uncertainties and alternative scenarios systematically. In the absence of a stochastic approach, a comprehensive, deterministic sensitivity study, which systematically examines the impact of different scenarios and parameter variations on the risk posed by the site, provides a similar examination of uncertainties and the impacts of assumptions, which is essential to decision making. Another informative approach is to examine results of a so-called practical worst case that would consider multiplicative effects in parameter variations in determining potential environmental exposures. The predictions based on sensitivity analysis are then used for the assessment of long-term impacts and risk analysis as part of the performance objective evaluation. Such methods do not necessarily improve the accuracy (realism) of model results, but they can more accurately represent the consequences of what is known about the systems.

As presented, a conceptual model is developed as a first step in the PA process using the available data. As more data become available through continuous monitoring, this initial conceptual model may require updating. This need would be apparent through regular recalibration and testing of predictions made earlier. Neuman and Wierenga (2003) encourage an iterative approach to modeling, whereby a preliminary conceptual-mathematical model is gradually refined. In addition to updating the conceptual model, well-designed and well-thought-out monitoring systems could provide data to improve the numerical model. A critical component of this iterative improvement of performance assessments is the external review of a performance assessment at its conception, and continuing external review as the performance assessment is updated. If done well, the performance assessment will be an evolving procedure that builds confidence in predictions and in its ability to carefully define the risks posed to public health and the environment by waste discharges on a site.

Presentation of the performance assessment is as important as any other step in the method because if the results are not communicated clearly to decision makers, the performance assessment has not done its job. The performance assessment is best documented in a comprehensive report that describes the basis for all of the assumptions that underpin the conceptual model, the development of simulation models, model calibration, and predictive analysis. In a good performance assessment, the basis for selection of compliance points is also clearly described. Review of a performance assessment is easier to complete if the most recent assessment is readily available as a single, updated document, rather than as a series of reports that may have been released over a number of years.

Appendix J

Relevant Maps of the Three Sites

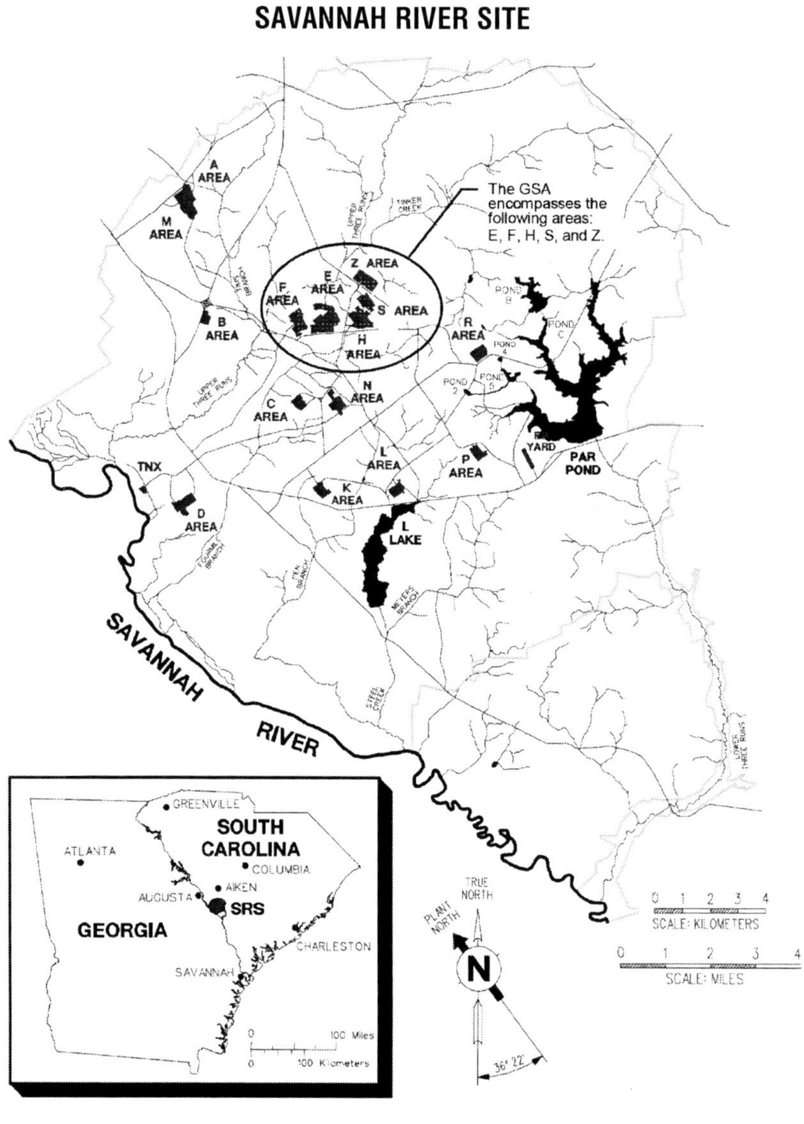

FIGURE J-1 Map of the Savannah River Site. The GSA is the General Separations Area. The area labeled F is the location of the F Canyon and F Tank Farm. E Area includes low-level waste disposal units. H Area is the location of the H Canyon and H Tank Farm. S Area is the Defense Waste Processing Facility. Z Area is the location of the Saltstone Production Facility and Saltstone Vaults. SOURCE: Buice et al., 2005.

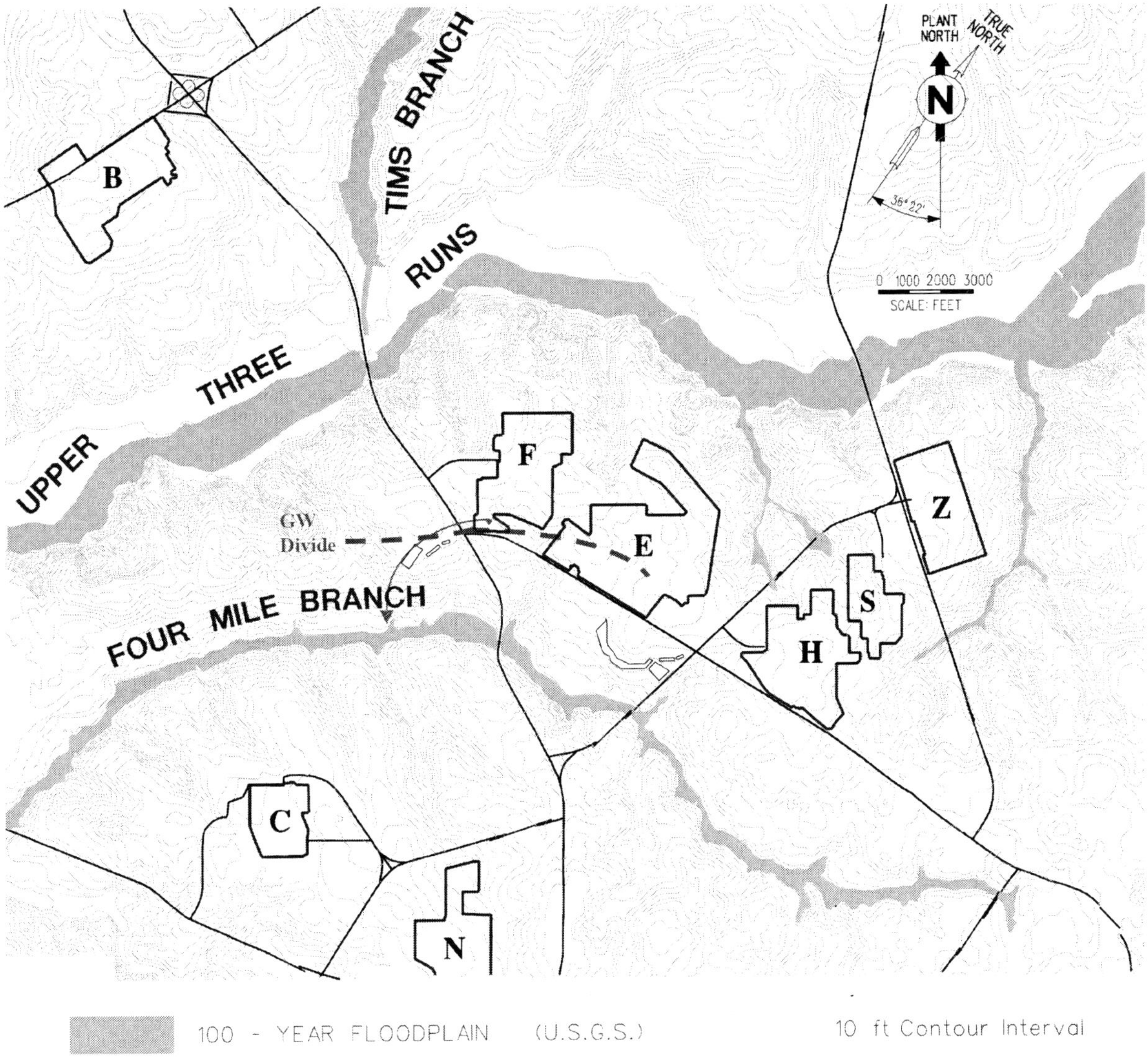

FIGURE J-2 Topographical map of the General Separations Area at the Savannah River Site. The area labeled F is the location of the F Canyon and F Tank Farm. E Area includes low-level waste disposal units. H Area is the location of the H Canyon and H Tank Farm. S Area is the Defense Waste Processing Facility. Z Area is the location of the Saltstone Production Facility and Saltstone Vaults. GW Divide refers to the line that separates groundwater that flows north to Upper Three Runs from groundwater that flows south to Four Mile Branch. SOURCE: Buice et al., 2005.

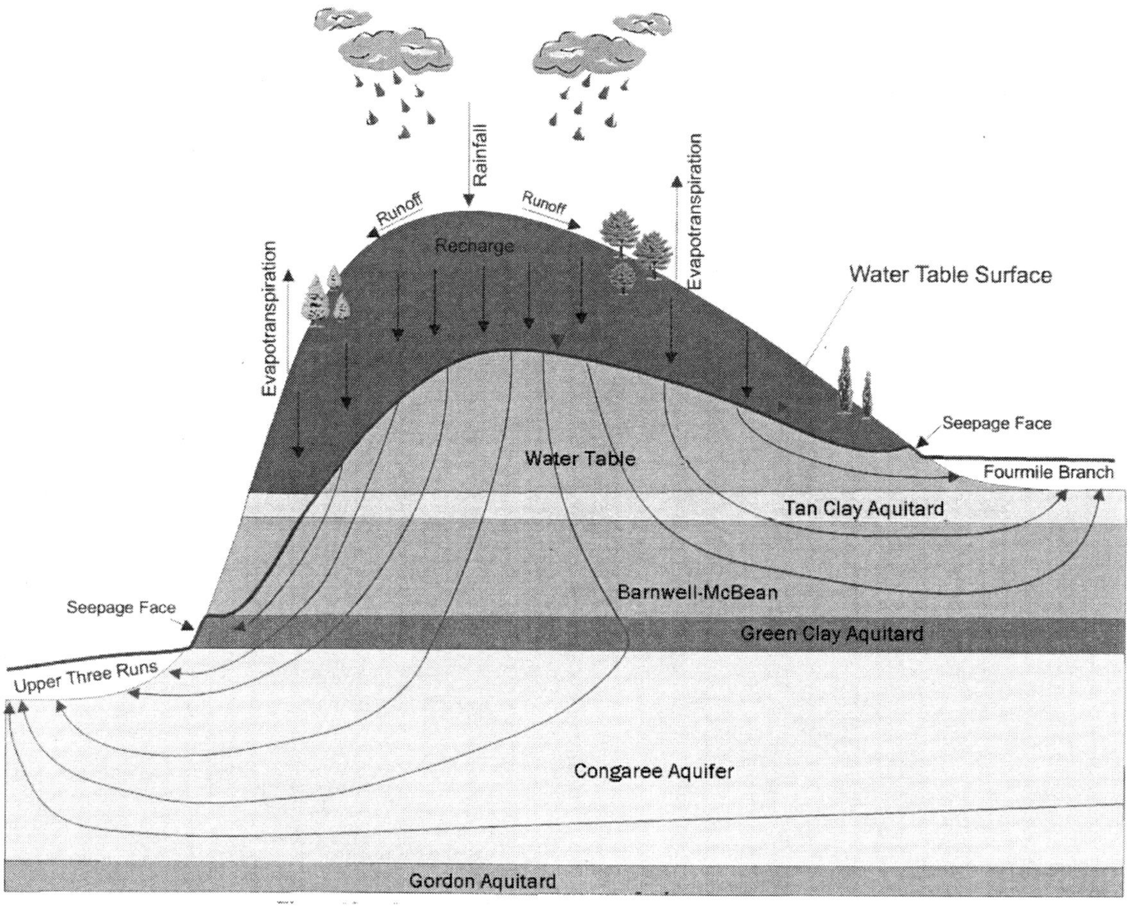

[NOT TO SCALE]

FIGURE J-3 Hydrologic units and flow directions in the General Separations Area. SOURCE: Buice et al., 2005.

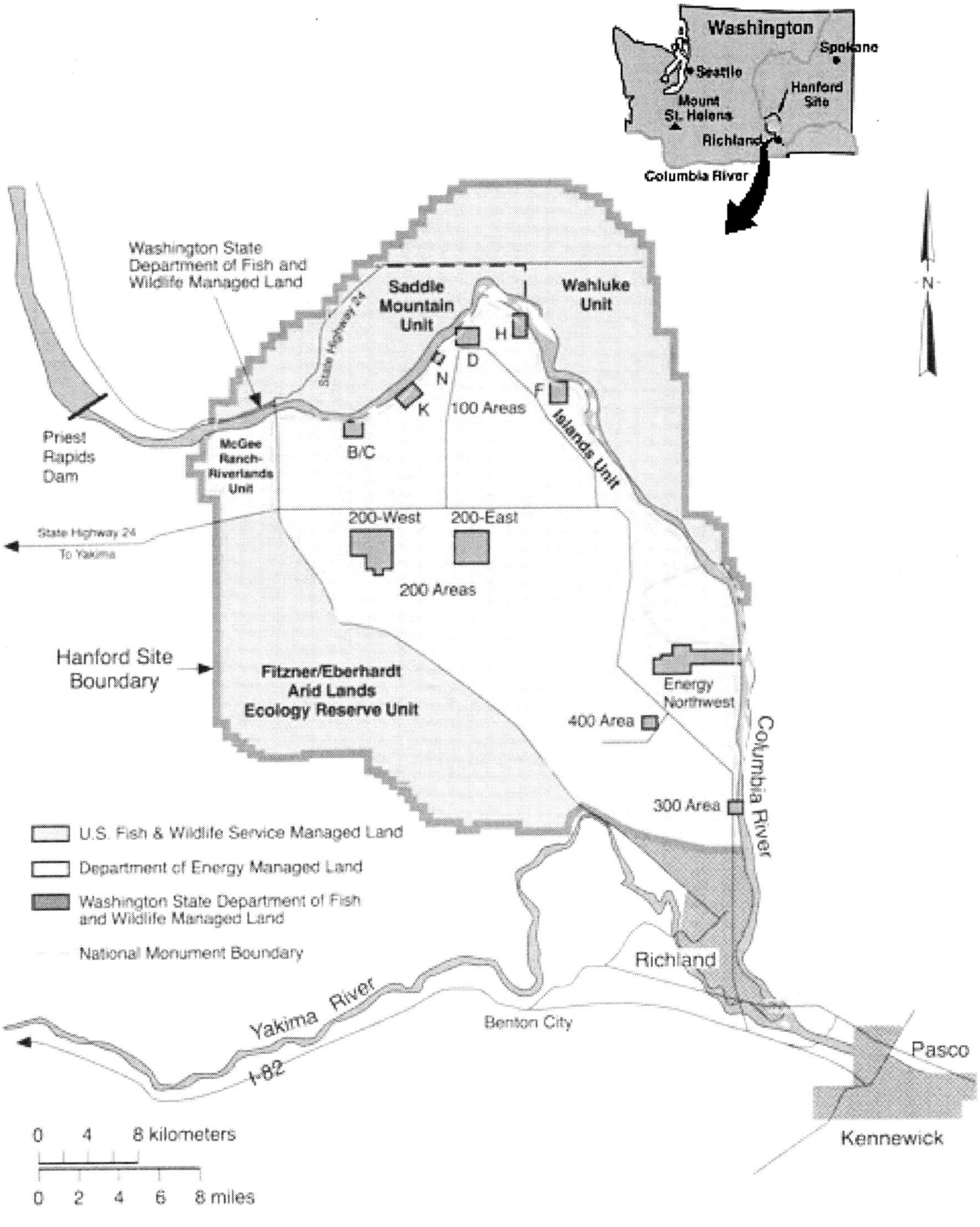

FIGURE J-4 The Hanford Site. SOURCE: Hanford Site, Washington, available at *http://www.pnl.gov/ecomon/Hsmap.HTML.*

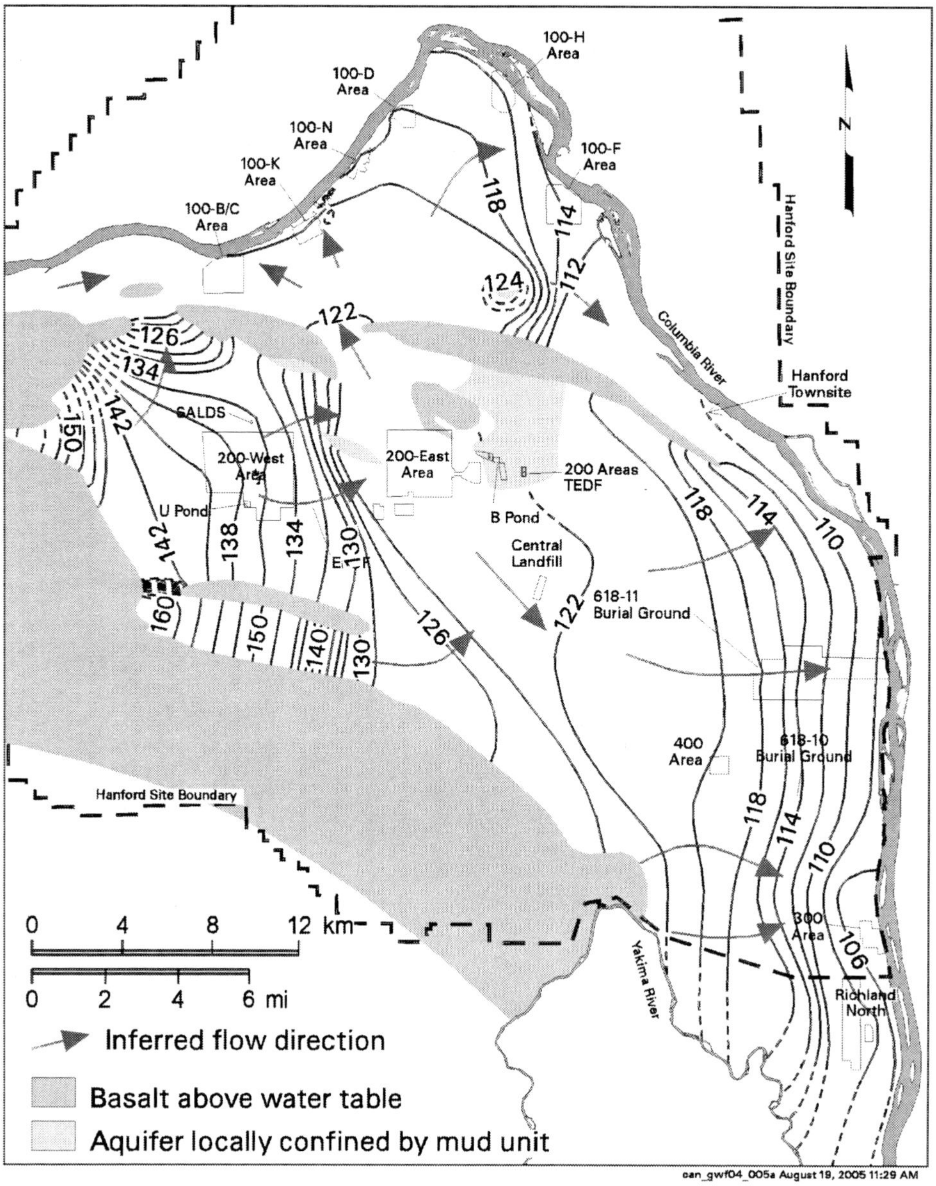

FIGURE J-5 Water Table Elevations in Meters (1m = 3.28 feet) and Inferred Groundwater Flow Directions for the Unconfined Aquifer at Hanford, Washington, March 2004. SOURCE: Hartman et al. 2005.

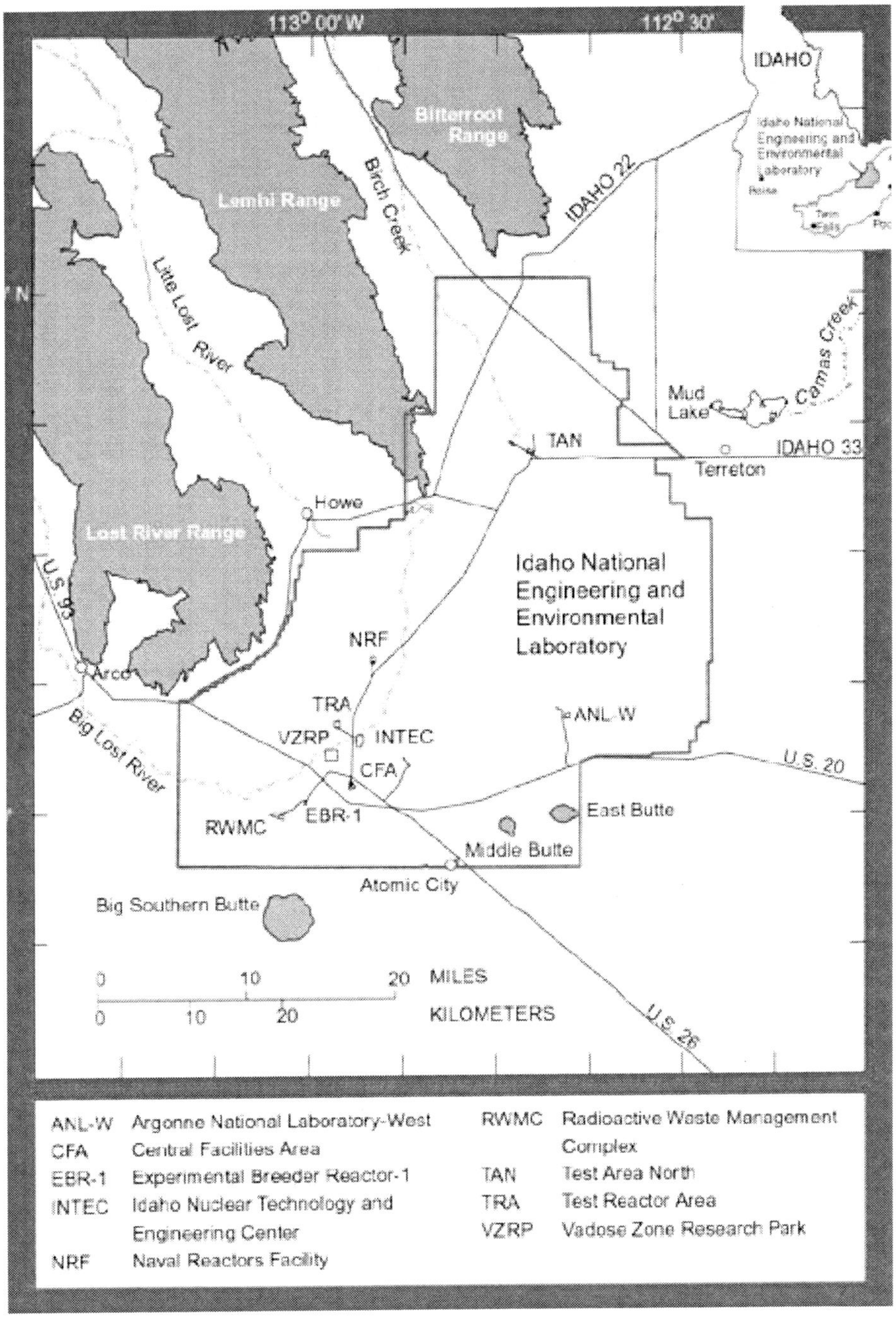

FIGURE J-6 Idaho National Laboratory. Source: United States Geological Service (USGS), Available at *http://wwwrcamnl.wr.usgs.gov/uzf/ineelmap.html*.

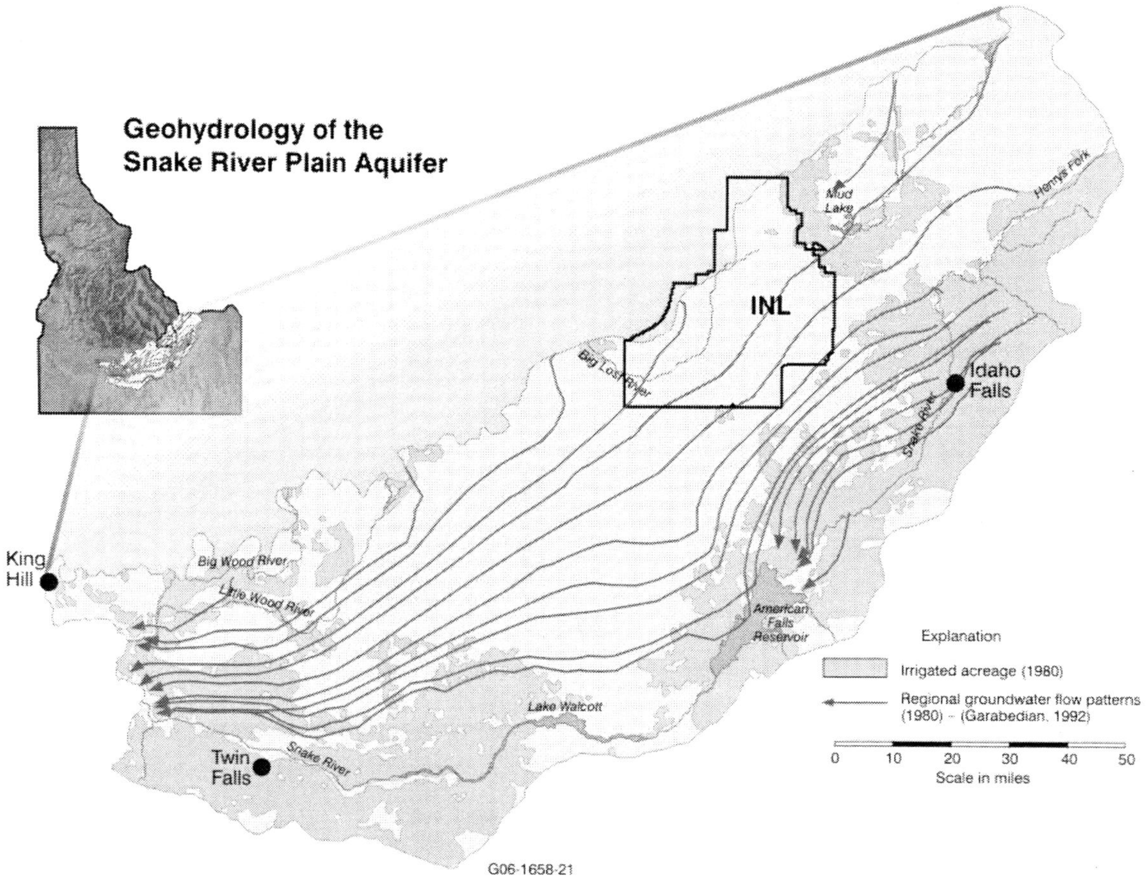

FIGURE J-7 Idaho National Laboratory water table. SOURCE: Lockie, 2006.

Appendix K

Glossary

ACTINIDE REMOVAL PROCESS (ARP). A chemical process for treating tank wastes in which monosodium titanate will be added to the waste tanks to sorb strontium and the actinides. To be used at the Savannah River Site to remove strontium, neptunium, uranium, and plutonium from salt solution.

ANNULUS. Annular space between underground tanks and the concrete vaults in which they sit.

AS LOW AS REASONABLY ACHIEVABLE (ALARA). Reduction of radiation doses or environmental releases as far below applicable limits as technical, social, economic, practical, and public policy considerations allow (NCRP, 2002).

ATOMIC ENERGY ACT (AEA). The purpose of the Atomic Energy Act (42 U.S.C. 2011 - 2259), passed in 1946 and revised in 1954, is to ensure the proper production and use of source, special nuclear, and by-product material for defense and peaceful purposes and for their regulation to protect public health and safety. The AEA and the statutes that amended it delegate the regulation of atomic energy defense, research, and development activities to the U.S. Department of Energy and the authority for licensing commercial nuclear activities to the U.S. Nuclear Regulatory Commission.

BINS. Vertical stainless steel tanks inside concrete vaults partially or wholly above ground, storing high-level waste in a granular form (calcine) at the Idaho site.

BISMUTH PHOSPHATE PROCESS. The bismuth phosphate ($BiPO_4$) process separated plutonium from uranium and other radionuclides in irradiated fuel. This process operated on an industrial scale at the Hanford Site from 1944 through 1956. The high-level from this process contained uranium and was very acidic. It was neutralized and alkalinized to minimize corrosion before being sent to the tank farms

BULK. Majority of the volume of waste in a tank or a bin.

CESIUM AND STRONTIUM CAPSULES. Metal cylinders containing intensely radioactive halide salts (cesium chloride and strontium fluoride) stored at the Hanford Site. In the late 1960s, strontium and cesium were separated from the waste in tanks filled by the PUREX Plant and later directly from the PUREX Plant waste stream. At the Hanford Site there are 2,217 cesium and strontium capsules.

CLASS C LIMIT. The concentration limits under U.S. Nuclear Regulatory Commission regulations for disposal of low-level radioactive waste in a near-surface facility.

CLASTIC DIKE. An intrusion of sediment into an overlying sedimentary rock.

CODE OF FEDERAL REGULATIONS (CFR). The codification of rules published in the *Federal Register* by the executive departments and agencies of the Federal Government.

COMPLIANCE AGREEMENT. An agreement reached to comply with decisions regarding a Federal Facility Agreement.

COMPLIANT TANK. Term used at the Savannah River Site to indicate tanks that have a full secondary containment system (i.e., a tank inside a tank). Most of the tanks at this site have a carbon steel inner wall and an outer vault wall constructed of concrete. Tanks that have a metal liner on the outer wall are said to have a secondary containment. If the outer liner rises only partway up the outer wall, it provides only partial secondary containment.

COMPREHENSIVE ENVIRONMENTAL RESPONSE, COMPENSATION, AND LIABILITY ACT (CERCLA). Law, also known as "Superfund," passed in 1980 and amended by the Superfund Amendments and Reauthorization Act of 1986, that established prohibitions and requirements concerning closed and abandoned hazardous waste sites; provided for liability of persons responsible for releases of hazardous waste at these sites; and established a trust fund to provide for cleanup when no responsible party could be identified.

CRIBS. Shallow, subsurface drainage structures for filtering liquid waste into soil.

CURIE. A unit of radioactivity equal to 37 billion decays per second.

DECAY PRODUCT. An atom resulting from the decay of a radioactive isotope.

DEFENSE NUCLEAR FACILITIES SAFETY BOARD (DNFSB). An independent federal agency established by Congress in 1988 under the Atomic Energy Act to provide safety oversight of the nuclear weapons complex.

DEFENSE WASTE PROCESSING FACILITY (DWPF). A facility that immobilized high-level radioactive waste in glass. Located at the Savannah River Site.

DELIQUIFICATION, DISSOLUTION, AND ADJUSTMENT (DDA). A process for removing salt waste from tanks at the Savannah River Site. The DDA process involves: (1) removing the supernate from above the saltcake; (2) extracting interstitial liquid within the saltcake matrix; (3) dissolving the saltcake and transferring the resulting salt solution to a settling tank; and (4) transferring the salt solution to the Saltstone Facility feed tank where, if required, the solution is aggregated with other tank farm waste to adjust batch chemistry. Chemical adjustment may be required to ensure that the salt solution feed stream meets processing parameters (e.g., sodium concentration, organic content, facility shielding limitations) for processing at SPF (DOE-SRS, 2005b, p. 12).

DEPARTMENT OF ENERGY (DOE). A federal agency established in 1977 as one of the successor agencies (see U.S. Nuclear Regulatory Commission) of the Atomic Energy Commission to advance the national, economic, and energy security of the United States; to promote scientific and technological innovation in support of that mission; and to ensure the environmental cleanup of the nation's nuclear weapons complex.

DIATOMACEOUS EARTH. A naturally occurring, soft, chalk-like, sedimentary rock mineral that is easily crumbled into a fine powder. Diatomaceous earth was used as waste sorbent material to immobilize residual supernatant liquid in leaking tanks when the liquid removal by pumping was not feasible. It was used in the following Hanford waste storage tanks: BX-102 (1971), SX-113 (1972), TX-116 (1970), TX-117 (1970), TY-106 (1972), and U-104 (1972) (Wagoner, 1995).

DOE ORDER 435.1: RADIOACTIVE WASTE MANAGEMENT. A self-imposed regulation intended to ensure that DOE radioactive waste is managed in a way that is protective of worker and public health and safety and of the environment.

ENVIRONMENTAL PROTECTION AGENCY (EPA). A federal agency established in 1970 to protect human health and the environment.

FEDERAL FACILITY AGREEMENT (FFA). An agreement among the U.S. Department of Energy, the U.S. Environmental Protection Agency, and the relevant state regulator for a DOE site that lays out how the site will comply with environmental laws and regulations. The agreement may contain the milestones for waste retrieval and tank closure.

FISSION PRODUCT. An atom resulting from the splitting or fission of a heavier atom.

GERMAN LOGS. An agreement between an agency of the Federal Republic of Germany and DOE resulted in the production of 34 isotopic heat sources ("German logs") in the mid-1980s. Stainless steel canisters were filled with radioactive borosilicate glass. The logs contain a total of approximately 8.3 Mci consisting mainly of cesium-137 and strontium-90, but also containing transuranic contamination. Originally the logs were intended to be transported to Germany for use in experimental programs associated with the development of underground storage facilities. Currently, the logs are managed as remote-handled transuranic waste (approximately 6 m^3) and are slated for disposal at the Waste Isolation Pilot Plant (WIPP) sometime after 2012.

GROUNDWATER TRAVEL TIME. The time for a contaminant to travel a given distance through groundwater.

GROUT. Term used in this report to mean a cementitious material used for waste immobilization or tank fill.

HALF-LIFE. The time required for half of the atoms of a radioactive isotope to undergo decay.

HANFORD. The Hanford Site along the Columbia River in south central Washington State was claimed and developed by the federal government to produce plutonium for nuclear weapons as part of the Manhattan Project. After 50 years of operation, the site is now primarily a cleanup site. Hanford has 149 single-shell and 28 double-shell tanks for storage of liquid wastes from reprocessing spent nuclear fuel.

HEEL. The waste remaining in the bottom of a tank after removal of the bulk and residual waste to the maximum extent practical. The heel may be liquid, loose or encrusted solids, or all of these.

HIGH-ACTIVITY WASTE. The fraction of tank waste that will be disposed in a deep-geologic repository for spent fuel and high-level waste.

HIGH-LEVEL RADIOACTIVE WASTE (HLW). High-level waste is defined in terms of its source as the primary waste product resulting from the reprocessing of spent nuclear fuel. This waste is (1) highly radioactive liquid, containing mainly fission products as well as some actinides, which is separated during chemical reprocessing of irradiated fuel (aqueous) waste from the first solvent extraction cycle and those waste streams combined with it; (2) spent reactor fuel, if it is declared a waste (NRC, 1996b).

IDAHO NATIONAL LABORATORY (INL). A large reservation near Idaho Falls, Idaho, that has been used for research and test reactors, operations to support the Naval Nuclear Propulsion Program, and other research. The site has 11 underground storage tanks for storage of liquid wastes from reprocessing spent nuclear fuel and seven bin sets of calcined wastes from reprocessing spent nuclear fuel, which are located below, or partially below the land surface in concrete vaults.

IDAHO NUCLEAR TECHNOLOGY AND ENGINEERING CENTER (INTEC). Facility that houses the reprocessing operation at Idaho National Laboratory, including chemical processing facilities, waste calciners, tank farms, and bin sets.

IMMOBILIZED. Bound up in a solid to isolate from environmental release or transport.

INTEGRATED DISPOSAL FACILITY (IDF). A single disposal facility for immobilized low-activity waste from tank cleanup and other waste, located at the Savannah River Site, Aiken, South Carolina.

ION EXCHANGE. a usually reversible exchange of one ion with another, either in a liquid, on a solid surface, or within a crystalline lattice. This mechanism is sometimes used to separate specific constituents, such as cesium, from waste.

IONIZING RADIATION. Any radiation capable of displacing electrons from atoms or molecules, thereby producing ions. Examples include alpha radiation, beta radiation, gamma radiation or X-rays, and cosmic rays. The minimum energy of ionizing radiation is a few electron-volts.

LONG-LIVED. In the context of waste disposal, having a half-life that is comparable to or longer than human history. For example, technetium-99 with its 212,000-year radioactive half-life is long-lived.

LOW-ACTIVITY waste (LAW). Radioactive waste that contains concentrations of radionuclides low enough that managing these wastes may not require all of the radiation protection measures necessary to manage higher-activity radioactive material to be fully protective of public health and the environment. Several classes of radioactive waste, including some low-level wastes contain low enough concentrations of radionuclides to be considered low-activity waste (EPA, available at *http://www.epa.gov/radiation/ larw/larw.htm*).

LOW-LEVEL RADIOACTIVE WASTE (LLW). "Radioactive waste that (A) is not high-level radioactive waste, spent nuclear fuel, transuranic waste, or byproduct material as defined in Section 11(e)(2) of the Atomic Energy Act, and (B) the U.S. Nuclear Regulatory Commission, consistent with existing law, classifies as low-level radioactive waste. The byproduct material referred to in Clause (A) essentially is uranium or thorium mill tailings" (NCRP, 2002).

MODULAR CAUSTIC-SIDE SOLVENT EXTRACTION UNIT (MCU). A chemical separation facility for waste treatment, located at the Savannah River Site. The MCU will use a solvent extraction process to recover cesium from the salt waste. It will be operated downstream of the actinide removal process (ARP) before the Salt Waste Processing Facility (SWPF) becomes operational (DOE-SRS, 2005b).

NATIONAL RESEARCH COUNCIL (NRC). Part of the National Academies which also comprise the National

Academy of Sciences, the National Academy of Engineering, and the Institute of Medicine.

NONCOMPLIANT TANK. See compliant tank.

NUCLEAR WASTE POLICY ACT OF 1982 AND ITS AMENDMENTS (NWPA). U.S. Code, Title 42, Chapter 108, Nuclear Waste Policy, Section 10101 establishes the federal government's responsibility to provide a place for the permanent disposal of high-level radioactive waste and spent nuclear fuel.

PERFORMANCE ASSESSMENT. An iterative process to support decisions about disposal facilities (i.e., regulatory compliance and design of the monitoring plan, the facility, and other controls) by evaluating risks based on modeling on interactions between the disposal facility, people, and the environment; contaminant release and transport; and exposures.

POINT OF COMPLIANCE. A location some distance away from the disposal facility boundary. Inside the point of compliance is the facility operations area, which is subject to institutional controls for at least 100 years after closure. DOE Order 435.1 and Nuclear Regulatory Commission Guidance set the point of compliance most commonly used at 100 m from the disposal facility boundary. Beyond the point of compliance, DOE must ensure that members of the public would not receive environmental exposures that exceed the dose limits in DOE Order 435.1 and 10 CFR 61.

PROGRAMMATIC RISK. The risk to cost, schedule, and technical performance of a program. It is associated with all uncertainties, including legal uncertainties that can result in delays, growth cost, and failure to reach the established goals.

PUREX PROCESS. A chemical process for separating plutonium and uranium from dissolved spent nuclear fuel.

RADIOACTIVITY. The property of an unstable atomic nucleus to spontaneously transform with the emission of energy in the form of radiation.

RADIONCULIDE. A naturally occurring or artificially produced radioactive element or isotope.

REDOX PROCESS. Process to separate both uranium and plutonium from other radionuclides in spent nuclear fuel. This process was developed after the bismuth phosphate process and before the PUREX process.

RESIDUAL. Waste remaining in a tank after bulk waste retrieval has been completed. Residual waste can be retrieved from the tanks. The material remaining is called the "heel."

RESOURCE CONSERVATION AND RECOVERY ACT (RCRA). Law enacted by Congress in 1976 as an amendment to the 1965 Solid Waste Disposal Act to regulate the handling, storage, treatment, transportation, and disposal of solid waste, particularly hazardous chemicals, to protect the public from harm. The goals of RCRA are to protect human health and the environment from the hazards posed by waste disposal, by encouraging the conservation of energy and natural resources through reuse and recycling, and the reduction or elimination of the amount of waste generated.

RISK ANALYSIS. A detailed examination, including risk assessment, risk characterization, risk communication, and risk management, performed to understand the nature of unwanted, negative consequences to human life, health, property, or the environment (Society for Risk Analysis, available at *http://www.sra.org/).*

RISK ASSESSMENT. Scientific evaluation of known or potential adverse health effects resulting from exposure to hazards. It is a process of establishing information regarding the acceptable levels of risk for an individual, group, society, or the environment. The process consists of the following steps: (1) hazard identification, (2) hazard characterization, (3) exposure assessment, and (4) risk characterization. The definition includes quantitative risk assessment and also qualitative expressions of risk, as well as an indication of the attendant uncertainties (NRC, 1983, 1986).

RISK-INFORMED APPROACH. An approach in which risk is the starting point but still only one among several factors in a decision process.

RONALD W. REAGAN NATIONAL DEFENSE AUTHORIZATION ACT OF FISCAL YEAR 2005 (NDAA). Public Law 108-375 authorizing funding for defense activities, military construction, and national security-related energy programs.

SALTCAKE. The crystalline salt that forms in high-level radioactive waste tanks and contains much of the cesium and some of the actinides in the waste.

SALTSTONE. The cementitious waste form used at the Savannah River Site to immobilize low-activity waste from the tank farms.

SALTSTONE DISPOSAL FACILITY (SDF). Located at that Savannah River Site.

SALTSTONE PROCESSING FACILITY (SPF). Located at the Savannah River Site, Aiken, South Carolina.

SALT WASTE PROCESSING FACILITY (SWPF). A facility for separating key radionuclides from retrieved tank waste under development at the Savannah River Site. The facility will use the chemical process of ARP/MCU (see ARP and MCU) but should have higher decontamination fractions and greater throughput.

SAVANNAH RIVER SITE (SRS). The nation's second site developed for the production of plutonium, SRS still carries out missions for the nuclear weapons program. Located in southern South Carolina, the site has 51 underground tanks for storage of liquid wastes from reprocessing spent nuclear fuel.

SCARIFIER. A device that uses rotating water jets for the purpose of breaking up agglomerated solids or for polishing or cleaning metal or concrete surfaces.

SHORT-LIVED. In the context of waste disposal, this means having half-life that is short compared to human history. For example, cesium-137 with its 30.2-year half-life is short-lived.

SLUDGE. Insoluble wetted particles.

"SMART" GROUT. Grout engineered to be pumped into a tank, flow to the tank walls, not segregate, self-consolidate, generate minimal heat of hydration, and provides the requisite high pH and low Eh to stabilize the radionuclides and toxic heavy metals in the waste.

SUPERNATE. The fluid above a sediment or precipitate.

TANK WASTE. Waste from reprocessing spent nuclear fuel stored in underground tanks.

TITLE 10 CODE OF FEDERAL REGULATIONS PART 61 (10 CFR 61). Through this regulation, entitled, "Licensing Requirements for Land Disposal of Radioactive Waste," the United States Regulatory Commission regulates near-surface disposal of commercial low-level waste.

TRANSURANIC ISOTOPE. Isotope of an element with more protons than uranium (i.e., atomic number greater than 92).

TRANSURANIC (TRU) waste. Waste containing more than 100 nCi of alpha-emitting TRU isotopes (atomic number greater than 92) per gram of waste, with half-lives greater than 20 years, except for

- High-level radioactive waste;
- Waste that the Secretary of Energy has determined, with the concurrence of the Administrator of the Environmental Protection Agency, does not need the degree of isolation required by the disposal regulations; and
- Waste that the USNRC has approved for disposal on a case-by-case basis in accordance with 10 CFR 61 (Public Law 102-579).

TRI-PARTY AGREEMENT. Alternative name for the Hanford Federal Facility Agreement and Consent order.

U.S. NUCLEAR REGULATORY COMMISSION (USNRC). An independent federal agency established in 1974 by the Energy Reorganization Act, as one of the successor agencies (see Department of Energy) of the Atomic Energy Commission. The USNRC's regulatory activities are focused on the oversight and licensing of reactors at commercial power plants, materials safety oversight and licensing for a variety of other activities, and waste management of both high-level waste and low-level waste.

VADOSE ZONE. The zone between the earth's surface and the top of the water table, also called the unsaturated zone.

VITRIFIED. Immobilized in glass.

WASTE INCIDENTAL TO REPROCESSING (WIR). Waste that is a by-product of reprocessing operations but not high-level waste.

WESTINGHOUSE SAVANNAH RIVER COMPANY (WSRC). The current contractor for the Department of Energy at the Savannah River Site.

ZEOLITES. A class of hydrated aluminosilicate minerals with an ability to "trap" cesium (and other cations) that was used at the Savannah River Site to remove cesium from the tank system. Zeolite particles in tanks are difficult to retrieve because they can agglomerate into chunks that are difficult to maintain suspended in a slurry.